Science and Risk Regulation in International Law

The regulation of risk is a preoccupation of contemporary global society and an increasingly important part of international law in areas ranging from environmental protection to international trade. This book examines a key aspect of international risk regulation – the way in which science and technical expertise are used in reaching decisions about how to assess and manage global risks. An interdisciplinary analysis is employed to illuminate how science has been used in international legal processes and global institutions such as the World Trade Organization. Case studies of risk regulation in international law are drawn from diverse fields including environmental treaty law, international trade law, food safety regulation and standard-setting, biosafety and chemicals regulation. The book also addresses the important question of the most appropriate balance between science and non-scientific inputs in different areas of international risk regulation.

DR JACQUELINE PEEL is an Associate Professor of Law at the Melbourne Law School, University of Melbourne, Australia. Her research focuses on areas of domestic and international environmental law including climate change, the intersections between law and science, and the precautionary principle.

CAMBRIDGE STUDIES IN INTERNATIONAL AND COMPARATIVE LAW

Established in 1946, this series produces high-quality scholarship in the fields of public and private international law and comparative law. Although these are distinct legal sub-disciplines, developments since 1946 confirm their interrelation.

Comparative law is increasingly used as a tool in the making of law at national, regional and international levels. Private international law is now often affected by international conventions, and the issues faced by classical conflicts rules are frequently dealt with by substantive harmonisation of law under international auspices. Mixed international arbitrations, especially those involving state economic activity, raise mixed questions of public and private international law, while in many fields (such as the protection of human rights and democratic standards, investment guarantees and international criminal law) international and national systems interact. National constitutional arrangements relating to 'foreign affairs', and to the implementation of international norms, are a focus of attention.

The Board welcomes works of a theoretical or interdisciplinary character, and those focusing on the new approaches to international or comparative law or conflicts of law. Studies of particular institutions or problems are equally welcome, as are translations of the best work published in other languages.

A list of books in the series can be found at the end of this volume.

Science and Risk Regulation in International Law

Jacqueline Peel

CAMBRIDGE
UNIVERSITY PRESS

CAMBRIDGE
UNIVERSITY PRESS

University Printing House, Cambridge CB2 8BS, United Kingdom

One Liberty Plaza, 20th Floor, New York, NY 10006, USA

477 Williamstown Road, Port Melbourne, VIC 3207, Australia

314-321, 3rd Floor, Plot 3, Splendor Forum, Jasola District Centre, New Delhi - 110025, India

79 Anson Road, #06-04/06, Singapore 079906

Cambridge University Press is part of the University of Cambridge.

It furthers the University's mission by disseminating knowledge in the pursuit of education, learning and research at the highest international levels of excellence.

www.cambridge.org
Information on this title: www.cambridge.org/9780521768634

First published 2010

A catalogue record for this publication is available from the British Library

Library of Congress Cataloging in Publication data
Peel, Jacqueline.
Science and risk regulation in international law / Jacqueline Peel.
p. cm. – (Cambridge studies in international and comparative law)
Includes bibliographical references and index.
ISBN 978-0-521-76863-4 (hardback)
1. Risk management–Law and legislation. 2. Science and law.
3. International law. I. Title.
K487.R57P44 2010
344.04′6–dc22
2010033026

ISBN 978-0-521-76863-4 Hardback

Contents

Preface

In our modern, industrialised world science is the way we have come to know and understand risks to our health and the environment. These risks range from the potential cancer-causing properties of traces of chemical contaminants in foodstuffs, to large-scale challenges, such as reducing the effects and managing the impacts of worldwide climate change. Most such risks are seen as having a global dimension, whether because of the potential for broad-ranging effects on human and environmental systems (as in the case of climate change) or because globalised processes of trade and commerce have the capacity to disseminate widely potential risks associated with particular products (for instance, genetically modified crops). In response to these factors, requirements for risk regulation have proliferated in international law and global governance, with science and expert processes of risk assessment as their basis.

As international law has come to take on a more important role in the governance of risk, it is exercising greater influence over key aspects of risk regulation, such as the role of science in assessing and managing health and environmental risks. Science is considered by most to be a necessary component of risk regulation, especially when dealing with risks to human health and the environment that would otherwise be difficult to perceive and comprehend. However, science and expertise hold less power than in the past to legitimate the exercise of governing authority. Greater acknowledgment of uncertainties in scientific knowledge and the gradual acceptance of social scientific research illustrating the potential for a diversity of perspectives on risk issues have led to questions over whether science should be the only (or primary) resource relied upon in international risk decision-making.

These questions – that have significant ramifications for international law in the area of risk regulation, as well as for states and for their citizens affected by international risk decision-making – are the focus of this book. It explores the appropriate role for science in risk regulation undertaken at the global level. This requires striking a delicate balance between the desire for credibility and the need for international law and institutions to be perceived as a legitimate source of risk governance – something that, in turn, often depends on recognising local risk perspectives and political concerns. While the book argues that science alone is rarely sufficient as a basis for credible and legitimate risk decision-making under international law, it by no means seeks to displace the role of science entirely. Ultimately the goal of the book is to find better ways to use science, and to prevent the misuse of science, in the international law of risk regulation.

The role of science in risk regulation, and the way this is influenced by domestic legal processes, has been, and continues to be a topic of intense policy debate and the basis for a substantial scholarship that crosses the disciplines of science, social science and law. In international law the topic of science and risk regulation rose to prominence with interstate disputes in the late 1990s such as the *Hormones* case over the application of the scientific evidence and risk assessment requirements of the World Trade Organization's Sanitary and Phytosanitary Measures Agreement (WTO SPS Agreement). Such disputes generated a thriving literature, but it is only recently that scholars have begun to explore more broadly the role of science in risk regulation in a range of areas of international law. The significant level of current interest in these questions among international legal and other scholars is attested by events such as the joint American and European Societies of International Law October 2009 forum on 'Science and International Law', and the publication of major edited collections on the issue such as *Uncertain Risks Regulated* (2009)[1].

This work builds upon, and also seeks to extend, the existing literature and practice concerning science and risk regulation in international law. To do so, the book adopts an interdisciplinary approach that combines the knowledge and findings of both international law and the sciences. This has allowed a fuller survey and treatment of the ways scientific evidence, and regulatory notions of 'sound science' and

[1] Michelle Everson and Ellen Vos (eds.), *Uncertain Risks Regulated* (Abingdon: Routledge-Cavendish, 2009)

precaution, are and might be addressed in international law. The book applies these insights to a range of case studies of the use of science in international risk decision-making processes. The most substantial case study addresses the now extensive practice of risk regulatory review by WTO bodies under the SPS Agreement. However, in line with the book's ambition to examine science and risk regulation in international law more broadly, other case studies are drawn from diverse areas, including international standard-setting in the food safety area, biosafety and the regulation of genetically modified organisms, risk assessment and management of pesticides and other potentially harmful chemicals, and global assessments of the risk of climate change, that inform international negotiations on the appropriate political and legal response.

Acknowledgements

This book originally began life as my Ph.D. thesis at the University of Melbourne and accordingly I owe an enormous debt of gratitude to my thesis supervisor, Professor Anne Orford, for her diligence, insights and thoughtful comments in guiding the initial articulation of this work. In reworking the thesis into its current book form I also derived much assistance and encouragement from the feedback of my examiners, Professors Rob Howse and Mary Footer, as well as the comments of two anonymous reviewers.

Over the course of writing the book I benefited greatly from opportunities to discuss my ideas and research with many colleagues, both in Australia and overseas. I am particularly grateful to Professor Joseph Weiler and the Jean Monnet Center at New York University who hosted me as a Hauser Research Fellow in 2003–4 while I was researching the SPS case study for this book. The book's discussion of the 'democratisation' of science in international risk decision-making benefited also from feedback from Professor José Alvarez and other international law colleagues during my participation in the International Law and Democratic Theory Symposium hosted under the auspices of the American, Australian/New Zealand, Canadian and Japanese International Law Societies in June 2006.

Back in Australia, at my home law school of Melbourne University, I was very fortunate to receive research support funds from the Law School that allowed me to employ fantastic research assistants such as Michael Power. In addition, many of my colleagues provided support for, and feedback on, my work. I would particularly like to mention in this regard David Morgan, Dr Jürgen Kurtz, Professor Lee Godden and Dr Jenny Beard.

I am also immensely grateful to international colleagues who have supported my work and professional development over many years. In particular I would like to thank Professor James Crawford and Professor Philippe Sands, both of whom have been wonderful mentors since I first met them as a NYU Masters student in 2000.

In bringing this book to fruition, I would like to acknowledge the diligent work and effort of editors and staff at Cambridge University Press. They include law commissioning editor, Nienke van Schaverbeke, Richard Woodham, Lorraine Slipper and the project manager for the book, Rob Wilkinson.

Last, but certainly by no means least, I would like to thank my family and friends who have given great support during the writing of the book. The production of my thesis and this book coincided with the birth of my children Aly (in July 2007) and Will Findlay (in August 2009), to whom I would like to dedicate this work, with love.

1 Introduction: science and risk regulation in international law

Introduction

Environmental and health risks are today a subject of great debate and concern in many countries, as well as at the global level. Risks of climate change, ozone depletion, the spread of disease and loss of species, among many others, have become central issues of policy and legal development preoccupying national governments and international organisations. The language of 'risk' is used in discussing these issues because, in many cases, available information is inadequate or incomplete.[1] Enough is known to suspect or predict that a threat exists, but the full outcomes for human health and the environment, including for future generations, may not be well understood. This uncertainty, together with the complexity of the ecological systems and processes at issue, encourages a proliferation of plausible perspectives on risk problems and the best way to manage them.[2] In this context, the regulatory and adjudicative systems of international law may be turned to as a forum for mediating between different risk perspectives and, indeed, for determining whether risks exist that should be the subject of legal intervention.

Where international legal disputes arise over the nature and extent of health and environmental risks – such as the 2006 World Trade Organization (WTO) dispute involving genetically modified organisms (GMOs) – typically an enormous amount of scientific material is

[1] The term 'risk' here is used in the sense of unknown dangers rather than in the more limited sense characteristic of scientific risk assessment exercises. As to the latter, see further Chapter 3.

[2] John Dryzek, *The Politics of the Earth: Environmental Discourses*, 2nd edn, (New York: Oxford University Press, 2005), p. 9.

gathered in order to substantiate (and contest) risk claims.[3] In the *GMO* case, for instance, the WTO panel involved consulted six independent experts on scientific and technical matters, who produced reports running to hundreds of pages. The parties to the dispute then hired their own scientists to digest and review the reports of the panel's independent experts, as well as the analyses of those reports produced by the opposing sides' experts. The result was reams of technical data and expert opinion regarding the health and environmental risks of GMOs intended to inform the legal findings made by the three non-scientifically trained members of the WTO panel.[4]

From the perspective of international law, the *GMO* case raises many important issues about the interaction of trade rules with health and environmental regulation.[5] One of the most critical questions posed by the dispute, and others like it, concerns the role of science in the evolving international legal system governing risk regulation. Given the uncertainties surrounding many risks and the multitude of different perspectives on health and environmental issues, we may ask whether it should be primarily to science that international law and decision-makers turn in understanding and managing such issues. Alternatively, if a broader information base for international risk regulation is seen to be appropriate (or at least in those risk situations where uncertainties abound or there is intense socio-political debate over potential harms), what additional sources should be consulted, and how might such views be integrated with scientific inputs?

These are the questions at the heart of this book, which addresses the role of science in risk regulation, and in the development and application of relevant areas of international law, such as international trade law. The book brings to this task an interdisciplinary perspective and analytical approach that allow a more comprehensive treatment of the

[3] See *European Communities – Measures Affecting the Approval and Marketing of Biotech Products*, Reports of the Panel, WTO Docs WT/DS291/R, WT/DS292/R, WT/DS293/R, 29 September 2006 (*GMO* case), [7.39], Annexes H–J.

[4] The members of the WTO panel in the *GMO* case were Christian Häberli (Head of International Affairs at the Swiss Federal Office for Agriculture), Mohan Kumar (India's Deputy High Commissioner in the Diplomatic Mission in Sri Lanka) and Akio Shimizu (Professor in Law at Waseda University, Tokyo, Japan).

[5] A comprehensive analysis of these issues is beyond the scope of this book. For a useful overview of the principal questions in the dispute see Simon Lested, 'International Decisions: European Communities – Measures Affecting the Approval and Marketing of Biotech Products. WT/DS291/R, WT/DS292/R, & WT/DS293/R', Am. J. Int'l L., 101 (2007), 453.

ways that scientific evidence and risk regulatory processes are, and might be, addressed in international law. The analysis reveals that the question of science's role in international risk regulation is one that has deep interconnections with a number of pivotal issues in current international legal scholarship and practice. These include debates over the legitimacy of international law, calls for greater democratic input into global governance, the desirability or otherwise of reducing fragmentation in international law, and the role of particular international institutions, such as the WTO, in shaping normative understandings and processes adopted in international law. While the book does not seek to deal definitively with all such issues, it situates the field of science and global risk regulation against this backdrop and demonstrates how the questions raised in the risk regulatory arena may illuminate broader discussions in the general field of international law.

Science-based regulation of global risks

As the *GMO* case illustrates, science increasingly occupies a central place in the risk decision-making processes of international organisations, such as the organs of the WTO dispute settlement system. In this respect, the WTO regime, established in 1995, appears to have played an important role through agreements such as the Sanitary and Phytosanitary Measures Agreement (SPS Agreement). This Agreement explicitly requires WTO members to ensure that national trade measures taken to protect human, animal or plant life or health have a basis in scientific evidence and risk assessment.[6] In the event that a dispute arises over trade-restrictive sanitary or phytosanitary (SPS) measures adopted by any member, the matter may be brought before the WTO dispute settlement system where decision-makers (with the help of experts)[7] review the scientific justification for the measures.

The SPS Agreement and disputes under it are of relatively recent origin, but the trend in international law towards science-based regulation of risk has its source in developments that go back over a century. These lie in the evolution of global legal rules in tandem with a culture of scientific rationality, and the emergence of future harms, in addition to

[6] Agreement on the Application of Sanitary and Phytosanitary Measures, Marrakesh, 15 April 1994, in force 1 January 1995, 1867 UNTS 493, Articles 2.2 and 5.1.

[7] Article 11.2 of the SPS Agreement directs panels, in disputes involving scientific or technical issues, to seek advice from experts chosen by the panel in consultation with the parties to the dispute.

present dangers, as a preoccupation of industrialised societies around the globe. In recent times rules developed at the international level are seen to have become more pervasive in their influence over nation states and the decisions governments take regarding the identification of, and response to, risks to the health of their populations and the environment. Together these factors – the growing importance of scientific knowledge to international regulatory processes and the transition to greater governance over risk issues exercised by international legal rules – have combined to position science at the heart of global debates and decision-making on matters of health and environmental concern.

Yet, at the same time as science has achieved such prominence in international risk regulation, there has been an improved understanding of its potential limitations to provide complete and accurate information about the threats posed by human activities to health and the environment, especially over the longer term.[8] In many fields scientific knowledge has developed to the point where certain risks are accepted to exist (for example, the risk of developing cancer as a result of exposure to asbestos fibres).[9] In other cases scientists are able to specify the possible adverse consequences of an activity with reasonable confidence, but recognise that the odds of occurrence of these events remain uncertain (for example, in respect of projections of the degree of future global warming and associated sea level rise).[10] Equally, though, there are many areas of science, particularly in the field of the environment, where there are significant unknowns – 'we don't know what we don't know'.[11] Greater understanding of the scope for scientific uncertainty in predicting threats of damage has led to the development of the precautionary principle in international law, which some see as a necessary counterweight to the proliferation of more narrowly science-based decision-making processes of expert risk assessment.[12]

[8] European Environment Agency, *Late Lessons from Early Warnings: the Precautionary Principle 1896–2000* (Luxembourg: European Union, 2001).

[9] General acceptance of the health risks posed by asbestos was evident in the rulings of the WTO Appellate Body in its decision in *European Communities – Measures Affecting Asbestos and Asbestos-Containing Products*, Report of the WTO Appellate Body, WT/DS135/AB/R, 12 March 2001.

[10] See Intergovernmental Panel on Climate Change, *Climate Change 2007: The Physical Science Basis – Summary for Policy-makers* (Geneva: IPCC, 2007), pp. 12–17.

[11] Brian Wynne, 'Uncertainty and Environmental Learning: Reconceiving Science and Policy in the Preventative Paradigm', *Global Environmental Change*, 2(2) (1992), 114.

[12] Andy Stirling and David Gee, 'Science, Precaution and Practice', *Public Health Reports*, 117(6) (2002), 525–6.

Undoubtedly there will be those who deplore the rise of science-based risk regulation as a triumph of neo-liberal conceptions of risk on the international stage or the technologising of global society.[13] Such concerns, however, have not prevented rapid growth in globally oriented, science-based decision-making processes that often give scientific forms of knowledge a key – if not privileged – place in risk regulation. International legal development of this kind makes pertinent the issue of whether it is appropriate for science to play such a central role in global legal structures dealing with the regulation of risk. This is particularly so in light of acknowledgement of the many uncertainties and limitations in scientific knowledge regarding risks, especially where complex human–environmental interactions or poorly understood ecosystems are at issue.

Rise of science in international risk governance

The first three chapters of the book address the questions of how and why science is becoming a fundamental organising principle in international legal regimes concerned with risk, particularly in the areas of health and environmental protection.

Chapter 2 traces the way in which questions of risk to human health and the environment – traditionally a matter over which national governments enjoyed virtually unlimited regulatory control – are now subject to substantial constraints dictated by global legal rules or other supranational regulations. Such rules and regulations are often to be found in governance arrangements of an administrative character, which operate at a level below the legislative processes of intergovernmental negotiation and agreement.[14] The far-reaching effects of rules generated by such structures of 'global governance',[15] and their remoteness from democratic mechanisms operating in many nation

[13] See, e.g., Daniel Kleinman and Abby Kinchy, 'Against the Neoliberal Steamroller? The Biosafety Protocol and the Social Regulation of Agricultural Biotechnologies', *Agriculture and Human Values*, 24(2) (2007), 195.

[14] Peter Lindseth, 'Democratic Legitimacy and the Administrative Character of Supranationalism: The Example of the European Community', Columbia L. Rev., 99 (1999), 632.

[15] The term 'governance' is used to signify the authoritative effects of these rules that yet do not originate from a particular government or governments. 'Global' is preferred to 'international' since many forms of governance originate from sources that are not strictly inter-national (in the sense of being collective decisions of national governments) but are rather supranational or trans-national in character.

states, have given rise to questions over their legitimacy. In the area of risk regulation, this has prompted governments and international organisations to turn to expertise as one possible means for legitimating the increasing reach of global rules into the daily lives of individuals, communities and businesses.

Chapter 3 investigates, in more detail, the concept of risk that has emerged as a central concern of regulation in the (world) 'risk society'.[16] The chapter explores how the determination of risks to health and the environment has come to be heavily reliant on science. This has led to the proliferation of procedures for science-based decision-making and risk assessment in international legal instruments, which in turn seek to provide legitimacy for the increasing transfer of decisions on risk issues from the national to the international level.

The chapter also discusses changes in international law's relationship with science over time as other disciplines, such as the social sciences, have brought to light the potential limitations of scientific knowledge as a reliable basis for predicting future risks. Such insights have exposed the inherently fuzzy boundaries between science and values and, indeed, the difficulty of drawing any firm line between the two in the context of regulating uncertain, complex risks.[17]

The permeability of the science/values boundary in risk regulation is the starting point for the analysis in Chapter 4 of the principal competing paradigms of risk regulation that have emerged in contemporary international law. These are encapsulated in the notion of sound science and the international legal principle known as the precautionary principle. Whereas proponents of sound science emphasise the importance of empirical, field-tested or peer-reviewed studies as a prerequisite for risk regulation, precautionary approaches advocate for action to address threats even in circumstances where the potential for harm is not well established by the available scientific evidence. These two regulatory paradigms are increasingly being brought into contact and conflict in diverse international settings, with indications that some of the potential breadth and flexibility of precautionary approaches are

[16] Ulrich Beck, *Risk Society: Towards a New Modernity* (London: SAGE Publications, 1992).

[17] While the book uses 'science' and 'values' (or 'politics') as key terms in the discussion, it is recognised that these refer to fluid, and eminently contestable, categories. Nevertheless, the distinction between science and values, albeit unfixed and permeable, serves a useful purpose in international risk regulation; namely that the form of knowledge that is generally labelled 'science' organises information in a useful way for the regulatory task of making decisions on health and environmental risk.

being eroded as a result.[18] This is particularly so in the key area of international trade law where precautionary approaches must navigate the widely held perception that the precautionary principle is often mere rhetoric masking protectionist motives.[19]

Science-based risk regulation in practice: the SPS Agreement

The emphasis on (sound) science-based regulation of risk in international law can be problematic where it overestimates the extent to which scientific evidence provides universally accepted, universally valid, guidance for risk policy. Elevating science to a privileged position in international risk regulation may often downplay the necessary role of non-scientific considerations in producing social – and also scientific – consensus on the importance of the risks posed by a given activity, especially in the face of unknowns.

This is well illustrated by science-based risk determination under the WTO SPS Agreement, which is the subject of a detailed case study in Chapter 5. The SPS Agreement is often put forward as a leading example of the adoption of a sound science decision-making model in international law.[20] It has been the forum for the adjudication of several interstate disputes, including that over GMOs. It has also been the focus of political discussion in the SPS Committee, a body 'which self-consciously aim[s] to bring together networks of like-minded regulators to discuss and elaborate norms of behaviour of particular relevance to the trade regime'.[21]

Applying the interdisciplinary understanding of science and risk regulation developed in the previous chapters, Chapter 5 analyses the approach which regulators and decision-makers have taken to the role of scientific evidence and risk assessment under the WTO SPS Agreement. This analysis is undertaken both in respect of the political

[18] John Applegate, 'The Taming of the Precautionary Principle', William & Mary Envtl L. & Policy Review, 27 (2002), 13.

[19] Sabrina Shaw and Risa Schwartz, *UNU-IAS Report: Trading Precaution – The Precautionary Principle and the WTO* (Tokyo: Institute of Advanced Studies, United Nations University, 2005).

[20] Warren H. Maruyama, 'A New Pillar of the WTO: Sound Science', *International Lawyer*, 32 (1998), 651.

[21] Andrew Lang, 'Some Sociological Perspectives on International Institutions and the Trading System', in Colin B. Picker, Isabella D. Burn and Douglas W. Arner (eds.), *International Economic Law: The State and Future of the Discipline*, (Portland: Hart Publishing, 2008), p. 79.

forum of the SPS Committee, and in the adjudicative arena of dispute settlement. In the latter area there has been an emphasis on the need for positive scientific evidence in order to establish risks justifying the introduction of trade measures (although the decision of the WTO Appellate Body in *Hormones II* suggests the pendulum may be swinging back towards a position that is more deferential to risk analysis undertaken by national authorities).[22] This approach effectively precludes reference to other, non-scientific considerations or values (such as those underlying policy decisions, consumer preferences, intuitive judgments, and ethical or socio-economic concerns) as a basis for risk regulation.

In a fragmented international legal environment, the relative institutional strength of the international trade rules and their associated dispute settlement procedures gives added importance to the treatment afforded scientific risk assessment in SPS law. There is thus the potential for the narrower approach to science-based decision-making that has been characteristic of the SPS area to exercise significant influence over the way in which science is used in other international legal fora concerned with risk regulation.

Alternatives to sound science in international risk regulation

The trend of strictly science-based decision-making emerging in SPS law illustrates the limitations of a one-dimensional over-reliance on sound science by global risk governance structures such as the WTO. Given the realities of international risk regulation as a value-laden process characterised by numerous contingencies, a broader approach would seem to be warranted. Yet this raises vexed questions as to available and reliable alternatives that might be looked to as the basis for international risk regulation. The discussion in Chapter 5 of the WTO Appellate Body's procedurally focused approach in the SPS case, *Hormones II*, provides an evaluation of one such alternative that purports to avoid searching international review of the science underlying particular risk regulatory measures.

Chapter 6 examines several other international legal contexts in which the role of science in risk regulation has emerged as a key issue. These include the settlement of health and environmental disputes

[22] United States – Continued Suspension of Obligations in the EC-Hormones Dispute, Report of the WTO Appellate Body, WT/DS320/AB/R, 16 October 2008 (Hormones II).

under the WTO's General Agreement on Tariffs and Trade;[23] consensus-seeking processes of the international organisation charged with developing global food safety standards, the Codex Alimentarius Commission; negotiations for the Cartagena Biosafety Protocol governing the transboundary movement of GMOs, which purports to adopt a precautionary approach;[24] and the operation of scientific assessment processes under multilateral environmental agreements, such as the convention regulating persistent organic pollutants,[25] and the international climate change regime.[26] These sites of international risk decision-making illustrate a variety of models for the use of science in global risk regulation. There is hence the potential for cross-institutional learning whereby elements of particular models could be incorporated into other international risk regulatory processes.

Another rich source of experience with science and risk regulation lies in the domestic systems established in many industrialised countries to assess and manage health and environmental risks. For instance, the United States of America (USA) has well-developed structures for the formulation and judicial review of risk regulatory measures on a range of health and environmental topics, which have been highly influential in the design of similar systems around the globe.[27] For those who look to domestic models as a guide for the appropriate role of science in international risk regulation, a common theme is the need for values and public views to inform determinations made about risk. A further, nascent thread of the literature looks at how democratic values (rather than domestic models of democracy) can be translated into a realisable institutional form for the purpose of

[23] General Agreement on Tariffs and Trade, Marrakesh, 15 April 1994, in force 1 January 1995, 55 UNTS 194, 1867 UNTS 187.

[24] Cartagena Protocol on Biosafety to the Convention on Biological Diversity, Montreal, 29 January 2000, in force 11 September 2003, 2226 UNTS 208.

[25] Convention on Persistent Organic Pollutants, Stockholm, 23 May 2001, in force 17 May 2004, (2001) 40 ILM 532.

[26] Two treaties currently make up this regime: the United Nations Framework Convention on Climate Change, Rio de Janeiro, 9 May 1992, in force 24 March 1994, 1771 UNTS 164 and the Kyoto Protocol to the United Nations Framework Convention on Climate Change, Kyoto, 11 December 1997, in force 16 February 2005, 2303 UNTS 148. The latter is supplemented by a detailed set of rules agreed by the parties known as the Marrakesh Accords: see Report of the Conference of the Parties on its Seventh Session, held at Marrakesh from 29 October to 10 November 2001, FCCC/CP/2001/13/Add.2. The Kyoto Protocol expires at the end of 2012 and international negotiations are underway with the aim of agreeing on post-2012 arrangements.

[27] See generally, Sheila Jasanoff, *Designs on Nature: Science and Democracy in Europe and the United States* (Princeton University Press, 2005).

designing or reforming global governance systems.[28] Chapter 7 critically reviews the potential for so-called 'democratisation' of global risk regulation through deference to national risk decision-making or the institution of transparency and participatory mechanisms that permit international decision-makers to take account of non-scientific inputs, alongside science. As in Chapter 5, the focus is upon science-based processes of review under the WTO SPS Agreement that have been the subject of a significant number of reform proposals in the international legal literature.

To the extent that such proposals allow for a more comprehensive appraisal of uncertainty concerns and conflicting values in processes of risk evaluation, they represent a means of reintroducing socio-political dimensions of risk lost where there is an insistence on narrowly science-focused assessments. Nonetheless, a continual obstacle that must be confronted in any attempt to translate accountability processes to the global level is the lack of conventional modes of democratic representation and underdeveloped structures for public participation in international law. This may not necessarily be a reason to abandon efforts for greater democratisation of international risk regulation (an outcome which could leave in place equally flawed, narrowly science-focused processes). However, it highlights the difficult trade-offs involved in seeking a broader basis for global risk governance: enhanced legitimacy may only come at the expense of decreasing the technical credibility of an assessment, at least for some audiences.[29]

What role for science in international risk regulation?

Emerging as a crucial issue for global risk regulation and governance is not whether science *or* values should triumph, but rather how scientific *and* non-scientific inputs might be blended in risk assessment in different settings to ensure a broadly acceptable balance of credibility and legitimacy concerns. In approaching this task an important prerequisite is a realistic understanding of the capacities of science to support risk assessment, as well as of those of international legal and governance structures to accommodate non-scientific inputs in a fair

[28] See, e.g., Gráinne de Búrca, 'Developing Democracy Beyond the State', Colum. J. Transnat'l L., 46 (2008), 221.

[29] Ronald B. Mitchell, William C. Clark and David W. Cash, 'Information and Influence', in Ronald B. Mitchell *et al.* (eds.), *Global Environmental Assessments: Information and Influence* (Cambridge, MA: MIT Press, 2006), p. 309.

and transparent manner. Narrowly focused science-based approaches to risk regulation, which have emerged in dominant international governance institutions such as the WTO, threaten such efforts by perpetuating a myth that complex risk questions can be reduced to matters of science.

Given the highly contextualised nature of risk and the potential for significant scientific uncertainty in many areas, questions about the appropriate role for science in international risk regulation are unlikely to yield a unitary response. In addition, the limitations of the international legal system constrain its capacity to accommodate a plurality of risk perspectives in relevant decision-making processes. Consequently, rather than a single defined role for science in global risk regulation, there is a menu of options and strategies available by which international legal structures can seek to determine the best possible balance between science and non-scientific perspectives in different circumstances. Concrete examples of how these mechanisms might operate in the particular institutional setting of WTO dispute settlement under the SPS Agreement are surveyed in the concluding section of Chapter 7.

Reforms to institute a more appropriate role for science in international risk regulation do not promise to be straightforward. Indeed, for some time to come we may have to accept a compromise based upon relatively imperfect attempts to democratise global risk decision-making. However, work to improve the use of science in international law governing risk regulation is not something that can or should be avoided. It is becoming clear that if global governance systems – and perhaps even science itself – are to play a long-term role in responding to issues of health and environmental risk, they will need to find ways to preserve space for value debates and political contestation alongside the evaluation of technical evidence of potential harms.

2 Global risk governance and its legitimacy

Introduction

One of the most debated topics in contemporary international law is the extent to which it constrains the regulatory autonomy of national governments. The interest that this issue presently attracts attests to the substantial changes that have taken place in international legal structures founded on the sovereignty of independent nation states. In a relatively short period of time, many issues traditionally conceived as ones of exclusively domestic concern – such as the health and safety of national populations and environmental protection – have come to be viewed as matters of global import, requiring systems of international regulation. The emergence of global risks, like that of climate change, as well as the processes of economic globalisation, have provided the impetus for the development of international rules that cover an increasingly wider range of activities and penetrate more deeply into national regulatory regimes.

Often the new constraints emerging at the international level do not take the form of specific obligations agreed by states in intergovernmental negotiation processes. Instead they may be the product of decisions taken about the implementation of ongoing multilateral regimes or supranational arrangements, which interpret the nature of governments' commitments or specify requisite modes of decision-making with implications both for the participating nation states and those that they govern.[1] These developments raise the prospect that, in more and more cases, decisions traditionally taken in the domestic sphere will be subject to systems of global regulation and governance.

[1] Gráinne de Búrca, 'Developing Democracy Beyond the State', Colum. J. Transnat'l L., 46 (2008), 235.

Hence the decision-making processes of international bodies, and the inputs that they allow, have assumed growing importance in a variety of regulatory fields, including that of risk regulation.

This chapter tracks the evolution of global rules that exert increasingly more wide-ranging control over the policies and activities of nation states, and explains their particular impact in the field of decision-making concerned with health and environmental risks. Playing a part in the narrative of a transition from nation state regulatory sovereignty to a significant role for global governance are international rules in the environmental field, as well as those in other areas of international law, such as in the sphere of global trade. In addition, the emergence of trans-governmental arrangements of various kinds has generated pressures for regulatory convergence, premised on the adoption of the environmental and other standards of 'advanced' nations by 'weaker states'.[2]

As internationally determined requirements have come to assume a more central role in the daily life of people and private entities, rather than simply affecting the relationships between governments, questions are emerging over their legitimacy. These questions have been asked particularly in (Western) democratic societies which are historically and culturally accustomed to the notion that any form of government should be accountable through the electoral process to those governed. However, the dominant view in international law has been that the rationale of democratic legitimacy, drawn from domestic analogies, has little traction in an international legal system with no identifiable global democratic community that could supply 'input-oriented legitimacy'[3] for supranationally determined rules.[4]

Rather than relying upon democratic support, the legitimacy of global governance has been seen to depend upon the capacity of international law to provide effective solutions to common problems, utilising standards or decision-making criteria that command widespread acceptance

[2] Kal Raustiala, 'The Architecture of International Cooperation: Transgovernmental Networks and the Future of International Law', Virginia J. Int'l L., 43(1) (2002), 7.

[3] Fritz Scharpf, *Governing in Europe: Effective and Democratic?* (Oxford University Press, 1999), p. 6.

[4] E.g. Benedict Kingsbury, Nico Krisch and Richard Stewart, 'The Emergence of Global Administrative Law', *Law and Contemporary Problems*, 68 (2005), 15; Daniel Esty, 'Good Governance at the Supranational Scale: Globalizing Administrative Law', Yale L.J., 115 (2006), 1490; J. H. H. Weiler, 'The Geology of International Law – Governance, Democracy and Legitimacy', *Zeitschrift fur auslandisches offentliches Recht und Volkerrecht*, 64 (2004), 547.

and deference.[5] In emerging areas of international law concerned with the regulation of risk, expertise based on scientific and technical knowledge is typically viewed as a plausible basis for legitimating the growing authority exercised by relevant international rules.

Sovereignty, interdependence and globalisation

Sovereignty and international law

In classical accounts of international law, a foundational principle is that of the sovereignty of independent nation states.[6] Although the origin of the sovereignty doctrine is generally traced back to the Peace of Westphalia, the concept is one that has evolved over time, moulding to accommodate different meanings in different contexts. While sovereignty was originally directed to the internal aspect of state affairs, in terms of recognising the authority of sovereigns as supreme legislators over their people, in modern times the notion came to be equated more with the external aspect of each state's independence vis-à-vis other nation states.[7] This change of focus in respect of the international understanding of sovereignty was promoted by the technological progress of the industrialisation age that 'brought about a near congruence between state and society: each "nation" state was now, or at least could conceivably be, its own society, considered complete … unto itself.'[8] Sovereignty in this modern sense thus signified 'the right of a state freely to exercise its power under international law without the permission of any other state', together with 'the right of a state to exclude from its territory the exercise of power by any other state.'[9]

In practical terms, sovereignty is often said to adhere in a state's exclusive control of its territory, which encompasses its position as the ultimate political authority exercising power over its resident

[5] See particularly Esty, 'Good Governance at the Supranational Scale'.

[6] Alfred von Staden and Hans Vollaard, 'The Erosion of State Sovereignty: Towards a Post-Territorial World?', in Gerard Kreijen *et al.* (eds.), *State, Sovereignty, and International Governance* (Oxford University Press, 2002), p. 165. This view of sovereignty is not uncontested; see particularly, Stephen D. Krasner, *Sovereignty: Organized Hypocrisy* (Princeton University Press, 1999).

[7] Malcolm Shaw, *International Law*, 5th edn, (New York: Cambridge University Press, 2003), p. 21.

[8] Andreas Osiander, 'Sovereignty, International Relations, and the Westphalian Myth', *International Organization*, 55(2) (2001), 282.

[9] Eli Lauterpacht, 'Sovereignty – Myth or Reality?', *International Affairs*, 73(1) 140.

population, its monopoly on the legitimate use of force within its territory, and its capacity to regulate the movement of persons and the flow of economic transactions across its borders.[10] Such notions of exclusivity of territorial control and non-interference were incorporated into the Charter of the United Nations concluded in the aftermath of the Second World War.[11] Article 2(1) of the Charter declares 'the sovereign equality' of member states, whereas Article 2(4) exhorts members to 'refrain in their international relations from the threat or use of force against the territorial integrity or political independence of any state'. In addition, Article 2(7) of the Charter preserves the independence of member states vis-à-vis the United Nations organisation, stating that nothing in the Charter 'authorise[s] the United Nations to intervene in matters which are essentially within the domestic jurisdiction of any state'.

In this setting, the role of international law has been described as one of ensuring the 'coexistence' of independent states, which exercise full powers within the ambit of their own jurisdictions but must also accord the same faculty to others by refraining from encroaching on their ambits.[12] The minimal international laws required for this purpose - such as the Charter prohibition on the use of force or the commitment to non-intervention in the domestic affairs of other states - are entered into on the basis of self-interest and reciprocity, and concern exclusively relations between states (domestic affairs being under the sole regulation of each state).

Crucial to this vision of international law is the principle that any diminution of the sovereign rights of states in favour of international rules must be based upon the express agreement of the states themselves. Hence, the sovereignty doctrine is traditionally underpinned by the requirement that states may only be bound by international law with their own consent.[13] Except where they have expressly agreed otherwise by way of a treaty or through their practices that give rise

[10] John Jackson, 'Sovereignty-Modern: A New Approach to an Outdated Concept', American J. Int'l L., 97 (2003), 786. See also Krasner, *Sovereignty: Organized Hypocrisy*, pp. 3–4, who distinguishes four separate uses of the term 'sovereignty': international legal sovereignty, Westphalian sovereignty, domestic sovereignty and interdependence sovereignty (the ability of public authorities to regulate flows of information, goods, people, pollutants or capital across the borders of the state).

[11] Charter of the United Nations, 26 June 1945, in force 24 October 1945, 1 UNTS XVI.

[12] Georges Abi-Saab, 'Whither the International Community?', European J. Int'l L., 9(2) (1998), 250–4.

[13] Shaw, *International Law*, p. 9.

to norms of customary international law, states, according to the conventional view, are considered to have complete autonomy to act as they choose without legal limitation by any superior entity.[14] Even where a binding international law is in existence that obliges states to limit their freedom in some respect, domestic implementation and any effects on private actors are considered a matter solely for each state. International law in this conception thus remained principally the domain of executive governments and national diplomats, remote from, and with very little penetration into, the daily lives of the citizenry of states.

Interdependence and international cooperation

This traditional picture of international law has attracted fewer and fewer adherents in the last few decades. At some point after the First World War, international law came to be seen by many of its practitioners as a (relatively) unified legal *system*, displacing the former international sphere of 'completely decentralised, disparate national rationalities, world-views and value-systems each claiming total control over a population and independence from others'.[15] Such views have been promoted by the development, over the course of the twentieth century, of an increasingly dense body of international rules premised on the idea of the interdependence of states and demands for international cooperation to meet common problems.

For some prominent international lawyers, such as Elihu Lauterpacht, one consequence is that the conventional sovereignty principle of international law must now 'be seen largely as myth – except when it is used as a word to describe a state's title to territory'.[16] Even those who are more sceptical about the demise of the decentralised system of sovereignty in international law nonetheless maintain that substantial changes have occurred in the last few decades that may be traced to the greater interdependence of states. Robert Jennings, for example, remarks that:

[14] This proposition was most famously declared by the Permanent Court of International Justice in *SS Lotus (Turkey* v. *France)*, PCIJ Series A-No. 10 (1927), p. 18.

[15] Martti Koskenniemi, *Global Legal Pluralism: Multiple Regimes and Multiple Modes of Thought* (Erik Castrén Institute of International Law and Human Rights, University of Helsinki, 2005), pp. 8–9. Available at www.helsinki.fi/eci/Publications/MKPluralism-Harvard-05d%5B1%5D.pdf.

[16] Lauterpacht, 'Sovereignty – Myth or Reality?', 149.

Most, if not indeed all, sovereign governments nowadays have very seriously limited choices in the exercise of their supposedly sovereign competence, because their theoretically important areas for decisions are much restricted and hemmed in by treaties, by customary international law and by the consequences, and especially the economic consequences, of the sheer interdependence of all sovereign states of today.[17]

Unlike the previous assumption of independence, *interdependence* signifies the perception of common interests shared by a number of states or the international community as a whole. In the environmental field, initial indications of the interdependence of state interests came with the intensification of industrial processes in the nineteenth and early twentieth centuries. Along with technological progress and the expansion of interstate economic activities, industrialisation brought with it trans-boundary externalities manifested in adverse effects on the populations and environment of other countries. This led to intergovernmental disputes over the use of shared resources such as fisheries, as well as to claims for compensation for environmental damage caused by trans-boundary pollution.[18]

In more recent times pollution problems have taken on a truly global dimension with the discovery of depletion of the ozone layer and concern over the risk of climate change. In his dissenting judgment in the International Court's *Advisory Opinion on the Legality of the Threat or Use of Nuclear Weapons*, Judge Weeramantry remarked that such instances of 'mutual interdependence' are the result of '[a] world order in which every sovereign state depends on the same global environment'.[19] According to the then President of the Court, Judge Bedjaoui, in the same case, interdependence also has important consequences for international law:

The resolutely positivist, voluntarist approach of international law still current at the beginning of the century ... has been replaced by an objective conception of international law, a law more readily seeking to reflect a collective juridical conscience and respond to the social necessities of States organized as a community.[20]

[17] Robert Jennings, 'Sovereignty and International Law', in Gerard Kreijen *et al.* (eds.), *State, Sovereignty, and International Governance* (Oxford University Press, 2002), p. 31.

[18] Philippe Sands, *Principles of International Environmental Law*, 2nd edn, (Cambridge University Press, 2003), pp. 29–30.

[19] Advisory Opinion on the Legality of the Threat or Use of Nuclear Weapons, (ICJ Reports, 1996), p. 505.

[20] *Ibid.*, pp. 270–1.

While the nature of the common values that might animate a more communitarian vision of international law (and indeed the very existence of a cohesive international community) remain a matter of considerable debate, the 'factual element' of greater interdependence of, and contact between states and peoples, is generally accepted.[21] Recognition of 'the growing necessity of international cooperation' has provided the impetus for 'an enlargement of the material scope of operation of international law' in areas where common interests or issues of global concern can be identified.[22]

Environmental treaties are a paradigmatic example of cooperative international laws that are intended to deal with issues that cannot be adequately addressed by one state acting alone. For example, the problem of climate change is presented as one requiring the concerted efforts of governments (as well as individuals and businesses) worldwide if the international legal objective of stabilising greenhouse gas emissions at safe levels is to be achieved.[23] Invocations of the international community and collective interests are not only the province of environmental law, as evidenced by the now numerous General Assembly, and even Security Council, resolutions that employ concepts of interdependence and community as a justification for restraints on the sovereignty of individual states.

Although, arguably, the term 'international community' often conceals the interests of a few powerful states (just as interdependence overlooks the growing *dependence* of many poorer countries on richer ones),[24] in the contemporary context it seems that these notions exercise a powerful rhetorical and practical influence over the actions of nation states. As Bruno Simma and Andreas Paulus conclude:

> [T]he world of the famous 'Lotus principle', according to which states are only bound by their express consent, seems to be gradually giving way to a more

[21] Bruno Simma and Andreas Paulus, 'The "International Community": Facing the Challenge of Globalization', European J. Int'l L., 9(2) (1998), 269.

[22] Pierre-Marie Dupuy, 'The Danger of Fragmentation or Unification of the International Legal System and the International Court of Justice', N.Y.U. J. Int'l L. Politics, 31 (1999), 795.

[23] United Nations Framework Convention on Climate Change (UNFCCC), Rio de Janeiro, 9 May 1992, in force 24 March 1994, 1771 UNTS 1664, Article 2.

[24] See further Don Greig, '"International Community", "Interdependence" and All That ... Rhetorical Correctness?', in Gerard Kreijen *et al.* (eds.), *State, Sovereignty, and International Governance* (Oxford University Press, 2002), p. 521.

communitarian, more highly institutionalized international law, in which states 'channel' the pursuit of most of their individual interests through multilateral institutions.[25]

Globalisation and its consequences

Interdependence, as a justification for international rules dealing with the interests of the international community, primarily affects what is often referred to as the external dimension of state sovereignty. This makes it increasingly difficult – even for the most powerful states – to accomplish their primary goals without entering into complex intergovernmental arrangements.[26]

At the same time as the external dimension of sovereignty is diminishing, its internal dimension also appears to be eroding as national borders become more permeable due to the effects of globalisation. In this regard, one of the main contributing factors is technological advances in the fields of transport and communication that have greatly increased the potential for exchanges between national populations. Globalisation is also generally associated with closer economic integration between states, evidenced only too well by the wide-ranging impacts of the global financial crisis.

In the economic area the Bretton Woods institutions established following the Second World War – the World Bank, the International Monetary Fund (IMF), and the General Agreement on Tariffs and Trade (GATT) (now the WTO) – have played an important part in fostering greater integration. For instance, pursuant to the GATT/WTO, the majority of states (including all the major economic powers) are committed to a rules-based multilateral framework for international trade, aimed at reducing or eliminating barriers to trade that previously restricted substantially the flow of foreign goods and services across national borders.[27] The early successes of the Bretton Woods institutions in facilitating greater economic integration led to their increasing focus on the internal regulatory structures of states, including issues of governance, transparency in public

[25] Simma and Paulus, 'The "International Community"', 276–7.

[26] See generally, Abram Chayes and Antonia Chayes, *The New Sovereignty: Compliance with International Regulatory Agreements* (Cambridge, MA: Harvard University Press, 1998).

[27] The WTO at present has 153 Members, including the major economic powers of the United States, the European Communities and Japan.

institutions and the external trade effects of public policy regulatory measures.[28]

The topic of globalisation has fostered a vibrant literature and social debate in which issues of the nature of the process itself and the desirability, or otherwise, of its consequences remain open.[29] Nonetheless, there is little disagreement that the technological and economic developments that underlie it are a matter of fact, nor that one of its primary effects is to increase the scale and scope of cross-border interactions.[30] Thus, an obscure European Union (EU) regulation on the labelling of sardines now affects Peruvian fishermen as much as European ones.[31] Likewise, the phasing-out of a potentially harmful fuel additive in California is a matter of interest to a Canadian producer of chemical feedstock, which can avail itself of treaty-based investor protections to seek compensation directly from the US government.[32]

As national borders become increasingly less effective as insulation from external influences, it is more difficult for states to maintain that they either do, or should, retain the monopoly to exercise regulatory power over a range of social activities within their territories. Rather, under the banner of globalisation many national policies – dealing with trade, foreign investment, protection of health and the prevention of environmental pollution – become the legitimate concern of other states (and their publics). These new constituents have begun

[28] Ngaire Woods and Amrita Narlikar, 'Governance and the Limits of Accountability: The WTO, the IMF, and the World Bank', *International Social Science Journal*, 53(170) (2001), 569.

[29] For discussion of this topic in the context of environmental issues see Arthur P. J. Mol, *Globalization and Environmental Reform: The Ecological Modernization of the Global Economy* (Cambridge, MA: MIT Press, 2001), pp. 71–94; Jerry Mander, 'Intrinsic Negative Effects of Economic Globalization on the Environment', in James Gustave Speth (ed.), *Worlds Apart: Globalization and the Environment* (Washington DC: Island Press, 2003), p. 109; Alan Scott, 'Globalization: Social Process or Political Rhetoric?', in Alan Scott (ed.), *The Limits of Globalization: Cases and Arguments* (London: Routledge, 1997), p. 1; Vandana Shiva, 'The Myths of Globalization Exposed: Advancing Toward Living Democracy', in James Gustave Speth (ed.), *Worlds Apart: Globalization and the Environment* (Washington DC: Island Press, 2003), p. 141.

[30] Oren Perez, Ecological Sensitivity and Global Legal Pluralism: Rethinking the Trade and Environment Conflict (Portland: Hart Publishing, 2004), p. 119.

[31] Leading to a WTO challenge by Peru in the case of *European Communities – Trade Description of Sardines*, Report of the WTO Appellate Body, WT/DS231/AB/R, 26 September 2002.

[32] *Methanex Corporation* v. *United States of America*, NAFTA Chapter 11 Arbitral Tribunal (2005) 44 ILM 1345. This dispute is discussed further in Chapter 7.

to demand a say in the way that countries go about regulating the behaviour of their citizens and businesses that could have implications beyond national borders. The response has generally been the strengthening of mechanisms and arrangements for international cooperation and regulation that tend to 'pull away' power or influence from national governments into the global arena.[33]

International law and global governance

Multilateral 'living' regimes

As globalised interdependency is seen to affect more and more areas of national regulatory activity, this has provided the impetus for an expansion of international law and regulation that shifts the locus of decision-making beyond state borders. This trend is most noticeable in fields such as security, health and environmental protection, trade in products and services, investment, financial regulation and the provision of developmental and financial assistance to developing countries.[34] Reflecting the growing concern with issues defined as global in nature, the number of international treaties grew substantially over the latter half of the twentieth century, more than tripling between 1970 and 1997.[35]

Many such treaties addressing collective interests are multilateral in nature and characteristically set up a 'living' regime that is intended to 'manage' a problem area over time.[36] For example, the climate change regime, established by the United Nations Framework Convention on Climate Change and its supplementary Kyoto Protocol, has as its ultimate, long-term objective the 'stabilization of greenhouse gas concentrations in the atmosphere at a level that would prevent dangerous anthropogenic interference with the climate system'.[37] In achieving this objective, parties to the Kyoto Protocol are obliged to meet their 'quantified emission and reduction limitation commitments' over the

[33] Anthony Giddens, *Runaway World: How Globalisation is Reshaping Our Lives* (London: Profile Books, 1999), p. 13.
[34] Kingsbury *et al.*, 'The Emergence of Global Administrative Law', 16.
[35] José Alvarez, *International Organizations as Law-makers* (Oxford University Press, 2005), p. 273.
[36] José Alvarez, 'The New Treaty Makers', Boston College Int'l & Comp. L. Rev., 25 (2002), 221–2.
[37] UNFCCC, Article 2.

first commitment period of 2008–12,[38] with additional commitment periods contemplated in the future.[39]

The participation of states in such regimes is increasingly difficult to rationalise on the traditional basis of state consent. Particularly in the case of economic regimes, such as the WTO, the IMF and the World Bank, that drive the processes of globalisation, states may feel compelled to accept these regimes or the conditions they impose because most of the rest of the world has done so or (in respect of many developing countries) they have little real choice.[40] Moreover, once parties to these regimes, states may be unable to exercise effective control over implementation of a regime. This may be the consequence of an evolving interpretation of obligations under the regime, developing practice under an agreement that shapes its scope of application, or voting rules and procedures that allow for a majority of members to impose a collective decision on dissenting parties.[41]

For example, an international institution such as the United Nations Security Council – dominated by the interests of its five permanent members – can make decisions binding on all United Nations members regardless of their specific consent to those decisions.[42] Other international organisations that exercise some degree of rule-making authority include the International Civil Aviation Organisation (which adopts international standards and recommended practices on a range of operational and safety requirements for aircraft), the International Maritime Organisation (which proposes conventions and standards concerning maritime safety, navigation and marine pollution that apply to more than 98 per cent of world merchant shipping tonnage) and the International Telecommunications Union (which establishes technical specifications for global telecommunications based on studies prepared by industry and expert groups).[43]

[38] Kyoto Protocol to the United Nations Framework Convention on Climate Change, 11 December 1997, in force 16 February 2005, 2303 UNTS 148, Article 3.

[39] Post-2012 arrangements are the subject of international negotiations under the auspices of the UNFCCC. For further details see www.unfccc.int

[40] Jackson, 'Sovereignty-Modern', 796. See also B. S. Chimni, 'International Institutions Today: An Imperial Global State in the Making', European J. Int'l L., 15(1) (2004), 1.

[41] Jackson, 'Sovereignty-Modern', 797.

[42] UN Charter, Articles 25, 48. A pertinent example is UN Security Council Resolution 1267 (1999) which establishes a mechanism for the blacklisting of persons and organisations suspected of funding terrorism.

[43] For further examples see Alvarez, *International Organizations as Law-makers*, pp. 109–22.

Where an international regime is established by a multilateral treaty, the treaty itself may allow for ongoing administration of the regime in accordance with the interests of most (but not all) of the participating states. A number of multilateral treaties thus permit decisions on amendments to technical annexes or lists of regulated activities or areas by way of a special majority vote.

A well-known example in the environmental field is the Montreal Protocol, which regulates state parties' production and consumption of ozone-depleting substances.[44] Article 2(9) of the Protocol allows for adjustments to the permitted levels of production and consumption of ozone-depleting substances that can be made binding on all the parties, as a last resort, by a qualified majority vote regardless of whether those parties expressly consented to the change. Successive adjustments of states' commitments pursuant to this provision are thought to be 'one of the decisive factors of the successful development of the Montreal Protocol' in terms of securing deep reductions in the manufacture and use of a number of ozone-depleting substances such as chlorofluorocarbons (CFCs).[45]

Pointing to the Montreal Protocol and other multilateral treaty regimes, as well as international organisations, José Alvarez contends that international law is witnessing the emergence of new 'law-makers'. He argues that these actors operate 'outside the constraints suggested by the traditional doctrine of sources' and 'need not find explicit delegations of power, are not confined to differentiated or closely circumscribed spheres of legal action, and are not limited to making "internal" institutional housekeeping rules'.[46]

The role of the WTO

While multilateral regimes are an important feature of areas of international law concerned with issues of collective interest, currently perhaps the most significant sites of international regulatory activity are those located within economic globalisation institutions. A focus

[44] Protocol on Substances that Deplete the Ozone Layer, opened for signature 16 September 1987, 1522 UNTS 3 (entered into force 1 January 1989).

[45] Johan Lammers, 'The Mechanism of Decision-making Under the Vienna Convention and the Montreal Protocol for the Protection of the Ozone Layer', in Gerard Kreijen *et al.* (eds.), *State, Sovereignty, and International Governance* (Oxford University Press, 2002) p. 413.

[46] Alvarez, *International Organizations as Law-makers*, p. 645.

of interest (and critique) in this regard are the institutional arrangements and agreements of the WTO that came into being following the successful conclusion of the Uruguay Round of trade negotiations. Unlike its predecessor, the GATT, the WTO is a fully-fledged international institution with broad coverage of trade-related issues and a binding dispute settlement system. Its underlying legal agreements are a package deal accepted by all members, reflecting the WTO's creation:

> on an all-or-nothing basis whereby countries had to commit to full membership in a 'single undertaking', binding themselves to a rule-based system, not just for the short-term periods of loans or negotiations.[47]

The commitments countries have made in their acceptance of WTO rules are often far from clear, giving rise to disputes that can be taken before three member dispute settlement panels and ultimately (on questions of law) the standing WTO Appellate Body. This system of dispute settlement is 'extraordinarily powerful' and 'basically unique in international law history'[48] since it combines virtual automaticity of members' acceptance of decision-makers' rulings with the availability of sanctions against members that remain in non-compliance.[49]

In this respect the WTO enjoys a substantial advantage over many other multilateral treaty regimes, which can often be plagued by problems concerning the implementation of obligations by states.[50] Moreover, for the international decision-makers involved in WTO dispute settlement, the autonomous operation of the process places them in the unique position of effective insulation from the direct political

[47] Woods and Narlikar, 'Governance and the Limits of Accountability', 570.

[48] Jackson, 'Sovereignty-Modern', 799. See also John H. Jackson, 'The WTO Dispute Settlement System after Ten Years: the First Decade's Promises and Challenges', in Yasuhei Taniguchi, Alan Yanovich and Jan Bohanes (eds.), *The WTO in the Twenty-First Century: Dispute Settlement, Negotiations, and Regionalism in Asia* (Cambridge University Press, 2007), pp. 31–2.

[49] The automaticity of the process is a function of the 'negative consensus' rule whereby a dispute settlement decision is deemed to be adopted unless *all* WTO Members (including the successful party in a dispute) vote against adoption. See Mary Footer, 'The Role of Consensus in GATT/WTO Decision-making', Northwestern J. Int'l L. & Business, 17 (1996), 363.

[50] In the environmental context see Michael Faure and Jürgen Lefevere, 'Compliance with Global Environmental Policy', in Regina Axelrod, David Downie and Stacey Van Deveer (eds.), *The Global Environment: Institutions, Law, and Policy* (Washington DC: CQ Press, 2011), p. 172.

control of the state membership of the organisation.[51] This gives decision-makers considerable freedom to interpret and evolve the rules in ways that, in some cases, WTO members may not have intended.[52]

Another, often noted, feature of the WTO Agreements is the extent to which they have shifted the attention of the trade regime from its traditional focus on border barriers (such as tariffs and quantitative restrictions) towards 'new categories of prohibited domestic regulation under the broad rubric of 'behind-the-border' barriers to trade'.[53] Such behind-the-border restrictions (or non-tariff barriers to trade) consist of trade measures put in place for a variety of non-economic, public policy purposes, such as health, quarantine, worker safety, consumer protection and environmental protection. The category of non-tariff trade barriers covers a substantial portion of national regulatory activity since most regulatory measures burden commercial activity, much of which is now international.[54] An underlying goal of WTO initiatives in this area is to reduce the trade-distorting effects of divergent national regulations by encouraging greater harmonisation.

For instance, in the case of both the SPS Agreement and the Technical Barriers to Trade Agreement (TBT Agreement), WTO members are urged to use international standards as a basis for their national regulatory measures.[55] The international standards referenced by the SPS and TBT Agreements are elaborated, not in inter-governmental negotiation processes, but instead under the auspices of hybrid public–private expert bodies such as the Codex Alimentarius Commission (responsible for adopting recommendations on food safety standards),[56] the

[51] Ernst-Ulrich Petersmann, 'WTO Dispute Settlement Practice 1995–2005: Lessons from the Past and Future Challenges', in Yasuhei Taniguchi, Alan Yanovich and Jan Bohanes (eds.), *The WTO in the Twenty-First Century: Dispute Settlement, Negotiations, and Regionalism in Asia* (Cambridge University Press, 2007), p. 38.

[52] *Ibid.* See also Philippe Sands, 'Turtles and Torturers: The Transformation of International Law', N.Y.U. J. Int'l L. & Pol., 33 (2001), 555–6 for a discussion of particular examples in WTO case law.

[53] Andrew T. F. Lang, 'Some Sociological Perspectives on International Institutions and the Trading System', in Colin B. Picker, Isabella D. Burn and Douglas W. Arner (eds.), *International Economic Law: The State and Future of the Discipline* (Portland: Hart Publishing, 2008), p. 82.

[54] David Driesen, 'What is Free Trade?: The Real Issue Lurking Behind the Trade and Environment Debate', *Virginia Journal of International Law*, 41 (2001), 283.

[55] Agreement on the Application of Sanitary and Phytosanitary Measures, Marrakesh, 15 April 1994, in force 1 January 1995, 1867 UNTS 493, Article 3.1; Agreement on Technical Barriers to Trade, Marrakesh, 15 April 1994, in force 1 January 1995, 1868 UNTS 120, Article 2.4.

[56] The standard-setting processes of Codex are discussed further in Chapter 6.

International Office for Epizootics (dealing with animal health matters), the International Plant Protection Convention (concerned with issues of plant health and disease), the International Electrochemical Commission and the International Organization for Standardisation. Although the standards promulgated by these organisations are not directly binding on WTO members, the WTO lends its (considerable) regulatory force to them by providing that conformity with the standards secures compliance with SPS and TBT commitments.[57]

In light of their recognition by relevant WTO agreements, the once rather arcane processes of international standard-setting bodies now also have vastly increased salience as potential sources of regulation. The standards they issue are 'measured against their market-opening potential and their costs and benefits for trade', which heightens the potential for division and places new stresses upon the previously favoured consensus-adoption approach.[58]

Nevertheless, even where the standards adopted by these organisations are not supported by all participating states, they are still effective for WTO purposes. In the case of *Sardines*, the WTO Appellate Body ruled that 'consensus is not required for standards adopted by the international standardizing community' in order to be relevant for the purposes of the TBT Agreement.[59] Although they may not have agreed to, or even participated in, the formulation of international standards, for many members of the WTO harmonising their regulatory measures in accordance with international standards may be the only feasible option to ensure SPS/TBT compatibility.[60] This will be particularly the case where members otherwise have insufficient technical,

[57] SPS Agreement, Article 3.2; TBT Agreement, Article 2.5.

[58] Doaa Motaal, 'The "Multilateral Scientific Consensus" and the World Trade Organization', J. World Trade, 38(5) (2004), 866.

[59] *European Communities – Trade Description of Sardines*, Report of the WTO Appellate Body, WT/DS231/AB/R, 26 September 2002, [222]. For discussion of the significance of this decision see Robert Howse, 'A New Device for Creating International Legal Normativity: The WTO Technical Barriers to Trade Agreement and "International Standards"', in Christian Joerges and Ernst-Ulrich Petersmann (eds.), *Constitutionalism, Multilevel Trade Governance and Social Regulation* (Portland: Hart, 2006), p. 383.

[60] On the particular problems facing developing countries in implementing WTO requirements see Mary Footer, 'The WTO, Developing Countries and Technical Assistance for Trade Law Reform', in Julio Faundez, Mary Footer and Joseph Norton (eds.), *Governance, Development and Globalization* (London: Blackstone Press, 2000), p. 353.

scientific or financial capacity to produce the required justification for a divergent measure.[61]

Transnational arrangements and regulatory convergence

The pressure for regulatory convergence increasingly comes not just from international organisations such as the WTO, IMF and World Bank, but also through a range of horizontal state-to-state mechanisms. These include transnational networks of governmental regulators, bilateral investment treaties (BITs), free trade agreements (FTAs) and mutual recognition arrangements (MRAs).

Trans-governmental networks

The phenomenon of trans-governmental networks has been a particular focus of research for international relations and international law scholars, such as Anne-Marie Slaughter. In her book *A New World Order*, Slaughter extensively discusses such networks, which she describes as collectives of national regulators (and sometimes legislators and judges).[62] In some cases the creation of trans-governmental networks has been facilitated by international institutions, such as the WTO. The Committee established under the SPS Agreement as a forum for information exchange and elaboration of SPS norms is one such example (see further Chapter 5).[63] According to Slaughter, trans-governmental networks employ 'soft power' (the power of persuasion and information) to promote convergence and improved enforcement of regulatory systems and standards among their participants in a manner that is more effective than traditional multilateral institutional processes.[64]

Examining case studies of trans-governmental regulatory networks in the fields of securities, competition and environmental regulation, Kal Raustiala concludes that such networks do indeed seem to promote 'regulatory export' from the major powers (invariably the US or EU) to 'weaker states'.[65] Raustiala describes the incentives that exist for 'weak

[61] Shyam Gujadhur, 'Influencing Market Standards: A Voice for Developing Countries', *International Trade Forum*, 2 (2003), 30.

[62] See Anne-Marie Slaughter, *A New World Order* (Princeton University Press, 2004).

[63] Joanne Scott, *The WTO Agreement on Sanitary and Phytosanitary Measures: A Commentary* (Oxford University Press, 2007), Chapter 2.

[64] Anne-Marie Slaughter, 'Disaggregated Sovereignty: Towards the Public Accountability of Global Government Networks', *Government and Opposition*, 39 (2004), 162.

[65] Raustiala, 'The Architecture of International Cooperation', 7.

jurisdictions' to import the regulatory approaches of the 'advanced' industrial democracies as follows:

> In a complex, uncertain economic environment, the strategy of adopting successful foreign models can markedly reduce regulatory costs. Importing jurisdictions do not bear the (often considerable) expense of creating the regulatory institutions they adopt. While these institutions 'may not match domestic conditions precisely ... [they] are ready-made, pre-tested, and provide international compatibility.' Foreign regulatory rules and systems also may come 'pre-interpreted' – with a body of case law and other decisions that have elaborated and improved the rules over time. Finally, technical assistance programs further ease the transition and enable regulators to learn from experienced practitioners.[66]

Trans-governmental networks can thus become 'sites of normative consensus-building and persuasion' that socialise members to accept certain norms of appropriate behaviour and regulatory concepts.[67]

Investment and free trade treaties

Also operating in a 'below-the-radar' manner to promote regulatory convergence are a range of bilateral and regional economic agreements concluded between states. Examples include proliferating BITs (under which countries commit to ensure fair treatment of foreign investors and provide protections against the expropriation of foreign investments) and FTAs (which provide a means for countries to agree to more far-reaching trade liberalisation commitments than exist under the WTO).

The ability for investors to bring claims for damages against a state, without the need to resort to customary requirements for the exhaustion of domestic remedies, is a unique feature of the many BITs and other similar investment treaties concluded particularly since the 1990s.[68] Gus Van Harten and Martin Loughlin argue that, as a result, such treaties 'subject the regulatory conduct of states to control through compulsory international adjudication to an unusual extent'.[69] For example, elaboration by international arbitral tribunals of treaty concepts of 'expropriation' of an investment – including whether they

[66] *Ibid.*, 59 (footnotes omitted).

[67] Lang, 'Some Sociological Perspectives on International Institutions and the Trading System', pp. 78–79.

[68] Gus Van Harten and Martin Loughlin, 'Investment Treaty Arbitration as a Species of Global Administrative Law', European J. Int'l Law, 17(1) (2006), 122.

[69] *Ibid.*

extend to losses in the value of an investment brought about by law and regulatory changes in a host country – has been seen as a substantial constraint on regulatory freedom in treaty parties hosting foreign investment.[70] In this way BITs may act as a force for regulatory convergence through facilitating 'internationally generated adjudicative norms and mechanisms' that come to exert 'a strong disciplinary influence over domestic administrative programmes'.[71]

In a similar fashion, FTAs may operate as a means for the harmonisation of laws and regulatory practices between states. FTAs (many of which incorporate chapters on investment) have been enthusiastically pursued by the USA.[72] Pursuant to the trade promotion authority granted by Congress for the negotiation of such agreements, certain standard objectives must be pursued, with the strongest agreement achieved to that time generally serving as a template for subsequent FTAs.[73] There is thus a large degree of similarity between the FTAs concluded by the USA and its various free trade partners, which cover a wide variety of regulatory areas with potential trade consequences, including SPS requirements and the enforcement of environmental laws.[74] Obligations undertaken pursuant to FTAs may be given teeth by binding dispute settlement arrangements, some of which (like the provisions of the North American Free Trade Agreement) also allow corporate investors to take direct action against governments for compensation where investment protection commitments are not met.[75]

[70] Michael J. Trebilcock and Robert Howse, *The Regulation of International Trade*, 3rd edn, (London: Routledge, 2005), pp. 463–4.

[71] Van Harten and Loughlin, 'Investment Treaty Arbitration as a Species of Global Administrative Law', 122.

[72] The United States has concluded FTAs with seventeen countries, with a further three awaiting Congressional approval, and has ongoing negotiations for such agreements with a dozen or so countries. European organisations, such as the European Union and the European Free Trade Association, have also concluded a large number of FTAs. See further, www.bilaterals.org.

[73] Daniel Esty, 'Economic Integration and Environmental Protection', in Regina Axelrod, David Downie and Stacy Van Deveer (eds.), *The Global Environment: Institutions, Law and Policy* (Washington DC: CQ Press, 2011), p. 164 (describing the binding negotiating objectives related to the environment agreed to by the Bush administration as part of the trade promotion authority granted in 2002).

[74] See, e.g., the Australia–United States Free Trade Agreement, 18 May 2004, in force 1 January 2005, [2005] ATS 1.

[75] North American Free Trade Agreement, 17 December 1992, (1993) 32 ILM 289, chapter 11.

While at a 'microscopic' level both FTAs and BITs are bilateral (or regional), contractual arrangements, 'telescopically, taken in aggregate they define a multilateral regime'.[76] As Joseph Weiler observes, where such agreements are offered by powerful 'vendor' states, such as the USA, they are essentially 'the international equivalent of domestic Standard Form contracts'.[77] In negotiations for such agreements, most 'buyer' states (often developing countries) are left with little choice but to sign on the dotted line.

Mutual recognition arrangements

Compared with BITs and FTAs, arrangements for mutual recognition of national regulatory standards are often viewed as a form of horizontal governance that allows for a more equal partnership between participants. Indeed, the use of MRAs has been endorsed by international economic institutions such as the WTO. Under the provisions of the SPS and TBT Agreements, WTO Members are encouraged to establish MRAs in order to recognise 'equivalent' SPS measures or 'conformity assessment procedures' for technical requirements maintained by their trading partners.[78] To their supporters, MRAs accommodate, rather than suppress, regulatory diversity by allowing states to agree on the circumstances and assessment procedures for treating each other's standards as sufficient for national regulatory purposes. Thus, Kalypso Nicolaidis and Gregory Shaffer argue that mutual recognition represents:

> a search for a more effective division of labor, not between a global center and the periphery (or a hegemonic state and peripheral states), but between regulators and lawmakers across countries through relatively more optimal combinations of home- and host-country control.[79]

Nevertheless, although MRAs require the express agreement of each participating state to treat the level of risk ensured by a particular regulatory standard or production process as acceptable, as a practical matter the ambit of 'proactive political choice' available to less economically powerful states in negotiations may be very limited.[80]

[76] Weiler, 'The Geology of International Law', 554.

[77] *Ibid.*, 554–5.

[78] SPS Agreement, Articles 4.1 and 4.2; TBT Agreement, Articles 6.1, 7 and 8.

[79] Kalypso Nicolaidis and Gregory Shaffer, 'Transnational Mutual Recognition Regimes: Governance without Global Government', Law & Contemp. Probs., 68 (2005), 268.

[80] *Ibid.*

Nicolaidis and Shaffer concede that where the EU is involved in MRA negotiations, it 'wields considerable market leverage in determining standards and regulatory structures required to implement mutual recognition policies ... because other countries' constituents desire access to the valuable and expanding E.U. market.'[81] This concern seems to be borne out in the SPS context, where a number of developing countries have complained that developed countries employ mutual recognition to require 'sameness' rather than 'equivalence' of SPS measures.[82]

Global governance

The collection of regulatory developments emerging at the international level, discussed in the previous sections, has led international lawyers and international relations scholars to 'a range of significantly different diagnoses'.[83] There are different views, for instance, as to whether the intensification of legal activity that has followed in the wake of growing interdependencies and processes of globalisation is leading to the disintegration of the nation state and the emergence of a 'new sovereignty',[84] reinforcing the importance of states as 'exclusive territorial communities',[85] or resulting in the disaggregation of the state into its individual functional elements.[86] Yet, the perception remains that the international legal order has changed (or is rapidly changing) to embrace new forms.[87] Such new forms challenge orthodox understandings and self-understandings of the field as simply a 'legal matrix for coexistence and community among and of States ensuring order and justice'.[88]

[81] *Ibid*, 311.

[82] Kevin Kennedy, 'Resolving International Sanitary and Phytosanitary Disputes in the WTO: Lessons and Future Directions', Food Drug L.J., 55 (2000), 88.

[83] Philip Alston, 'The Myopia of the Handmaidens: International Lawyers and Globalization', European J. Int'l Law, 3 (1997), 447.

[84] See Chayes and Chayes, *The New Sovereignty*.

[85] Michael Reisman, 'Designing and Managing the Future of the State', European J. Int'l Law, 3 (1997), 413 emphasising that this is occurring despite (or indeed perhaps because of) an intensification of transnational regulatory activity. See also Oscar Schachter, 'The Decline of the Nation-State and its Implications for International Law', Colum. J. Transnat'l L., 36 (1998), 23, who argues that reduced state autonomy does not portend the demise of the nation state.

[86] Slaughter, 'Disaggregated Sovereignty'.

[87] Benedict Kingsbury, 'The Administrative Law Frontier in Global Governance', Am. Society Int'l L. Proc., 99 (2005), 147–8.

[88] Weiler, 'The Geology of International Law', 547.

This sense of change is often given expression through the use of the term 'global governance'. In a domestic context, 'governance' signifies the disjuncture of functions of regulation from the formal institutions of national government, and thus 'lies in the conceptual gray zone between electoral politics and administrative rule-making'.[89] Likewise, where the terminology of global governance is applied, it is used to refer to rules, regulations and practices, extending beyond the law of states or of state-controlled institutions, which are directed to the management of interdependence.[90] The diverse sources of global governance include non-consensual rules generated by bodies under multilateral treaties, regulations and standards promulgated by international institutions or their dispute settlement organs, or the activities of groupings of bureaucrats or non-governmental actors charged with functions of norm elaboration and dissemination.[91]

Characteristic of global governance, as opposed to orthodox ideas of international law, is its behind-the-border focus, which entails significant direct and indirect effects on private individuals and businesses within states.[92] It is these actors, ultimately, 'who have to alter their behaviour in order say to reduce CO_2 or CFC emissions'.[93] As a consequence, the daily lives of citizenry are conditioned to a far greater extent by the rule-making activities of international institutions.[94] In turn, the interstitial existence of global governance between the realms of national governments and their governed populations tends to dissolve the classic dichotomy between the domestic and international spheres, and between the roles of public and private actors.[95]

One reason for the deep penetration of global governance into national systems is that such rules typically have an administrative

[89] Marybeth Martello and Sheila Jasanoff, 'Introduction: Globalization and Environmental Governance', in Sheila Jasanoff and Marybeth Martello (eds.), *Earthly Politics: Local and Global in Environmental Governance* (Cambridge, MA: MIT Press, 2004), p. 2.

[90] Esty, 'Good Governance at the Supranational Scale', 1500.

[91] *Ibid.*, 1497–8 for a comprehensive list of potential sources of global governance.

[92] Weiler, 'The Geology of International Law', 550.

[93] Michael Zürn, 'Global Governance and Legitimacy Problems', *Government and Opposition*, 39(2) (2004), 268.

[94] John Braithwaite and Peter Drahos, *Global Business Regulation* (Cambridge University Press, 2000), p. 488.

[95] Nico Krisch and Benedict Kingsbury, 'Global Governance and Global Administrative Law in the International Legal Order', European J. Int'l Law, 17(1) (2006), 1, 3; Serge Sur, 'The State between Fragmentation and Globalization', European J. Int'l Law, 3 (1997), 422.

or regulatory character, dealing with issues once within the exclusive domain of states and their domestic administrations.[96] These rules, moreover, often impose positive obligations that go not just to the results to be obtained, but also to the specific processes to be employed in seeking to achieve those results.[97] This has been accompanied by the creation of more detailed and effective mechanisms under multilateral treaties and within international institutions that serve a range of bureaucratic functions, such as auditing state performance with treaty commitments,[98] monitoring compliance[99] and conducting quasi-judicial review of regulatory action taken by states.[100]

Given the administrative orientation of much of what is now labelled 'global governance', some have argued that we are in fact witnessing the emergence of a new body of global administrative law.[101] For instance, researchers involved with the innovative Global Administrative Law Project at New York University Law School point to amassing evidence of a multifaceted 'global administrative space'[102] dominated by the activities of various 'global administrative bodies'. These include:

> formal intergovernmental regulatory bodies, informal intergovernmental regulatory networks and coordination arrangements, national regulatory bodies operating with reference to an international intergovernmental regime, hybrid public-private regulatory bodies, and some private regulatory bodies exercising transnational governance functions of particular public significance.[103]

Adoption of a framework of administrative law has led authors in this vein to look for ways in which global administrative action might be made more transparent, participatory and accountable by extending

96 Weiler, 'The Geology of International Law', 550.

97 *Ibid.*, 559.

98 E.g., the expert review teams established under the Kyoto Protocol, Article 8.

99 E.g., the non-compliance mechanism of the Kyoto Protocol. See Conference of the Parties, *Procedures and Mechanisms Relating to Compliance under the Kyoto Protocol*, Decision 24/CP.7, 7th sess, 8th plen mtg, s XV, UN Doc FCCC/CP/2001/13/Add.3 (21 January 2002), available at http://unfccc.int/resource/docs/cop7/13a03.pdf.

100 Krisch and Kingsbury, 'Global Governance and Global Administrative Law in the International Legal Order', 3, note that WTO dispute settlement can be viewed in this light.

101 See, particularly, Kingsbury *et al.*, 'The Emergence of Global Administrative Law'.

102 Krisch and Kingsbury, 'Global Governance and Global Administrative Law in the International Legal Order', 1.

103 Kingsbury *et al.*, 'The Emergence of Global Administrative Law', 17.

principles of domestic administrative law and politics to the global level.[104]

While recognition of the growing importance and authority of global governance need not entail acceptance of all the premises underpinning notions of global administrative law,[105] it does suggest the need for a rethinking of conventional understandings and functions of international legal arrangements. In particular, as international law evolves towards a system of governance – mimicking national regulation in its scope and effects – a demand arises for new kinds of global structures that are equipped to deal with the complexities, uncertainties and constantly evolving knowledge basis characteristic of contemporary regulatory problems.[106]

Like their counterparts in advanced regulatory states, these structures will need to be flexible and adaptable to allow for ongoing implementation efforts associated with the interpretation of regulatory requirements, and the enforcement and verification of compliance. There is also an increasing call for global governance structures to be transparent and accessible to a wider range of actors other than states, not only because such structures often need to draw on information provided by private entities, but also because of the potentially far-reaching effects of international regulation for citizens and businesses.[107]

Global governance and risk regulation

The trend in international law towards global governance, and its wide-ranging effects behind national borders, is particularly noticeable in regulatory fields concerned with issues of risk to human health or the environment. Risk, as is further discussed in the next chapter, has become a central concern of advanced regulatory states over the last few decades. As global governance takes on more and more of the tasks of national governments, it is unsurprising to find that risk regulation is an important function of many global governance structures

[104] This exercise is not unproblematic as leading figures in the global administrative law movement acknowledge: see Nico Krisch, 'The Pluralism of Global Administrative Law', European J. Int'l Law, 17(1) (2006), 247.

[105] See Krisch and Kingsbury, 'Global Governance and Global Administrative Law in the International Legal Order', for an elucidation of these.

[106] Zürn, 'Global Governance and Legitimacy Problems', 269.

[107] This latter point is explored further below in the discussion of democratic legitimacy, international law and global governance.

in the health and environmental field. Indeed, the content of the new 'regulatory layer' emerging in international law might be regarded as being particularly addressed to 'issues associated with the risk society in which we live'.[108]

Risks to health and the environment are arguably well suited to international regulation and governance in an interdependent, globalising world.[109] These processes contribute both to the production of new risks (from novel diseases such as SARS (severe acute respiratory syndrome) to environmental problems such as ozone depletion) and the ability to disseminate information about them.[110] In addition, many risk-producing activities tend to be undertaken by diffuse private, as well as public, entities, making it more difficult for one state to control the effects of 'misbehaviour' on the health and environmental well-being of the citizens of other states.[111] In the case of health risks, like those posed by mad cow disease or H1N1 influenza, global dissemination of a disease is promoted by the permeability of national borders to trade, travel and migrations of species that may thwart national efforts to prevent transmission. In the environmental sphere, risks from greenhouse gases and atmospheric and marine pollution are made more diffuse by processes of global mixing that spread the consequences of an activity in one location to many other areas around the world.

Even in the case of risks which are relatively contained in the sense of being associated with a particular product (for example, building materials including asbestos fibres), a country's capacity to manage the risk through national means will be reduced in an environment of economic integration. Products giving rise to risk concerns will often have been produced overseas, beyond the regulatory authority of the nation to whose consumers the product is sold.

National governments also face difficulties where the risks of concern to their populations are not simply those associated with the riskiness of a particular product, but rather those deriving from the process by which it was produced. For example, concerns that manufacturing processes in one country are unsustainable in terms of their resource

[108] Weiler, 'The Geology of International Law', 550.

[109] Carlo Jaeger *et al.*, *Risk, Uncertainty, and Rational Action* (London: Earthscan Publications Ltd, 2001), p. 13.

[110] Perez, *Ecological Sensitivity and Global Legal Pluralism*, p. 119.

[111] Christian Joerges, 'Law, Science and the Management of Risks to Health at the National, European and International Level – Stories on Baby Dummies, Mad Cows and Hormones in Beef', Colum. J. Eur. L., 7 (2001), 9.

use or environmental impacts will be difficult for (most) nations to address through recourse to the limited, bilateral arrangements of traditional international law.[112] The need for some kind of global regulatory structure to achieve common interest goals will be even more acute in circumstances where the risks of concern result from complex causes or necessitate reference to a broad and evolving knowledge base for their management.[113]

These features of health and environmental risks, which lend themselves to international regulation and governance rather than exclusive national control, have supported the development of global structures that seek solutions that cannot be adequately devised at the domestic level.[114] In the case of global health risks, for example, it is the World Health Organization (WHO), rather than individual national governments, that undertakes assessments of these risks and issues warnings.[115] In the environmental sphere, the major multilateral treaties dealing with resource management and environmental threats are structured and operated as ongoing regimes which set standards, establish obligations, allocate shared resources and establish prohibitions for their constituent parties.[116]

On the other side of the equation, the uncertainty surrounding many health and environmental risks makes national risk regulation more susceptible to charges that it has been put in place to keep out the products of foreign competitors or, if not intentionally protectionist, that it has been designed with insufficient regard to the effects on non-citizens or is more trade restrictive than necessary to achieve legitimate public policy objectives. Where the country concerned is a WTO member, the organisation's dispute settlement system offers a means for international review of the country's risk measures to be sought by other WTO members. To avoid the risk of such review (and any trade

[112] Unilateralism has generally been a tool only of the powerful in the environmental field: see Laurence Boisson de Chazournes, 'Unilateralism and Environmental Protection: Issues of Perception and Reality of Issues', European J. Int'l L., 11(2) (2000), 315.

[113] Zürn, 'Global Governance and Legitimacy Problems', 269.

[114] The development of global risk regulatory structures is discussed in more detail in the next chapter.

[115] Eric Stein, 'International Integration and Democracy: No Love at First Sight', *AJIL*, 95 (2001), 489, 497–8. This role was exemplified in the organisation's response to the SARS crisis between 2003 and 2005: for discussion see Esty, 'Good Governance at the Supranational Scale', 1551.

[116] Von Staden and Vollaard, 'The Erosion of State Sovereignty', p. 176.

sanctions that may follow), the country concerned may find it is easier simply to adopt off-the-shelf standards supplied by international organisations or the major economic players that have the financial and technical wherewithal to conduct the justificatory risk assessments required by WTO rules.[117]

In many respects then, decisions about risk, like those in a variety of other areas traditionally the exclusive domain of states, 'are highly, probably increasingly, dependent on actors and actions that are outside the country's boundaries and not directly subject to its government'.[118] Moreover, the perception that matters of risk assessment and management are 'low politics' has facilitated their delegation to global bodies of a regulatory or administrative character that are subject to less direct oversight by participating states.[119] The transition to global governance in the area of health and environmental risk regulation not only decreases the credibility of arguments that such rules are justified by the express consent of sovereign states, but also raises questions over whether states can indeed give effective consent where the rules that result may have far-reaching effects on private actors.

As international law increasingly 'goes public', these questions become more pressing and are framed in new ways.[120] Thus, when the WTO Appellate Body overrules a European Communities' (EC) ban on imports of hormone-containing beef on the basis that purported health risks are not scientifically established, European consumers want to know why they must abide by this decision and who they should hold to account if they disagree. Such questions reflect the expectation of those in (Western) democratic states that where there is the assertion of authority to govern, this exercise of power should be legitimate and governors accountable to those governed. It is expectations of this kind, raised by the penetration of global governance into fields traditionally regulated by governments, which has prompted increasing attention to the question of legitimacy in international law.

117 Gujadhur, 'Influencing Market Standards'.

118 Robert Dahl, *Democracy and its Critics* (New Haven: Yale University Press, 1989), p. 319.

119 Weiler, 'The Geology of International Law', 550 noting that global governance rules are typically not about 'high politics' issues such as security or human rights.

120 Philippe Sands, *Lawless World: America and the Making and Breaking of Global Rules* (London: Allen Lane, 2005), p. 15.

Legitimacy of global governance

Questions over the legitimacy of international law and international institutions – as with the governance arrangements they establish – are of relatively recent origin. Reflecting this novelty, issues and debates surrounding the legitimacy of international law and global governance are far from settled. Indeed, the *significance* of particular global governance structures possessing legitimacy, and the *kind* of legitimacy they require, may vary depending upon perceptions of an institution's nature and the extent of its authority.[121] Taking the WTO as an example, this institution may require a very strong form of legitimacy if it is viewed as an incipient global economic constitution balancing competing public values such as free trade versus environmental protection.[122] On the other hand, to the extent that the WTO displays deference to domestic regulatory choices and the rules of other international regimes reflecting non-trade values, it 'need not have the kind of legitimacy that it would require if it was to act as the final authority in the prioritization of diverse human and social values'.[123]

Complicating debates over the legitimacy of international law and institutions is the fact that the concept of legitimacy is one that eludes easy definition.[124] In general, the function served by the perceived legitimacy of a system of governance is said to be its capacity to persuade subjects of the order's 'worthiness to be recognised'.[125] Even so, a variety of notions of legitimacy exist, the majority of which focus on explaining how authority is legitimated in a domestic context.[126]

[121] See also Stein, 'International Integration and Democracy', 493 and Esty, 'Good Governance at the Supranational Scale', 1511.

[122] For examples of such views see Ernst-Ulrich Petersmann, 'From "Member-Driven Governance" to Constitutionally Limited "Mutlilevel Trade Governance" in the WTO', in Giorgio Sacerdoti, Alan Yanovich and Jan Bohanes (eds.), *The WTO at Ten: the Contribution of the Dispute Settlement System* (Cambridge University Press, 2006), p. 86; Thomas Cottier, 'Limits to International Trade: the Constitutional Challenge', in American Society of International Law (ed.), *International Law in Ferment: A New Vision for Theory and Practice, Proceedings of the 94th Annual Meeting, April 5–8, 2000* (Washington DC: American Society of International Law, 2000), p. 220.

[123] Robert Howse and Kalypso Nicolaidis, 'Enhancing WTO Legitimacy: Constitutional ization or Global Subsidiarity?', *Governance*, 16(1) (2003), 75.

[124] For discussion see Jens Steffek, 'The Legitimation of International Governance: A Discourse Approach', European J. Int'l Relations, 9(2) (2003), 251–2.

[125] Jürgen Habermas, *Communication and the Evolution of Society* (London: Heinemann, 1979), p. 178.

[126] Gráinne de Búrca, 'The Quest for Legitimacy in the European Union', Modern Law Review, 59(3) (1996), 349.

Most such formulations focus on the 'social legitimacy' of the exercise of authority connoted by 'a broad, empirically determined societal acceptance of the system'.[127] Essentially, the 'more positive the public's attitudes about an institution's right to govern', the greater the institution's social legitimacy.[128] In addition, legitimacy is generally seen to have a normative dimension, which 'refers to the validity of political decisions and political orders and their claim to legitimacy'.[129] In this sense, 'legitimacy occurs when the government process displays a commitment to, and actively guarantees, values that are part of the general political culture, such as justice, freedom, and general welfare.'[130]

The issue of legitimacy is considered to be particularly acute for global governance (as compared with national decision-making) given the increased distance between those exercising authority and the affected public (or publics).[131] For example, a Codex food standard determining a particular additive to be 'safe' may govern – via the workings of the WTO's Agreements and its dispute settlement process – what imported products can be lawfully excluded from a state's territory. However, individual consumers and local non-governmental organisations within the affected state may take a different view of the risks associated with the food additive at issue. The legitimacy of such instances of risk-related governance for those effectively governed by them will thus turn on whether affected individuals, private entities and states are willing to substitute the decisions of global bodies for their own evaluation of the situation.[132]

Legitimacy challenge in international law

In the past, the question of international law's legitimacy was not one that greatly exercised the minds of scholars and practitioners in the field. One reason for this may be that international institutions were previously thought to be relatively weak in terms of their capacity to exercise authority over powerful states, meaning that the issue of the legitimacy of their authority did not really arise.[133] In the years

[127] J. H. H. Weiler, 'The Transformation of Europe', Yale L.J., 100 (1991), 2469.
[128] Daniel Bodansky, 'The Legitimacy of International Governance: A Coming Challenge for International Environmental Law?', American J. Int'l L., 93 (1999), 601.
[129] Zürn, 'Global Governance and Legitimacy Problems', 260.
[130] Weiler, 'The Transformation of Europe', 2469.
[131] Esty, 'Good Governance at the Supranational Scale', 1505.
[132] Bodansky, 'The Legitimacy of International Governance', 602.
[133] *Ibid.*, 597.

following the Second World War, international organisations and multilateral treaties proliferated but without generating significant debates over their legitimacy. Over this period consent remained a sufficient rationale for states' acceptance of these regimes, perhaps drawing on the memory of earlier times when international law was more limited in terms of its scope and participants.

The same perception of change in international law that has driven discussions about its role in global governance has underpinned a growing concern among international legal writers and others that legitimacy is a critical challenge facing global institutions and rules. The expansion of the international community and its increasing interdependence is seen to be eroding the 'consensual underpinnings' of international law in many areas as the significant implications of international rules for non-state actors become clear.[134]

However, the legitimacy of international law and global governance is becoming a concern not just of scholars but also of individuals, non-governmental organisations and other private entities, as illustrated by the anti-globalisation protests that first came to international prominence at the Seattle WTO Ministerial meeting in 1999.[135] Where previously people looked to hold their national governments to account for policies they did not agree with, there is a growing appreciation of the fact that decisions shaping everyday life may well now be taken at the international level, out of reach of the direct control of local authorities.[136] Even where decisions are not socially contested (for instance, because decision-making processes are not transparent nor their results widely known), the 'public-oriented' character of many of the products of global governance arguably still raises questions over the source and justification for their authority.[137]

Some loss of national control over policy-making and regulation is inevitable and not necessarily undesirable in a context of interdependence where the nature of problems requiring legal intervention is such that more effective solutions may be designed and achieved by pooling resources and knowledge at a higher level of integration.[138] In addition, with increasing (economic) integration, national governments arguably

[134] *Ibid.*

[135] Mary Kaldor, '"Civilising" Globalisation? The Implications of the "Battle in Seattle"', *Millennium*, 29(1) (2002), 105.

[136] Dahl, *Democracy and its Critics*, p. 319.

[137] de Búrca, 'Developing Democracy Beyond the State', 224.

[138] Weiler, 'The Transformation of Europe', 2471–2.

should be accountable beyond their borders 'where one jurisdiction's decisions can have significant impact on outsiders, whether because the national system may regulate too little, too much, or simply take account of the interests of its own constituents before those of affected outsiders'.[139] Nevertheless, as Gráinne de Búrca points out, the 'felt need' for regulation at the supranational level is not itself sufficient to legitimate such governance.[140]

For many then, the critical legitimacy challenge faced by international law and global governance is also a paradoxical one. On the one hand, there is the 'logic' of global governance arising from 'the presence of issues that spill across national borders and the need to manage the interdependence generated by this intertwining of fates'.[141] On the other hand, the resulting proliferation of supranational and transnational rules has not (yet) been accompanied by the development of a genuine global public that sees itself in those terms. The addressees of much of the regulatory activity of global governance largely remain 'nationally enclosed' in their sense of polity, seeing international institutions and decision-making processes as remote and inaccessible.[142]

Importantly this trend is one affecting not only those countries which are tied to the implementation of administrative reforms and global regulatory standards by the requirements of international loans and development assistance arrangements, but also industrialised democracies. If international decision-making is seen to lack validity in the eyes of the societies of the world's most powerful and influential nations, the task of global governance could become one fraught with difficulty. Indeed, some see resolution of the international legitimacy challenge as crucial because '[i]f an answer is not found to this, the huge gains attained in the systemic evolution of law making and law enforcement may be normatively and even politically nullified.'[143]

Democratic legitimacy, international law and global governance

Given the widely held view that global governance – if perceived as illegitimate by the global populace – will face increasing resistance from those it seeks to govern, the search for an adequate source of legitimacy

[139] Nicolaidis and Shaffer, 'Transnational Mutual Recognition Regimes', 299.
[140] de Búrca, 'Developing Democracy Beyond the State', 228.
[141] Esty, 'Good Governance at the Supranational Scale', 1500.
[142] Zürn, 'Global Governance and Legitimacy Problems', 275.
[143] Weiler, 'The Geology of International Law', 561.

to sustain global governance rules has become an important focus of international legal scholarship and related, broader fields of research.

Older forms of legitimation, such as religion and tradition, are generally thought to have lost their influence in the face of secularisation and modernisation in many parts of the world. As societies have come to see the exercise of power as a matter of decision-making choice and discretion – rather than as a matter pre-determined by fate or constrained by long-accepted custom – direct input into decision-making, or at least decision-making that is reflective of the will of the majority, has emerged as a widely accepted basis for the legitimation of governmental authority.[144] Accordingly, democracy for many has become 'the *sine qua non* of legitimacy'.[145] At the empirical level, this is supported by the fact that, in an increasing number of nation states, authority derives its legitimacy from being exercised by representative governments accountable to their people through regular and fair elections.[146]

'Democratic deficit' of global governance

The exercise of authority by global governance bodies, in the absence of democratic support from the world's peoples, is often described as giving rise to a 'democratic deficit'. The language of democratic deficit, and concerns over the legitimacy crisis this may precipitate, draw on the experience of the most developed system of supranational regulation in existence today, namely that of the EU. This system is frequently looked to as a model for the development of global governance, as well as an instructive example of the legitimacy challenges it may face as its rules gain more authority and penetrate deeper into national regulatory systems. The supranational governance systems of the EU and questions over their legitimacy have generated their own

[144] 'Democracy' means 'rule by the people' but there are many different views as to the form this should take. In his influential work on democracy, David Held identifies nine models of democracy: David Held, *Models of Democracy*, 3rd edn (Cambridge: Polity Press, 2006).

[145] Esty, 'Good Governance at the Supranational Scale', 1515; Thomas Franck, 'The Emerging Right to Democratic Governance', American J. Int'l L., 86 (1992), 46. See also UNGA res. Doc. A/RES/60/1, 24 October 2005, [135], declaring democracy to be 'a universal value'.

[146] In its 2009 report, Freedom House found that 62 per cent of the world's states were electoral democracies, compared with 40 per cent in 1987: Freedom House, *Freedom in the World 2009* (Freedom House, 2009) available at www.freedomhouse.org/template.cfm?page=351&ana_page=352&year=2009.

rich literature, to which it is impossible to do justice here.[147] For our purposes, however, a much briefer treatment will suffice.

In the initial, relatively successful, phase of integration in the various European Communities, the work of the 'European project', like that of early international institutions, was of low profile and the 'largely elite-driven and technocratic' processes involved were a matter of little popular concern.[148] The very success of those processes in achieving greater economic integration encouraged Community institutions to direct their attention to various non-tariff barriers to trade deriving from members' regulatory systems in the fields of environmental, health and safety, and consumer protection. Maintaining social regulatory programmes and acceptable levels of health and environmental protection in the face of the deregulatory pressures that this created for member states spurred the development of European-level standards covering a wide range of issues.[149] However, the occurrence of several health scares in the EU – culminating in the crisis over the management of the risks of mad cow disease in the late 1990s[150] – exposed serious deficiencies in the adequacy of European regulatory standards. Europeans became much more aware of the activities of the Community institutions with their direct impacts on the citizens of member states, generating widespread public malaise with European-level governance.[151]

Subsequently there have been significant efforts made to improve the transparency and inclusiveness of EU regulatory systems and to strengthen the role of member states and representative bodies such as the European Parliament in the rule-making process.[152] Despite

[147] For a selection see Kalypso Nicolaidis and Robert Howse (eds.), *The Federal Vision: Legitimacy and Levels of Governance in the United States and the European Union* (Oxford University Press, 2001); Christian Joerges and Renaud Dehousse (eds.), *Good Governance in Europe's Integrated Market* (Oxford University Press, 2002); Gráinne de Búrca and Joanne Scott (eds.), *Law and New Governance in the EU and the US* (Portland: Hart Publishing, 2006).

[148] de Búrca, 'The Quest for Legitimacy in the European Union', 350.

[149] Renaud Dehousse, 'Misfits: EU Law and the Transformation of European Governance', in Christian Joerges and Renaud Dehousse (eds.), *Good Governance in Europe's Integrated Market* (Oxford University Press, 2002), p. 207.

[150] The ramifications of this crisis for European governance and law in the risk regulatory area are discussed further in Chapter 4.

[151] Howse and Nicolaides, 'Enhancing WTO Legitimacy', 85.

[152] See, particularly, European Commission, 'European Governance: A White Paper, COM(2001) 428 final' (European Union, 2001).

this, questions over the legitimacy of the EU are still pervasive.[153] The 'crux' of the ongoing EU legitimacy debate remains the difficulty of reconciling the organisation's structure and mode of governance - which are neither those of a state nor a typical inter-governmental organisation - with the functions that it exercises. These replicate, and in some cases, displace those of national governments.[154]

The parallels between the EU's experience and that of international governance institutions such as the WTO have brought the debate over democratic legitimacy into the broader international domain.[155] Not all would agree, though, that this is a worthwhile debate as regards global governance.[156] For those who deny the existence of an international democratic deficit, such as Andrew Moravcsik, the remoteness of global governance rules from the citizens of states is unproblematic as it is not that different from the position that exists in many contemporary democracies where there are widespread trends of delegation to administrative agencies and insulation of rule-making authority.[157] This position, however, is belied by evidence of growing public opposition to the activities of particular global governance bodies, including the WTO, IMF and World Bank, as well as high-profile standardisation organisations such as Codex.[158] Arguably, it also underestimates the extent to which domestic agencies are embedded within an institutional and political framework of checks and balances that provides a substitute for the agencies themselves to satisfy democratic requirements. A similar framework is largely absent when we move to the global level.[159]

[153] Such tensions were brought to a head during the national processes for ratification of the proposed Treaty Establishing a Constitution for Europe. Rejection of the Constitution by voters in France and the Netherlands during 2005, and the subsequent abandonment or postponement of ratification processes by several other member states, led to its abandonment in favour of a new treaty - the Treaty of Lisbon.

[154] de Búrca, 'The Quest for Legitimacy in the European Union', 352.

[155] Stein, 'International Integration and Democracy', 489.

[156] For a succinct overview of the views of those taking 'denial approaches' see de Búrca, 'Developing Democracy Beyond the State', 238.

[157] Andrew Moravcsik, 'Is There a "Democratic Deficit" in World Politics? A Framework for Analysis', *Government and Opposition*, 39 (2004), 336.

[158] See Zürn, 'Global Governance and Legitimacy Problems', 279–85 for a discussion of examples of national and transnational resistance to the authority of global governance.

[159] de Búrca, 'Developing Democracy Beyond the State', 232–3.

No democratic legitimacy without a global democratic community?

The more common approach in the literature when it comes to questions of legitimacy in global governance sees the democracy problem as one which is crucial and unanswered; indeed, perhaps even unanswerable.[160] For authors in this vein, the key sticking point is the lack of an identifiable world political community:[161] 'a certain community of a common good and common expectations of the people that bridges the cultural differences to the extent necessary to sustain community institutions as their powers increase.'[162]

Joseph Weiler puts the argument thus:

> The international system form of governance with government and without demos means that there is no purchase, no handle whereby we can graft democracy as we understand it from Statal settings on to the international arena.[163]

This view is clearly premised on an understanding of democracy that is 'closely tied to the context of the nation state, and to the idea of a bounded political community'.[164] Nonetheless, it seems that the reference to models (or at least the values) of democracy drawn from national systems is justified given the 'empirical reality' of a pervasive

160 Kingsbury *et al.*, 'The Emergence of Global Administrative Law', 49–50. Prospects for democratic legitimacy at the international level are regarded as bleaker than for the EU where the European Parliament provides at least some basis for establishing a democratic link to EU citizens (or might do so with further expansion of its powers).

161 Another, albeit more controversial, term used to convey this idea is that of the *demos*.

162 Stein, 'International Integration and Democracy', 527. See also J. H. H. Weiler, *The Constitution of Europe: Do the New Clothes Have an Emperor and Other Essays on European Integration* (Cambridge University Press, 1999), p. 337.

163 Weiler, 'The Geology of International Law', 560.

164 de Búrca, 'Developing Democracy Beyond the State', 225. This can be contrasted with other democratic governance theories which are non-hierarchical and deliberative in nature. See, e.g., Oliver Gerstenberg and Charles Sabel, 'Directly-Deliberative Polyarchy: An Institutional Ideal for Europe?', in Christian Joerges and Renaud Dehousse (eds.), *Good Governance in Europe's Integrated Market* (Oxford University Press, 2002), p. 289. Consensus decision-making in the WTO institutions and committees (such as the SPS Committee considered in Chapter 5) may offer a possible model here: see Footer, 'The Role of Consensus in GATT/WTO Decision-making'.

and strong socio-political attachment to the idea of democracy as it is practiced in nation states.[165]

Despite some proposals for a 'world parliament',[166] development of representative global democracy along the lines familiar in nation states looks to be a remote, even an undesirable, possibility.[167] In addition, attempts to develop alternative theories of democratic international governance that do not rely on the existence of a global political community have been roundly criticised as having 'hardly resolved the problems of defining "the public" that is supposed to govern or be represented globally, or of designing the mechanisms by which global participation or deliberation can indeed occur'.[168] Robert Keohane and Ruth Grant argue, for example, that assigning participatory rights in any international decision-making process to 'those affected' is too diffuse and imprecise a criterion in the absence of global political structures that could define these boundaries.[169]

Absent the capacity to achieve legitimacy through representative or participatory democratic forms, some commentators see a stark choice between sacrificing the achievements of multilateral institutions and international cooperative efforts for the sake of democracy, or accepting rule by non-democratic, unaccountable international bodies.[170] A body of literature and international practice is emerging that challenges this view, positing that it is still possible to establish, or at least strive towards, the democratisation of global governance without the need to establish representative government institutions as exist in many nation states. Such proposals are explored in the context of the democratisation of global risk governance in Chapter 7.

In general, however, the dominant position remains one that accepts the impossibility of global governance achieving democratic legitimacy or suggests that questions of democracy be bracketed for the present.[171]

[165] Peter Lindseth, '"Delegation is Dead, Long Live Delegation": Managing the Democratic Disconnect in the European Market Policy', in Christian Joerges and Renaud Dehousse (eds.), *Good Governance in Europe's Integrated Market* (Oxford University Press, 2002), p. 139

[166] See, e.g., Richard Falk and Andrew Strauss, 'Towards Global Parliament', *Foreign Affairs*, 80 (2001), 212.

[167] Kingsbury *et al.*, 'The Emergence of Global Administrative Law', 49.

[168] *Ibid.*

[169] Ruth Grant and Robert Keohane, 'Accountability and Abuses of Power in World Politics', *American Political Science Review*, 99(1) (2005), 33.

[170] *Ibid.*

[171] Kingsbury *et al.*, 'The Emergence of Global Administrative Law', 51.

Instead such authors argue for 'new pragmatic approaches' for securing effective legitimacy at the international level, 'approaches that do not depend on the existence of a clearly defined global public'.[172] These approaches typically draw on a number of 'good governance' values, some of which (such as transparency, accountability and deliberation) are democratic in nature, albeit more readily detached from a notion of political community. Other good governance values have no necessary link to democracy, such as the qualities of independence, efficiency and expertise.

Alternative sources of legitimacy for global governance

In her recent article discussing democracy 'beyond the state', Gráinne de Búrca provides a useful categorisation of the arguments that have been advanced for the legitimacy of global governance, which do not rely on the establishment of representative democratic forms at the international level. The three general groupings she identifies (which are not necessarily mutually exclusive) focus either upon (a) the merits of the decision-makers; (b) the qualities of the process of decision-making; or (c) the outcomes or impact of the governance process.[173] The latter category of arguments, concentrated on the outputs of governance, emphasises factors of efficiency of decision-making or general acceptance of the norms that have been generated by the process. For example, Fritz Scharpf has argued that the supranational governance regime of the EU relies largely upon 'output-oriented' legitimacy derived from its capacity to tackle effectively problems requiring collective solutions.[174]

By contrast, arguments focused on the decision-making process suggest that certain qualities of the processes themselves can render global governance (more) legitimate. At the simplest level are calls for greater transparency in the governance process, for example, so that decision-making fora are open to the general public,[175] or to ensure that decisions are readily accessible, say via their publication on the

[172] Grant and Keohane, 'Accountability and Abuses of Power in World Politics', 34.

[173] de Búrca, 'Developing Democracy Beyond the State', 242.

[174] Scharpf, *Governing in Europe*, pp. 11–12.

[175] A long-standing criticism of the GATT/WTO dispute settlement process points to its operation behind closed doors: e.g., J. H. H. Weiler, 'The Rule of Lawyers and the Ethos of Diplomats: Reflections on WTO Dispute Settlement', in Roger B. Porter *et al.* (eds.), *Efficiency, Equity, and Legitimacy: the Multilateral Trading System at the Millennium* (Washington DC: Brookings Institution Press, 2001), p. 334.

Internet. More ambitious proposals in this vein call for some degree of accountability of global governance organs to those affected by their decisions through mechanisms such as reason-giving or the availability of administrative review.

Another variant of the process-oriented approach stresses deliberative governance procedures as a guarantor of the quality of decision-making that may compensate for its otherwise non-democratic nature. For instance, Christian Joerges and Jürgen Neyer have described the operation of 'deliberative technocratic' processes in the EU's regulatory committee procedures, known as comitology, which involve the interaction of a variety of experts, interest groups and member state representatives in the decision-making process.[176] Although not fully inclusive in their membership, Joerges and Neyer argue that, as a result of the deliberative processes of reasoning and debate that go on in the committees:

> delegates not only learn to reduce differences between national legal provisions but also to develop converging definitions of problems and philosophies for their solution. They slowly proceed from being representatives of national interests to being representatives of a Europeanized inter-administrative discourse characterized by mutual learning and an understanding of each other's difficulties in the implementation of specific solutions.[177]

The final category of legitimacy rationales, focused on the merits of the decision-makers, largely claims support for the regulatory activities of governance bodies on the basis that certain decisions require expert judgments and that decision-makers have, or can draw on, the requisite knowledge to make these judgments.[178] Justifying the issue of authoritative rules and decisions in this instance is said to be 'the special qualifications, knowledge and proficiency of expert actors, rather

[176] Christian Joerges and Jürgen Neyer, 'Transforming Strategic Interaction into Deliberative Problem-Solving: European Comitology in the Foodstuffs Sector', J. European Public Policy, 4(4) (1997), 609. Some question, however, whether such processes indeed promote solutions to common problems on a deliberative democratic, rather than a purely technocratic, basis: see Martin Shapiro, '"Deliberative", "Independent" Technocracy v. Democratic Politics: Will the Globe Echo the E.U.?', IILJ Working Paper 2004/5 (Global Administrative Law Series, 2004). Available at www.iilj.org.

[177] Joerges and Neyer, 'Transforming Strategic Interaction into Deliberative Problem-Solving', 620.

[178] Robert Baldwin, 'Regulatory Legitimacy in the European Context: the British Health and Safety Executive', in Giandomenico Majone (ed.), *Regulating Europe* (London; Routledge, 1996), pp. 90–1.

than the nature and breadth of participation in decision-making'.[179] For advocates of this approach, the administrative tasks of rule-making gain legitimacy through the use of expertise deployed to achieve effective policy outcomes, sometimes with a form of legal control – such as judicial review – providing a check on the exercise of discretion.[180]

The appeal to expertise reflects the perceived difficulty of leaving the formulation of effective solutions to problems in certain technical areas to processes of self-interested inter-governmental bargaining, together with the simple fact that decision-makers cannot regulate what they do not understand.[181] For instance, Daniel Esty remarks that '[w]hen a matter is largely scientific or technical, having designated supranational experts address the problem may be uncontroversial'.[182] Consequently, legitimacy based on the expertise of decision-makers, or that of their advisors, is often put forward as a sufficient justification for governance decisions taken about issues of a scientific or technical nature, especially where this is combined with an argument that rational decisions informed by expertise lead to 'good outcomes'.[183]

Global risk governance and expertise-based legitimacy

Risk regulation has long been seen as an activity of particular complexity, demanding high levels of scientific and technical capacity. At the same time, issues surrounding the assessment of risk have generally been designated scientific matters, separable from questions of politics and values.[184] In the next chapter we explore and critique the long-standing association between science and risk, and the idea that science can be cleanly separated from values in areas of risk regulation. Notwithstanding this critique, it remains the case that science is now an indispensable component of the regulation of matters of global environmental risk or risks to health, necessitating the input of knowledge and technical expertise by specialists in relevant fields.[185] Indeed,

[179] de Búrca, 'Developing Democracy Beyond the State', 242.

[180] Peter Lindseth, 'Democratic Legitimacy and the Administrative Character of Supranationalism: The Example of the European Community', Columbia L. Rev., 99 (1999), 632.

[181] Shapiro, '"Deliberative", "Independent" Technocracy v. Democratic Politics'.

[182] Esty, 'Good Governance at the Supranational Scale', 1511.

[183] *Ibid.*, 1517.

[184] As detailed in Dan M. Kahan *et al.*, 'Fear of Democracy: A Cultural Evaluation of Sunstein on Risk', Harvard L. Rev., 119 (2006), 1075–6.

[185] Bodansky, 'The Legitimacy of International Governance', 622–3.

the dependence of risk assessment on expert judgment has become all the more critical because many of the risks of contemporary concern are largely imperceptible to lay people or involve extended timeframes that require predictions extrapolating from current data.

The conventional view of risk regulation as a primarily technical activity has supported an approach that sees expertise-based legitimacy as being of particular importance for global governance bodies engaged in such tasks.[186] Consequently, virtually all international regimes dealing with risk issues in the health and environmental field involve scientific experts in the regulatory process.[187] Prominent examples include the scientific advisory committees at the heart of the activities of regional fisheries organisations,[188] the Joint Food and Agriculture Organization (FAO)/World Health Organization Expert Committee on Food Additives that supplies independent scientific expert advice to Codex,[189] and the Intergovernmental Panel on Climate Change (IPCC) operating in conjunction with the climate change regime. The latter is charged with the broad mandate of assessing on a comprehensive, objective, open and transparent basis the scientific, technical and socio-economic information relevant to understanding the scientific basis of the risk of human-induced climate change, its potential impacts and options for adaptation and mitigation.[190]

Scientific expertise has also been required by international regimes as a justification for national risk measures that affect flows of trade or investment across borders. The WTO SPS Agreement, for example, deems illegitimate any national SPS measures that do not conform to science-based international standards and lack a basis in 'sufficient scientific evidence'.[191] In negotiations for the Agreement, participants' acceptance of this structure reflected the view that '[h]armonization of health and

[186] Joerges, 'Law, Science and the Management of Risks to Health at the National, European and International Level', 2–3.

[187] Steinar Andresen *et al.*, *Science and Politics in International Environmental Regimes: Between Integrity and Involvement* (Manchester University Press, 2000), pp. 182–3.

[188] For example, based on scientific advice and relevant scientific evidence, the Commission under the Convention on the Conservation of Southern Bluefin Tuna, 10 May 1993, in force 20 May 1994, [1994] ATS 16, Article 8 is tasked with establishing an annual total allowable catch for southern bluefin tuna and allocating it among the Convention parties.

[189] This body is discussed further in Chapter 6.

[190] The scientific risk assessment process of the IPCC is discussed further in Chapter 6.

[191] SPS Agreement, Articles 2.2, 3.1 and 3.2.

sanitary regulations, that are based on objective scientific review, could facilitate trade'.[192] They agreed that an objective scientific basis for SPS measures could be demonstrated, either via a risk assessment or by relying on the standards of organisations such as Codex that were seen to have 'international scientific reputation and credibility'.[193]

In the event of a dispute over the particular SPS measures adopted by a member, the Agreement also allows WTO dispute settlement panels to seek independent expert advice (including from relevant international scientific organisations) to support conclusions about the extent of any proffered scientific justification. In this case, '[t]he 'epistemic authority' of experts' is seen to give 'expert-based WTO decisions their extra legitimacy'.[194]

Expertise-based legitimacy in domestic regulatory systems

Support for relying on expertise to found the legitimacy of global risk governance is often drawn from the experience of domestic regulatory systems. In many domestic systems – most prominently in the modern American administrative state – implementation of broad legislative mandates for health and environmental protection is delegated to agencies established as independent, expert bodies. The underlying rationale is that such bodies offer 'the possibility of achieving expertness in the treatment of special problems, relative freedom from the exigencies of party politics in their consideration, and expeditiousness in their disposition'.[195] Accordingly, expertise, combined with an argument that agency decision-making delivers optimal results, underpins the claims of administrative agencies to policy-making and decision-making legitimacy on basis that they perform a primarily technical role requiring expert knowledge and judgment.[196] Government regulators have traditionally fostered this perception, seeking 'legitimacy for their decisions by wrapping them in a cloak of scientific respectability'.[197]

192 'A Discussion Paper on Issues Related to the Negotiations Submitted by the United States', Negotiating Group on Agriculture, MTN.GNG/NG5/W/44, 22 February 1988.

193 *Ibid.*

194 Joost Pauwelyn, 'The Use of Experts in WTO Dispute Settlement', *International and Comparative Law Quarterly*, 51 (2002), 330.

195 Giandomenico Majone, 'Regulation and Its Modes', in Giandomenico Majone (ed.), *Regulating Europe* (London: Routledge, 1996), p. 16.

196 Esty, 'Good Governance at the Supranational Scale', 1517

197 Giandomenico Majone, 'Science and Trans-Science in Standard Setting', *Science, Technology and Human Values* (1984) 9(1), 15.

Indeed, for some, problems with present risk regulatory systems in Western countries, such as the USA, lie in dilution of the expertise of administrative agencies as a result of calls for, and reforms to institute, greater pluralism in policy-making.[198] A leading proponent of this view, Stephen Breyer, argued in a book published shortly before his elevation to the US Supreme Court that politicisation of the activities of American risk regulatory agencies was creating a vicious circle, leading to an ineffective and inefficient management of risks.[199] His recommended solution was the creation of a new centralised bureaucratic group, insulated from both politics and public opinion, with broad authority to bring uniformity and rationality in highly technical areas of decision-making by employing specialised expertise.[200]

In the field of global governance, reliance upon independent expert bodies, rather than participatory mechanisms, to found decision-making has also attracted support in the literature. For instance, echoes of this approach are evident in Andrew Moravcsik's arguments against the existence of a democratic deficit in world politics, drawing on analogies between regulatory decision-making at the national and international level. In respect of the former he states:

> Citizens delegate to assemble more efficient decision-making in areas where expertise is required. Involvement in the full range of government policies would impose costs beyond the willingness of any modern citizen to bear. Whether the area is environmental policy, medical drug authorization, or criminal law, we do not expect complex medical, legal, or technical decisions to be made by direct popular vote.[201]

Likewise, in conditions of 'social complexity, political uncertainty, and underlying differentials in social power', Moravcsik argues that 'under many circumstances more insulated and delegated authority of *global* governance structures might be thought of as more "representative" of citizen concerns precisely because they are less directly "democratic"'.[202]

198 For discussion of this debate see Edward L. Rubin, 'Getting Past Democracy', U. Pennsylvania L. Rev., 149 (2000–2001), 781–2.

199 Stephen Breyer, *Breaking the Vicious Circle: Toward Effective Risk Regulation* (Cambridge, MA: Harvard University Press, 1993).

200 *Ibid.*, pp. 60–1. Another, more recent, and well-developed articulation of this approach can be found in Cass Sunstein, *Laws of Fear: Beyond the Precautionary Principle* (Cambridge University Press, 2005), see e.g., p. 126.

201 Moravcsik, 'Is There a "Democratic Deficit" in World Politics?', 344.

202 *Ibid.*, 346–7 (emphasis added). For criticisms of drawing a direct analogy between domestic and global governance agencies in such circumstances, see de Búrca, 'Developing Democracy Beyond the State', 232.

Writing in the European context, Giandomenico Majone does not go so far as to claim general democratic credentials for expert regulatory bodies, but rather argues that the growing importance of non-majoritian institutions in Western democratic countries demonstrates that for many purposes 'reliance upon qualities such as expertise, professional discretion, policy consistency, fairness and independence of judgment is considered to be more important than reliance upon direct political accountability'.[203] While Majone believes that decisions involving a 'significant redistribution of resources from one social group to another cannot be legitimately taken by independent experts',[204] he contends that so long as decision-making tasks are concerned with 'problem-solving', and are precisely and narrowly defined, sources of legitimacy such as 'expertise, procedural rationality, transparency [and] accountability by results' should provide a sufficient justification for regulatory activity.[205]

The domestic model of legitimate risk regulation is one that has proved attractive for international organisations such as Codex, as well as supranational regulatory agencies such as the European Food Safety Agency. Decision-making structures in such bodies delegate scientific judgments about the existence of a risk (or the review of the scientific justification for risk measures) to expert-based decision-making processes, while reserving policy-oriented, value-based risk management decisions to nation states.[206] For instance, Codex's Working Principles for Risk Analysis prescribe a 'functional separation of risk assessment and risk management, to the extent practicable, in order to ensure the scientific integrity of the risk assessment, to avoid confusion over the functions to be performed by risk assessors and risk managers and to reduce any conflict of interest'.[207]

Expertise as necessary but not sufficient rationale

For all but its most fervent proponents, expertise has the potential to take global risk governance only some of the way down the path

203 Giandomenico Majone, 'Regulatory Legitimacy', in Giandomenico Majone (ed.), *Regulating Europe* (London: Routledge, 1996), p. 286.

204 *Ibid.*, p. 295. 205 *Ibid.*, p. 299.

206 Christian Joerges, '"Deliberative Supranationalism" – Two Defences', *European Law Journal*, 8(1) (2002), 140.

207 Codex Alimentarius Commission, 'Working Principles for Risk Analysis for Application in the Framework of the Codex Alimentarius', in Codex Secretariat (ed.), *15th Procedural Manual* (Rome: Food and Agriculture Organisation, 2005), p. 101.

towards establishing legitimacy for its far-reaching effects on peoples within states. Most recognise that expert knowledge is limited in its capacity to legitimise public authority, even in areas treated as scientific and technical in nature. Daniel Esty has sought to convey this idea by ranging issues dealt with by global governance along a spectrum running from 'scientific/technical issues' at one end to 'political/value-laden issues' at the other.[208] While Esty sees the application of international expertise to address matters that are 'largely' science/technical as unlikely to generate controversy, he acknowledges that:

> As an issue becomes more political or normatively charged, ... delegation to those lacking electoral legitimacy becomes increasingly problematic. The more sharply values diverge, the more intense will be the stress on the decision-making process.[209]

A similar conclusion was reached by the European Court of First Instance in its 2002 judgment in the case of *Pfizer Animal Health* v. *Council*.[210] The decision concerned a challenge to a Council Regulation removing a particular antibiotic growth promoter from the list of authorised additives permitted in animal feedstuffs on the grounds of its potential animal and human health risks. A scientific advisory body consulted about the regulation by the European Commission had advised that, based on current scientific evidence, the antibiotic in question did not constitute an immediate risk to public health.

Nonetheless, the Court of First Instance found that the Commission (and Council) was entitled to disregard the conclusions drawn by scientific advisors, noting that the latter did not possess 'political responsibilities and democratic legitimacy of the Commission'. 'Scientific legitimacy', the Court opined 'is not a sufficient basis for the exercise of public authority'.[211] By same token, though, the Court was not prepared to sanction the risk governance activities of Community institutions carried out without any reference to pertinent scientific information. Thus it emphasised that 'the competent public authority must ensure that any measures that it takes, even preventive

[208] Esty, 'Good Governance at the Supranational Scale', 1511.

[209] *Ibid.*, 1511–2. See also Bodansky, 'The Legitimacy of International Governance', 622–3.

[210] Case T-13/99 *Pfizer Animal Health SA* v. *Council of the European Union* [2002] ECR II-3305 (*Pfizer*). Fuller discussion of this case appears in Chapter 4.

[211] *Ibid.*, [201].

measures, are based on as thorough a scientific risk assessment as possible, account being taken of the particular circumstances of the case at issue'.[212]

The judgment of the Court of First Instance in *Pfizer* uncovers a tension that exists more generally in global risk governance between recognition of the desirability of scientific input and the inability to answer the questions involved in purely scientific terms.[213] This points to the need for the expertise deployed in international risk regulation to be bolstered by other mechanisms in order to be legitimate. Democratic mechanisms and values re-enter the debate at this point as a potentially necessary component of legitimate global risk governance; a topic we will revisit in Chapter 7.

As will we see further in the next chapter, however, the reality of global risk governance is of a highly 'scientised' regulatory space in which 'the topics and issues that are the stuff of political debate, conflict and action – and are expressed in political discourses – are more and more generated or discovered in and through science'.[214] Expertise based in scientific and technical knowledge thus retains significant power to persuade decision-makers and others as to correctness of a particular regulatory approach. Indeed, where it is alleged that a decision under an international regime or a particular government measure with external effects has failed to draw on the best available scientific expertise, it is likely to face legitimacy-based challenges.[215] For this reason when questions over risk arise in international law, a central, if not *the* central question, is whether there exists sufficient scientific evidence that, in the view of relevant experts, supports a particular regulatory measure.

Conclusion

In Chapter 1, the canvassing of expert opinion in the WTO *GMO* case decided under the SPS Agreement was highlighted as an example of what has emerged as a more widespread trend in international law;

212 *Ibid.*, [162].
213 Bodansky, 'The Legitimacy of International Governance', 622.
214 Tom Burns, 'Parliamentary Democracy in Crisis: The Challenges and Potentialities of High Science and Technology', (Paper presented at the European Parliamentary Technical Assessment, Annual Conference, 2 December 2000).
215 Bodansky, 'The Legitimacy of International Governance', 622.

namely, the central role of science in underpinning global rules and decision-making concerned with issues of health and environmental risk. This chapter sought to situate these developments against the broader narrative of change (or perceived change) in international law. In this narrative the regulatory sovereignty of nation states is giving way in favour of international rules and global organisations, which increasingly exercise governing authority over entities and peoples within states, albeit in a manner disassociated from the formal institutions of government. The broad-ranging effects of such global governance have made the issue of legitimacy a central concern of international law, especially the question of the potential for democratic legitimacy to be achieved at the global level.

For the area of risk regulation – one where the logic of global governance has considerable purchase – the increasing transition from a paradigm of nation state regulatory sovereignty to one of global administration has facilitated a heavy reliance on science and expert knowledge in relevant international legal structures. The turn to expertise can be explained, in part, by the perceived need for new systems of governance to have a legitimate basis. With international democratic legitimacy generally viewed as an unrealisable or unrealistic goal, expertise (sometimes in combination with other legitimation rationales that do not rely on the identification of a global political community) would seem to have substantial power to convince decision-makers and the broader public of the legitimacy of international regulatory efforts that draw on such expertise.

Underpinning the credibility of expertise-based legitimacy for global risk governance is the widespread perception of the technical nature of the regulatory tasks involved, and hence the belief that expertise can supply a neutral basis for any standards adopted. As discussed in the next chapter, this potential is linked to the positioning of science as a supplier of authoritative knowledge in modern society, as well as the strong, reflexive association that has developed between notions of risk and a basis in scientific evidence.

Nevertheless, the growth of science-based international decision-making processes exists alongside increasing acknowledgement on the part of decision-makers and others that science, albeit a necessary element of international risk regulation, will not always be sufficient to establish broad legitimacy for the exercise of public authority. Underlying such concerns is a growing appreciation of the impact of uncertainties and contingencies in science for the reliability of scientific knowledge, as well as the scope they may allow for different value judgments to affect decisions about risk assessment and risk management. These matters are taken up in more detail in the following chapter, which addresses the rise of, and challenges to, the culture of science in global society and its manifestation in international legal structures concerned with risk regulation.

3 Scientific rationality and risk in international law

Introduction

In areas of international law concerned with issues of risk, rules and processes of global governance with significant behind-the-border impacts are of growing importance. The potential reach of global risk governance into the domestic regulatory sphere of sovereign states has made questions over legitimacy increasingly pertinent for the broader acceptance of international risk regulation. This is particularly so in the context of an emerging (world) 'risk society' that posits the control of risk as a central concern of modern government.[1]

Drawing on analogies with national risk regulation, expertise in the form of scientific knowledge is often looked to as a means for strengthening the legitimacy claims of global decision-making processes dealing with highly technical and complex matters of health and environmental risk. Acceptance of science as a sound foundation for international risk regulation is underpinned by the perception that scientific knowledge offers an objective and universally applicable basis for rational decision-making, as well as the close association between notions of risk, and the scientific understanding of them, that has developed in contemporary times. Indeed, science and technology can be said to have provided the foundations of 'a dynamic, homogenizing global culture'; one that now creates 'continuously intense levels of interaction between and across territorial communities'.[2]

[1] Ulrich Beck, *Risk Society: Towards a New Modernity* (London: SAGE Publications, 1992); Ulrich Beck, *World Risk Society* (Cambridge: Polity Press, 1999).

[2] Michael Reisman, 'Designing and Managing the Future of the State', European J. Int'l Law, 3 (1997), 410 and 415.

This chapter undertakes an interdisciplinary, critical examination of the cultural authority of science and expertise in the international legal context, and the 'scientisation' of notions of risk that has occurred in societies preoccupied with risk regulation and control.[3] The first two sections of the chapter follow the history of the association between science, expertise and international law. This reveals the increasing importance placed on scientific knowledge and independent expertise in international legal processes, with a particularly significant role played in the policy-making and advisory processes of multilateral agreements concerned with potential health or environmental harms.

The next part of the chapter takes up the notion of risk. Its development is traced from original conceptions as a matter of chance, good or bad, to the current understanding as threats of future harm which are not only a central concern of (Western) societies and their governments, but also necessitate reference to science for their comprehension, diagnosis and remedy. As risk issues grow in importance but can no longer be effectively addressed by nation states acting alone, there are an increasing number of global instruments concerned with risk regulation.[4] In their ongoing implementation and administration, these instruments call for scientific evidence and expertise-based procedures, such as risk assessment.

However, at the same time as science is assuming a central role in the risk-related decision-making procedures of global governance institutions, public confidence in science and its technological products is declining in many parts of the world. There is a growing appreciation that technology lies at the heart of some of the world's most pressing risk problems, such as the spread of diseases and invasive pest species, chemical pollution, ozone depletion and climate change. Moreover, contrary to the faith displayed by societies in the early industrial era, science has often been unable to predict the occurrence of harmful events or to prevent their undesirable effects.[5] Thus, in a remarkably

[3] In interdisciplinary work 'an issue is approached from a range of disciplinary perspectives [e.g. law, science, social science] integrated to provide a systemic outcome': Roderick J. Lawrence and Carole Després, 'Futures of Transdisciplinarity', *Futures*, 36 (2004), 400.

[4] This chapter gives an outline of such instruments. A more detailed analysis of particular risk regulatory processes under international agreements and institutional arrangements is found in Chapter 6.

[5] European Environment Agency, *Late Lessons from Early Warnings: the Precautionary Principle 1896–2000* (Luxembourg: European Union, 2001).

short space of time, 'confidence about the physical world has turned into doubt; [o]nce the source of safety, science and technology have become the source of risk'.[6]

Public misgivings about science and technology find an echo in the increasingly critical literature dealing with the philosophy and sociology of science and its application in risk regulation.[7] There is now a substantial body of research in the social science field, reviewed in the final part of the chapter, which questions the authority of scientific knowledge and the objectivity of risk assessment by exposing the uncertainties and contingencies in various areas of science dealing with health and the environment.[8]

International law and global governance have not been immune from these developments, which have been productive of growing unease in their relationship with science. While scientific knowledge continues to be seen as essential for providing legitimacy in debates about risk that are invariably conducted in scientific terms, nevertheless acknowledged uncertainties in science, and the presence of subjective value judgments in risk assessment, open the way for significant political disagreement.[9] The potential for conflict is greatest, as disputes like the *GMO* case illustrate, when stark differences exist as to the basic values at stake and appropriate management goals. In these circumstances, science as 'knowledge, speaking to everyone else as power'[10] may be less effective as a means for achieving some degree of international consensus on risk than global institutions and decision-making processes with the capacity to integrate scientific and

[6] Mary Douglas and Aaron Wildavsky, *Risk and Culture: An Essay on the Selection of Technical and Environmental Dangers* (Berkeley: University of California Press, 1982), 10.

[7] Within this body of research itself there is a great diversity of views: see Tom Horlick-Jones and Jonathan Sime, 'Living on the Border: Knowledge, Risk and Transdisciplinarity', *Futures*, 36 (2004), 447–50.

[8] Parallels can be drawn with quantitative cost–benefit analysis which, along with scientific risk assessment, is often prescribed as an essential element of efficient and effective risk regulation: see, e.g., Cass Sunstein, *Risk and Reason: Safety, Law, and the Environment* (Cambridge University Press, 2002). Similar critiques have been mounted in recent years to claims of objectivity applied to cost–benefit analysis. While this debate lies outside the scope of this book, for an introduction to the critical literature in this area see David M. Driesen, 'Is Cost-Benefit Analysis Neutral', Uni. Colorado L. Rev., 77 (2006), 335.

[9] Peter Haas, 'Science Policy for Multilateral Environmental Governance', in Norichika Kanie and Peter Haas (eds.), *Emerging Forces in Environmental Governance* (Tokyo: United Nations University, 2004), p. 116.

[10] David Kennedy, 'The Politics of the Invisible College: International Governance and the Politics of Expertise', *European Human Rights Law Review*, 5 (2001), 463, 472.

non-scientific considerations to produce a more broadly acceptable, 'serviceable truth'.[11]

A global culture of scientific rationality

Rational scientific method

Science has been described as 'a cornerstone of modernity',[12] the 'archetype' of all knowledge which, in modern Western cultures, is privileged over both God and tradition.[13] It was the Enlightenment period of the seventeenth century that saw the rise of rationalistic, scientific thinking as a new, dominant force in Western societies. Enlightenment-era philosophies posited reason and experimental observation as the primary means for understanding nature and mastering its resources. The successes of the physical sciences, achieved through the work of the likes of Galileo and Newton, were celebrated as a triumph of reason and logic, promoting 'an optimistic feeling that everything of importance could be understood by the systematic application of rational thought'.[14]

For its leading seventeenth-century proponents, such as Francis Bacon, science was the means to free men from the illusions and myths of the past, including those of theological orthodoxy and the political structures of feudalism.[15] Conceived of in this way, Enlightenment-era science was a deeply normative project, but this is not the image of science that pertains today. Rather Baconian induction (and the modern traditions of science that it fathered) are renowned for introducing the *method* of science; one based on the progressive accumulation of knowledge about the natural world through objective observation. Ultimately this method had its source in the British empiricist tradition which held that we 'know' nature through the senses, through actually experiencing the natural world and thereby perceiving its

11 Sheila Jasanoff, *The Fifth Branch: Science Advisors as Policymakers* (Cambridge, MA: Harvard University Press, 1990), p. 250.

12 Karen T. Litfin, *Ozone Discourses: Science and Politics in Global Environmental Cooperation* (New York: Columbia University Press, 1994), p. 8.

13 Margaret Davies, *Asking the Law Question: the Dissolution of Legal Theory* (Sydney: Lawbook Co., 2002), p. 115.

14 Philip S. Baringer, 'Introduction: the "Science Wars"', in Keith M. Ashman and Philip S. Baringer (eds.), *After the Science Wars* (London: Routledge, 2001), pp. 1, 4.

15 Anthony O'Hear, *Introduction to the Philosophy of Science* (Oxford: Clarendon Press, 1989), p. 14.

essential order.[16] The empirical scientific method was later refined and developed by leading modern philosophers of science, such as Karl Popper, who proposed that scientific discovery proceeds through the rigorous testing and elimination of theories revealed to be incorrect on the basis of observational data.[17]

Contemporary practitioners of science continue to view their discipline in methodological terms. One widely cited definition, for instance, depicts science as 'a set of methods designed to describe and interpret observed or inferred phenomena, past or present, and aimed at building a testable body of knowledge open to rejection or confirmation'.[18] This understanding of science has clearly also influenced lay perceptions, as is illustrated by comments of the WTO Appellate Body in its *Hormones* decision. In that case, the Appellate Body approved 'ordinary', dictionary-derived meanings of 'scientific' as including:

> 'of, relating to, or used in science', 'broadly, having or appearing to have an exact, objective, factual, systematic or methodological basis', 'of, relating to, or exhibiting the methods or principles of science' and 'of, pertaining to, using or based on the methodology of science'.[19]

The modern emphasis on the methodology of science – and particularly its underlying rationality and empiricism – founds claims for the objectivity of scientific knowledge and its superiority as a source of information about the natural world. As Anthony O'Hear remarks, today the popular perception is that 'science and science alone gives access to the ultimate truth about man and the world'.[20] Science has thus become 'something for the modern Western culture to believe in, a world-view taking the place of religion'.[21] Indeed O'Hear goes further, presenting science as 'a mythology, perhaps the prevailing mythology of our time'.[22]

16 Lee Godden, 'Preserving Natural Heritage: Nature as Other', Melb. Uni. L. Rev., 22 (1998), 725–6.

17 Karl Popper, *The Logic of Scientific Discovery*, 4th edn, (London: Hutchinson, 1980), often referred to as the falsification model.

18 Michael Shermer, *Why People Believe Weird Things: Pseudoscience, Superstition, and Other Confusions of Our Time*, 2nd edn, (New York: W. H. Freeman, 1997), p. 17.

19 *EC – Measures Concerning Meat and Meat Products*, Report of the WTO Appellate Body, WT/DS26/AB/R & WT/DS48/AB/R, 16 January 1998 (*Hormones*), [187], footnote 172.

20 O'Hear, *Introduction to the Philosophy of Science*, p. 204.

21 Davies, *Asking the Law Question*, p. 116.

22 O'Hear, *Introduction to the Philosophy of Science*, p. 203.

Global spread of the scientific world-view

Increasingly, the scientific world-view is a truly worldwide view, albeit one that may have achieved its pre-eminence through the appropriation and suppression of the scientific cultures of non-Western civilisations.[23] Prominent sociologists of science, such as Robert K. Merton, described universalism as one of the core norms of science.[24] In a world often rent by political and ideological conflicts, such universalism has a comforting ring of assurance that scientists will 'see the world the same way whether they live in Japan, India, Brazil or the United States'.[25] Moreover, for international law and global governance it holds out the prospect that in debates and disputes over technical issues science can act as a unifying force.[26]

It is not only its potential universality that makes science an appealing partner for international law and global governance, but also the apparent progressiveness of scientific discovery in creating a better world 'for everyone'.[27] The narrative of progress is fundamental to standard accounts of the history of Western science that represent the growth of knowledge as a linear or cumulative process stretching from Galileo to Newton to Einstein and beyond. To be sure, dominant models of scientific methodology, such as Popper's falsification approach, are not always entirely clear as to how the elimination of *incorrect* theories actually *advances* knowledge. Nonetheless, it is generally regarded as 'a truism to say that scientists today know more than scientists in the past'.[28]

The progressivism of science was enthusiastically embraced by international law in the years following the Second World War with the establishment of organisations such as the United Nations Educational, Scientific and Cultural Organization (UNESCO).[29] This organisation was

23 Ziauddin Sardar, 'Above, Beyond, and at the Center of the Science Wars', in Keith M. Ashman and Philip S. Baringer (eds.), *After the Science Wars* (London: Routledge, 2001), p. 120.

24 Robert K. Merton, *The Sociology of Science: Theoretical and Empirical Investigations* (University of Chicago Press, 1973), pp. 267–8.

25 Sheila Jasanoff, 'The Dilemma of Environmental Democracy', *Issues in Science and Technology*, 13(1) (1996), 64.

26 Litfin, *Ozone Discourses*, p. 40.

27 Sandra Harding, *The Science Question in Feminism* (Milton Keynes: Open University Press, 1986), p. 231.

28 Steven Goldberg, *Culture Clash: Law and Science in America* (New York University Press, 1994), p. 7.

29 Evan Schofer, 'Science Associations in the International Sphere, 1875–1990: the Rationalisation of Science and the Scientisation of Society', in John Boli and

particularly influential in disseminating an international principle that held that the coordination and direction of science are necessary tasks of the modern state and require the institution of science bureaucracies. With UNESCO's aid, many states (both developed and developing countries) created domestic scientific policy institutions over the course of the 1950s and 1960s in order 'to comply with the new [global] norm about states' responsibility for science'.[30]

The international concept of development that emerged around the same time also drew links between science and the capacity for a state to progress socially and economically. Underlying the distinction between developed and developing countries 'was the belief that science and technology could transform ... countries into carbon copies of European industrialised states; [s]cience was seen as something that had to be acquired from the West, and technology had to be "transferred"'.[31] Jawaharlal Nehru, the first prime minister of post-colonial India, clearly sensed the perils for developing countries of ignoring science, observing prophetically: '[t]he future belongs to science and those who make friends with science.'[32]

Science and objectivity

Alongside progress and universality, another important feature of the modern, global conception of science is the emphasis on scientific neutrality and objectivity. In her discussion of the 1975 World Heritage Convention (sponsored and administered by UNESCO), Lee Godden catalogues how 'objective' scientific understandings of the natural environment came to be seen as a culturally and political neutral basis for the identification of natural heritage areas of 'outstanding universal value'.[33] The prominence given to objective scientific criteria in listing decisions also supports the Convention's distinction between 'natural' and 'cultural' areas of world heritage. In recent years this division has come under increasing pressure from indigenous peoples and developing countries who refute the Western scientific approach of separating

George M. Thomas (eds.), *Constructing World Culture: International Non-Governmental Organisations since 1875* (Stanford University Press, 1999), p. 264.

30 Martha Finnemore, 'International Organizations as Teachers of Norms: the United Nations Educational, Scientific, and Cultural Organization and Science Policy', *International Organization*, 47(4) (1993), 565.

31 Sardar, 'Above, Beyond, and at the Center of the Science Wars', p. 126.

32 Quoted in Max F. Perutz, *Is Science Necessary?: Essays on Science and Scientists* (New York: E. P. Dutton, 1989), p. vii.

33 Godden, 'Preserving Natural Heritage'.

'natural' and 'cultural' elements of the environment. Nonetheless, scientific views of nature have remained paramount in the Convention's decision-making processes.

The idea that science is value-free and objective posits that processes of scientific observation are independent of the theories within which the resultant data is rationalised. In addition, it is assumed that scientists carrying out their work are (relatively) free from bias that might distort interpretation of the results of studies. This underpins claims that science, unlike other forms of knowledge, 'does cut through political ideology, because its theories are about nature, and made true or false by a non-partisan nature, whatever the race or beliefs of their inventor, and however they conform or fail to conform to political or religious opinion'.[34]

The view that science produces positive knowledge free from the myths and distortions of religion and metaphysics was something argued by August Comte over a century and a half ago.[35] However, the critical importance of establishing science as a purely impartial and objective observer of facts became even more evident following the Second World War, when the nuclear bombs dropped on the Japanese cities of Hiroshima and Nagasaki 'shattered not just bodies and buildings but also the myth that scientists can remain detached from the uses of their knowledge'.[36] A speech given by one of the bomb's main scientific architects, J. Robert Oppenheimer, to the Association of Los Alamos Scientists on 2 November 1945 is illustrative of what has now become an important refrain in global scientific research:

> But when you come right down to it the reason that we did this job is because it was an organic necessity. If you are a scientist you cannot stop such a thing. If you are a scientist you believe that it is good to find out how the world works; that it is good to find out what the realities are; that it is good to turn over to mankind at large the greatest possible power to control the world and to deal with it according to its lights and its values.[37]

According to this conception, science is about finding out 'how the world works' and 'what the realities are'; what the world chooses to do with that knowledge is to be decided 'according to its lights and

[34] O'Hear, *Introduction to the Philosophy of Science*, p. 2.

[35] *Ibid.*, p. 202.

[36] Jasanoff, *The Fifth Branch*, p. v.

[37] Alice Kimball Smith and Charles Weiner (eds.), *Robert Oppenheimer: Letters and Recollections* (Cambridge, MA: Harvard University Press, 1980), p. 317.

its values', not those of scientists. Almost all contemporary working scientists (as well as many in the general public) would still identify with this positivist view of science.[38] Indeed, the facts/values distinction that it creates is crucial to maintaining the 'sacredness' of the scientific world-view,[39] and the legitimacy of scientific knowledge as a neutral arbiter in a divided world.

International law and expertise

The role of science and scientific experts in international law is not a topic that has traditionally attracted the attention of legal scholars.[40] International lawyers are more often preoccupied with 'foreground' issues of interstate politics, law-making and institutional development than with the 'background' activities and decisions of people other than sovereigns and legislators.[41] Recently, however, some scholars have begun to investigate more closely the part played in international law and governance by experts and expert knowledge (of which scientific expertise is a prominent example).

Martti Koskenniemi, for instance, has remarked on the growing 'managerialism' evident in international law that manifests in the creation of separate regimes covering different areas of international activity (such as environmental protection, trade, human rights or security). In this 'mindset', according to Koskenniemi, disputes appear as 'management problems' for which the proper response 'is always technical or economic'.[42] Recast as problems of expert knowledge, vast areas of decision-making are thus left to the various legal and technical experts appointed to the supervisory organs of the regimes.[43]

In an article entitled 'Challenging Expert Rule: The Politics of Global Governance', David Kennedy takes such arguments even further.

[38] Stephen Cole, *Making Science: Between Nature and Society* (Cambridge, MA: Harvard University Press, 1992), p. 5.

[39] Harding, *The Science Question in Feminism*, pp. 39–40.

[40] It is, however, a growing area of interest as evidenced by innovations such as the joint American and European Societies of International Law research forum on 'Changing Futures: Science and International Law', 2–3 October 2009, Helsinki.

[41] David Kennedy, 'Challenging Expert Rule: The Politics of Global Governance', *Sydney Law Review*, 27(1) (2005), 8–9.

[42] Martti Koskenniemi, 'The Fate of Public International Law: Between Technique and Politics', *Modern Law Review*, 70(1) (2007), 14.

[43] *Ibid.*, 4.

Starting from the proposition that power in our world 'lies in the capillaries of social and economic life' rather than in 'interstate diplomacy', Kennedy proposes that the dense network of global rules now governing international affairs means that 'we are increasingly governed by experts'.[44] He describes the role of experts in global governance as one of 'interpreting and enforcing the background norms and institutions which structure activity in the market, in the state, in the family'.[45] Kennedy gives as one example the WTO dispute settlement system, which he sees as 'a mechanism for settling disputes between nations each asserting that *their* background rule is normal and that their trading partner is imposing unfair costs or offering unfair advantages'.[46] In this system it is the decisions of legal and other experts (including scientific experts in health and environmental cases) that are crucial in determining 'globally tolerated levels of differentiation' among national regulatory standards.[47]

Kennedy notes that a key aspect of the background work performed by experts in international law and governance structures is the deployment of 'vocabularies of advice, implementation, technique, know-how' that represent the resultant decisions as technical and help to locate issues of politics 'elsewhere'.[48] Yet despite this prevalent self-image, Kennedy asserts that the role of experts is more than one of simply providing advice to states or inter-governmental organisations. Rather, he argues, '[e]xpertise can shape how problems are defined and narrow the range of solutions considered.'[49]

Interesting empirical work is now beginning to emerge in the international legal literature that lends credence to this idea, relying upon a close examination of actual practice in different areas of international law. For instance, Andrew Lang tracks the symbiotic relationship that has developed under the General Agreement on Trade in Services between the relevant WTO law and expert knowledge in shaping the understanding and dynamics of the global services economy.[50] Such work builds on a variety of other multidisciplinary work investigating the close association between expert knowledge and international institutions such as the International Monetary Fund, the Food and

44 Kennedy, 'Challenging Expert Rule', 6–7.

45 *Ibid.*, 6. 46 *Ibid.*, 12. 47 *Ibid.* 48 *Ibid.*, 15. 49 *Ibid.*, 17.

50 Andrew T. F. Lang, 'Legal Regimes and Regimes of Knowledge: Governing Global Services Trade', LSE Law, Society, Economy Working Papers 15/2009, available at http://ssrn.com/abstract=1423538.

Agriculture Organization and the Intergovernmental Panel on Climate Change (IPCC).[51]

As Koskenniemi, Kennedy and Lang all note, an important consequence of the increasing prominence of experts in global governance is whether the resulting decisions can be politically contested, despite their portrayal as technical matters of expert knowledge. This is a highly relevant question in the field of global risk regulation where, as we explore later in the chapter, the scientific and objective presentation of decision-making is belied by the inherent value dimension of choices about risk identification and control.

Science, expertise and international environmental law

Perhaps in no other field of international law does the rationale of science and scientific expertise as an impartial 'interpreter of reality'[52] hold greater sway than in that concerned with the protection of nature and the prevention of environmental pollution. Scientific research, monitoring and advice are widely regarded as necessary for the functioning of environmental treaties and other international environmental institutions. Scientific experts from a growing array of disciplinary areas people many global environmental regulatory structures, such as standard-setting bodies, risk assessment panels and advisory bodies to international courts. Their role according to one leading international environmental law text is not to take policy decisions 'that are ultimately the responsibility of politicians', but rather '"to refine problem definition and to identify and expand the range of response options", setting out uncertainties, assumptions, and the probable consequences of action or inaction'.[53]

The current predominance of science and scientific experts in international environmental law arguably has its heritage in the influential role played by the Western scientific tradition in shaping notions of nature and the environment since the Enlightenment period. While the exploitation and subjugation of nature was the early focus of Western science, environmental protection emerged as an issue of widespread concern as the consequences of large-scale industrialisation

[51] In relation to the IPCC see further Chapter 6.

[52] Litfin, *Ozone Discourses*, p. 29.

[53] Patricia Birnie and Alan Boyle, *International Law and the Environment*, 3rd edn, (Oxford University Press, 2009), p. 99 citing Lee A. Kimball, *Treaty Implementation: Scientific and Technical Advice Enters a New Stage* (1996), p. 7.

and intensive development became evident in the late nineteenth century.[54] Initial legal developments to safeguard natural resources (particularly forests and watercourses) were closely tied to the research efforts of eighteenth- and early nineteenth-century scientists whose work revealed links between deforestation and problems such as flooding, siltation and erosion.[55] The earliest fauna protection treaties, such as the 1902 Convention to Protect Birds Useful to Agriculture,[56] and the 1911 Convention for the Preservation and Protection of Fur Seals,[57] also followed in the wake of better understanding of the effects of destruction of habitat and over-harvesting on species.

However, it was particularly in uncovering the health and environmental risks of transboundary pollution posed by industrialisation that scientific expertise came to the fore. In the first major international dispute over atmospheric pollution, the *Trail Smelter Arbitration* (1938–41), the Arbitral Tribunal received advice from independent scientists and heard from a variety of experts regarding the links between sulphur dioxide emissions from a zinc smelter at Trail in British Columbia and damage to crops, forests, soil and waterways across the border in Washington State.[58] On this basis, the Tribunal made findings about the mechanism of atmospheric distribution of the sulphur dioxide gas and the extent of its detrimental effects on property and the environment in Washington State.

The *Trail Smelter Arbitration* was a harbinger of the kinds of environmental problems of greatest contemporary concern – those where the causes are imperceptible to the untrained eye, requiring study and the 'sensory organs' of science for detection.[59] Indeed, many of the environmental issues currently on the international agenda, such as ozone depletion and climate change, were only recognised as problems

[54] Philippe Sands, *Principles of International Environmental Law*, 2nd edn, (Cambridge University Press, 2003), pp. 26–30.

[55] *Ibid.*, pp. 26–7.

[56] Convention to Protect Birds Useful to Agriculture, 19 March 1902, 191 Consol. T. S. 91.

[57] Convention on the Preservation and Protection of Fur Seals, 7 July 1911, in force 14 December 1911, 214 Consol. T. S. 80.

[58] The two awards issued by the Arbitral Tribunal are reproduced in Am. J. Int'l L., 33 (1939), 182 and Am. J. Int'l L., 35 (1941), 684. Interestingly, the Tribunal was little impressed by the evidence presented by the parties' experts given their conflicting views and the potential for bias in an adversarial setting: Am. J. Int'l L., 33 (1939), 195–6.

[59] Beck, *Risk Society*, p. 27.

through inputs from research.[60] Improvements in scientific techniques and instruments have allowed greater precision and sophistication in scientific understanding of the natural world, even at the same time as revealing problems to be ones resulting from technological development.[61]

As a practical matter, the interrelationship between science and knowledge of a global environmental issue has thus had the effect of bringing 'diplomats and international lawyers together with the scientific community in ways not often seen in other areas of international law'.[62] In many cases the ability to produce credible scientific evidence that a lack of action by the international community could give rise to significant adverse effects will enhance '[t]he ease with which an international lawyer is able to present a cogent case for international legislation'.[63]

In modern international environmental law – that which emerged following the watershed 1972 Stockholm Conference on the Human Environment – the close association between scientific and legal developments is attested to in the main soft law documents that articulate the field's underlying norms and principles.[64] The Stockholm Declaration, issued at the 1972 Conference, endorsed science and technology as the means for 'identification, avoidance and control of environmental risks and the solution of environmental problems'.[65] A decade later at the United Nations Conference on Environment and Development (UNCED), governments observed that 'the sciences are increasingly being understood as an essential component in the search for feasible pathways toward sustainable development.'[66] Agenda 21, the policy framework for achieving sustainable development formulated at UNCED, devoted an entire chapter to the topic of science for sustainable development, stating that:

60 Steinar Andresen *et al.*, *Science and Politics in International Environmental Regimes: Between Integrity and Involvement* (Manchester University Press, 2000), p. 3.

61 Carlo Jaeger *et al.*, *Risk, Uncertainty, and Rational Action* (London: Earthscan Publications Ltd, 2001), p. 9.

62 Sands, *Principles of International Environmental Law*, p. 6.

63 *Ibid.*

64 Patricia Birnie and Alan Boyle, *International Law and the Environment*, 2nd edn, (Oxford University Press, 2002), pp. 24–7.

65 Declaration of the United Nations Conference on the Human Environment, U.N. Doc. A/CONF.48/14 (June 16, 1972), Principle 18.

66 UNCED, Agenda 21: Programme of Action for Sustainable Development (1993), [35.2].

One role of the sciences should be to provide information to better enable formulation and selection of environment and development policies in the decision-making process. In order to fulfil this requirement, it will be essential to enhance scientific understanding, improve long-term scientific assessments, strengthen scientific capacities in all countries and to ensure that the sciences are responsive to emerging needs.[67]

Most recently, the state participants at the 2002 World Summit on Sustainable Development (WSSD) reiterated the importance of science and technology for sustainable development in the Summit's Plan of Implementation. The Plan called for the improvement of policy-making and decision-making through urgent action at all levels to:

- increase the use of scientific knowledge and technology;
- make greater use of integrated scientific assessments, risk assessments and inter-disciplinary and inter-sectoral approaches;
- continue to support and collaborate with international scientific assessments supporting decision-making;
- assist developing countries in developing and implementing science and technology policies;
- establish partnerships between scientific, public and private institutions by integrating scientists' advice into decision-making bodies in order to ensure a greater role for science, technology development and engineering sectors; and
- promote and improve science-based decision-making and reaffirm the precautionary approach.[68]

Today it is rare to find an international instrument or multilateral treaty dealing with matters of health or environmental risk that does not make reference to science or technical considerations.[69] At the very least, most international environmental treaties provide for the transfer of necessary technologies to developing countries, the exchange of scientific information and the allocation of funds for technical assistance and capacity building.[70] A significant number of multilateral

[67] *Ibid.*, [35.1]. See also Rio Declaration on Environment and Development, A/CONF.151/26 (Vol. I) (1992), Principle 9.

[68] Plan of Implementation of the World Summit on Sustainable Development, A/CONF/199/20, (1992) [103].

[69] House of Lords Science and Technology Committee, *Science and Treaties* (2004, HL 110-I, London). See also the catalogue of institutions 'for the performance of science policy functions within MEAs' in Haas, 'Science Policy for Multilateral Environmental Governance', appendix.

[70] Examples include the Convention on Wetlands of International Importance especially as Waterfowl Habitat, Ramsar, 2 February 1971, in force 17 December 1975 996 UNTS 245, Article 4(3); Convention for the Protection of the World Cultural

environmental treaties, as well as less institutionalised forms of governance, such as the free trade agreements discussed in the previous chapter, also have standing scientific or technical advisory bodies made up of experts from a variety of disciplines who act independently of the governments appointing them.[71] Key functions performed by such bodies include the receipt and audit of national reports, the review of parties' compliance with technical obligations, the issue of guidelines for activities such as environmental monitoring, and performing assessments of

and Natural Heritage, Stockholm, 16 November 1972, in force 17 December 1975, 1037 UNTS 151, Article 22; Convention on the Prevention of Marine Pollution by Dumping of Wastes and Other Matter, London, 13 November 1972, in force 30 August 1975, 1046 UNTS 120, Article IX; Convention on the Conservation of Antarctic Marine Living Resources, Canberra, 5 May 1980, in force 7 April 1982, 1329 UNTS 48 (CCAMLR) Article XX; Vienna Convention for the Protection of the Ozone Layer, Vienna, 22 March 1985, in force 22 September 1988, 1513 UNTS 293, arts. 3, 4; Montreal Protocol on Substances that Deplete the Ozone Layer, Montreal, 16 September 1987, in force 1 January 1989, 1522 UNTS 3, arts. 9, 10, 10A; Basel Convention on the Control of Transboundary Movements of Hazardous Wastes and their Disposal, Basel, 23 March 1989, in force 5 May 1992, 1673 UNTS 57, arts. 10(2), 14(1); Protocol on Environmental Protection to the Antarctic Treaty, Madrid, 4 October 1991, in force 14 January 1998, 30 ILM 1455, Article 6(1); United Nations Framework Convention on Climate Change, Rio De Janeiro, 9 May 1992, in force 24 March 1994, 1771 UNTS 164 (UNFCCC), arts. 4, 11; Convention on Biological Diversity, Rio De Janeiro, 5 June 1992, in force 29 December 1993, 1760 UNTS 79, (CBD), arts. 12, 16–18, 21; United Nations Convention to Combat Desertification in those Countries Experiencing Serious Drought and/or Desertification, Particularly in Africa, Paris, 14 October 1994, in force 16 December 1996, 1954 UNTS 3, arts. 12, 16–18, 19; Agreement for the Implementation of the Provisions of the United Nations Convention on the Law of the Sea of 10 December 1982 relating to the Conservation and Management of Straddling Stocks and Highly Migratory Fish Stocks, New York, 4 December 1995, in force 11 December 2001, 2167 UNTS 3, arts. 10, 14, 25; Kyoto Protocol to the United Nations Framework Convention on Climate Change, Kyoto, 11 December 1997, in force 16 February 2005, 2303 UNTS 148, Article 10(d); Rotterdam Convention on Prior Informed Consent Procedure for Certain Hazardous Chemicals and Pesticides in International Trade, Rotterdam, 11 September 1998, in force 24 February 2004, 2244 UNTS 337, arts. 14(1)(a), 16; Cartagena Protocol on Biosafety to the Convention on Biological Diversity, Cartagena, 29 January 2000, in force 11 September 2003, 2226 UNTS 208, arts. 20, 22; Convention on Persistent Organic Pollutants, Stockholm, 23 May 2001, in force 17 May 2004, (2001) 40 ILM 532 (POPs Convention), arts. 11(2)(b), 12, 13.

[71] Examples include the Convention on the Conservation of Migratory Species of Wild Animals, Bonn, 23 June 1979, in force 1 November 1983, 1651 UNTS 333, Article VIII; CCAMLR, arts. XIV, XV; Ozone Convention, Decision VCI/6; Antarctic Treaty Protocol, arts. 11, 12; UNFCCC, article 9; CBD, Article 12; Desertification Convention, Article 24; Rotterdam Convention, Article 18(6); POPs Convention, arts. 8, 19(6); Australia–United States Free Trade Agreement, signed 18 May 2004, in force 1 January 2005 [2005] ATS 1, Annex 7-A.

environmental health or quality. In turn, treaty-based scientific bodies may have links with non-governmental or inter-governmental scientific organisations, such as the International Council for Exploration of the Seas, the Group of Experts on the Scientific Aspects of Marine Pollution, the Intergovernmental Oceanographic Commission, the International Council for Science and the IPCC.[72]

Through such mechanisms, the political institutions in a treaty regime have access to expert advice and analysis, including, in some cases, assessments of the state of scientific knowledge on a particular topic, information regarding available technologies and know-how relevant to addressing a particular environmental problem, and reviews of the effectiveness of environmental measures taken under the treaty. Thus, scientific research is recognised in international environmental regimes as 'a major supplier of relevant knowledge', which is drawn on by decision-makers 'for problem identification and diagnosis, and in some cases also for explicit policy advice'.[73]

Science in global environmental governance

An important question raised by the extensive presence of science and expertise in international legal arrangements governing health and environmental matters concerns what influence scientific knowledge exercises over the political institutions of treaty regimes and the behaviour of participating states. After all, if the role played by science is marginal, then the political contestability of expert decision-making becomes less salient (indeed, the concern might then be that insufficient attention is being paid to the scientific evidence). The interaction of science and global risk governance has, once again, not been the subject of much analysis in the international legal literature. One of the few areas where the relationship between scientific evidence and international law has emerged as a major focus of debate and scholarship is the adjudication of disputes under the SPS Agreement, discussed in detail in Chapter 5.

By contrast, the place and influence of science in global environmental governance has been the subject of several major studies in the social scientific literature.[74] Often the focus of this research is on how

[72] Birnie and Boyle, *International Law and the Environment*, 3rd edn, pp. 99–100.

[73] Andresen *et al.*, *Science and Politics in International Environmental Regimes*, pp. 182–3.

[74] For details see William C. Clark, Ronald B. Mitchell and David W. Cash, 'Evaluating the Influence of Global Environmental Assessments', in William C. Clark *et al.* (eds.),

to make 'science policy' resonate in international treaties and decision-making processes. This reflects a more general concern that in high-profile processes of international negotiation scientific knowledge may play a subsidiary role to political concerns, ultimately reducing the effectiveness of global environmental agreements.[75]

Science in international legislative and administrative processes

Empirical research tends to support the conclusion that non-scientific factors can play an influential role during the legislative phase of treaty negotiation and establishment of a new environmental regime, although not necessarily the implication that this has detrimental consequences for the regime's effectiveness. For example, Karen Litfin's analysis of negotiations for the Montreal Protocol suggested the 'crucial' importance of two non-scientific factors in mediating the capacity of the available scientific evidence to facilitate global cooperation regarding controls on the production and consumption of ozone-depleting substances. These factors were first, the ways in which ozone research was framed and interpreted by a group of ecologically minded policy officials, or 'knowledge brokers', within American and United Nations environmental agencies, and second, the discovery, part way through the negotiations, of the Antarctic ozone hole which enhanced the political acceptability of a precautionary approach in the face of scientific uncertainty.[76]

Another, oft-cited, example where an interactive relationship between science and politics seems to have been crucial in generating sufficient international agreement regarding the reality and seriousness of environmental risk is with regard to global action on climate change. The climate change field has been described as one 'born in politics' given the links between greenhouse gas production and economically important energy and land-use sectors in many nations.[77] The initial global assessment of climate change science produced by

Global Environmental Assessments: Information and Influence (Cambridge, MA: MIT Press, 2006), p. 1.

[75] Lawrence E. Susskind, *Environmental Diplomacy: Negotiating More Effective Global Agreements* (New York: Oxford University Press, 2994), p. 63; Gareth Porter, Janet Welsh-Brown and Pamela S. Chasek, *Global Environmental Politics*, 3rd edn), (Boulder, CO: Westview Press, 2000), pp. 18–19.

[76] Litfin, *Ozone Discourses*, p. 187.

[77] Shardul Agrawala, 'Context and Early Origins of the Intergovernmental Panel on Climate Change', *Climatic Change*, 39 (1998), 614.

the IPCC in 1990 was, by many accounts, influential in catalysing the decision-making process that led to the conclusion of the United Nations Framework Convention on Climate Change (UNFCCC) at UNCED in 1992. This occurred despite the uncertainties in the knowledge base at the time and the high political stakes of the issue.[78]

Overall, however, the IPCC's influence over international climate change policy and law has primarily been attributed to the organisation's unique institutional structure that involves both scientists and government officials in the process by which scientific assessments are translated into policy recommendations.[79] In particular, conclusions which appear in the IPCC's all-important *Summary for Policy-makers* are produced through a process of negotiation between government officials and the lead authors of the scientific assessments. One former chairman of the IPCC, Robert Watson, has commented that this process has proved much more powerful than one producing scientific findings in isolation from policy concerns. Although participants 'may not all agree with the outcome', he observed that 'if they're all part of designing the process in the beginning, they'll be more willing to let the chips fall where they may.'[80]

Once the intensity of the legislative phase is past, a different science policy dynamic often takes hold within regimes, particularly where the decisions necessary for the ongoing implementation of state obligations are seen to be of a technical character. This is consistent with the emergence of global governance as a new 'regulatory layer' in international law, focused on 'low politics' issues of regime administration (or in Kennedy's terms, 'background' work). The preoccupation of decision-making in this governance mode is frequently questions of regime adjustment and implementation, such as the need to keep treaty measures up to date with evolving scientific and regulatory knowledge.[81] Although such matters – for instance, whether catch levels for a particular fish species are sustainable or if a country's greenhouse emissions accounting system meets international guidelines – appear technical in nature, they often have significant political ramifications.

[78] *Ibid.*, 633.

[79] The processes of the IPCC are discussed in more detail in Chapter 6.

[80] Quoted in Andresen *et al.*, *Science and Politics in International Environmental Regimes*, p. 174.

[81] House of Lords Science and Technology Committee, *Science and Treaties*, [6.10].

Given the presentation of the work of global environmental governance as technical, even mundane, those involved in the ongoing implementation of international environmental regimes are generally government officials (rather than political representatives) and scientific experts who specialise in the area concerned. The kinds of tasks undertaken by these actors include the articulation of policy-relevant scientific knowledge, the drafting of reports and the preparation of responses to questions put by the secretariat or the state members of the treaty.[82] The separate existence and operation of different international legal regimes (for instance, different treaties deal with climate change, marine pollution and biodiversity despite the interrelationship in practice between these environmental issues), means that distinct networks of actors are mobilised for each regime. Research into the activities of such networks under multilateral environmental governance arrangements suggests that they are capable of generating scientific consensus, which in turn may play an influential role in policy-making and decision-making.

Epistemic communities and global environmental governance

This view of the work of experts in global environmental governance is presented most clearly in the writings of the social scientist Peter Haas. Haas coined the term 'epistemic communities' to describe the networks of experts and officials operating within multilateral treaty bodies.[83] According to Haas, epistemic communities disseminate knowledge and influence the policy process through their interpersonal networks, helping to shape the interests of state participants in a regime.[84] The defining feature of epistemic communities is their commitment to a shared set of normative and principled beliefs, shared causal beliefs, shared notions of validity and a common policy enterprise.[85] As Martin Shapiro puts it, a 'French nuclear engineer and a Greek nuclear engineer are far more likely to see eye to eye than a French and a Greek politician, and the eye they see with is likely to be nuclear engineering'.[86] Haas' work

[82] Haas, 'Science Policy for Multilateral Environmental Governance', p. 130.

[83] This idea was first presented in Haas' seminal work, Peter M. Haas, *Saving the Mediterranean: the Politics of International Environmental Cooperation* (New York: Columbia University Press, 1990).

[84] Peter Haas, 'Epistemic Communities and International Policy Coordination', *International Organization*, 46(1) (1992), 1.

[85] *Ibid.*, 3.

[86] Martin Shapiro, '"Deliberative", "Independent" Technocracy v. Democratic Politics: Will the Globe Echo the E.U.?', IILJ Working Paper 2004/5 (Global Administrative Law Series, 2004), 9, available at www.iilj.org.

suggests that the expertise and claims advanced by epistemic communities achieve their greatest influence when they are 'developed behind a politically insulated wall'.[87] However, as we shall see in Chapter 6, representation of the work of epistemic communities as technical in nature is belied in practice by the intervention of political factors that are often necessary to give scientific assessments broader legitimacy and policy salience.

Forums conducive to the formation and consolidation of the influence of epistemic communities exist under the administrative structures of a number of multilateral environmental agreements (MEAs), as well as under other global risk governance arrangements. For example, subsidiary scientific bodies established under major MEAs, such as the regimes in place governing long-range transboundary pollution,[88] ozone depletion,[89] climate change[90] and biodiversity reduction,[91] bring together experts on a regular basis to undertake tasks such as reviewing treaty-based regulatory controls in light of available scientific, environmental, technical or economic information. Similar standing expert bodies dealing with matters of sanitary and phytosanitary (SPS) risk exist under several free trade agreements between the USA and other nations.[92] In the Australian Senate inquiry into the Australia-United States Free Trade Agreement, the function of the agreement's standing body on SPS matters, was described to the reviewing committee in the following terms:

87 Haas, 'Science Policy for Multilateral Environmental Governance'.

88 Scientific support to the regime established by the 1979 Convention on Long-Range Transboundary Pollution is provided by the Steering Body to the Cooperative Programme for Monitoring and Evaluation of the Long-range Transmission of Air Pollutants in Europe (EMEP). For details of this body's work see www.emep.int/.

89 The Technology and Economic Assessment Panel, working from within the Montreal Protocol Secretariat, provides technical information on technological responses and alternatives to ozone-depleting substances, at the parties' request: see http://ozone.unep.org/teap/.

90 In addition to the IPCC, the UNFCCC has a subsidiary scientific body – the Subsidiary Body for Scientific and Technological Advice – which deals with methodological and scientific questions arising under the UNFCCC: see http://unfccc.int/essential_background/convention/convention_bodies/items/2629.php

91 The CBD established the Subsidiary Body on Scientific, Technical and Technological Advice to advise the Conference of the Parties on the technical implementation of the Convention: www.cbd.int/sbstta/.

92 E.g., Australia–US FTA, Annex 7-A; United States-Chile Free Trade Agreement, 6 June 2003, in force 1 January 2004, Article 6.3; Dominican Republic-Central America-United States Free Trade Agreement, 5 August 2004, in force 1 January 2009, Article 6.3; United States-Morocco Free Trade Agreement, 2 March 2004, in force 1 January 2006, art 19.2; United States-Morocco Joint Statement on Sanitary and Phytosanitary (SPS) Cooperation, 15 June 2004.

scientists on either side will attempt to achieve a meeting of their scientific minds and resolve ... to their mutual satisfaction, any of [the] kinds of issues which are germane in an import risk analysis or which may not be related to a specific import analysis but may be alive in international debate somehow.[93]

Some international scientific subsidiary bodies provide more than just a forum for discussion and review. A prominent example is the expert committee under the Persistent Organic Pollutants Convention (POPs Convention), which prepares risk assessments and risk management evaluations for chemicals proposed for international listing.[94] Structures of this kind offer the potential for epistemic communities to 'institutionalise' their influence within a regime and to 'insinuate' their particular world-views into broader international politics.[95] By this means, questions that may have had high political saliency during the legislative phase of regime creation are recast as matters of technical assessment, informed by the professional norms and advice of experts.

The supranational experience of the EU with the use of expert advisory committees in regulatory processes suggests there is indeed significant potential for the development of epistemic communities under governance arrangements to facilitate a highly technocratic mode of decision-making.[96] The EU's comitology committees that operate in various areas of risk management assist the Commission to articulate detailed standards and rules giving effect to the often broadly worded EU regulations issued by the Community's political institutions.[97] Scientific committees, which provide advice to a standing body of government representatives in such procedures, comprise scientists who are drawn from EU member states but appointed on the basis of their expertise.[98]

[93] Australian Senate Select Committee, *Final Report on the Free Trade Agreement Between Australia and the USA* (Parliament of Australia, 2004), p. 44.

[94] POPs Convention, Article 8. The scientific assessment processes of the POPs committee are discussed in Chapter 6.

[95] Haas, 'Epistemic Communities and International Policy Coordination', 4.

[96] Shapiro, '"Deliberative", "Independent" Technocracy v. Democratic Politics'.

[97] Concerns over the operation of the comitology procedures led to reforms in 2006 to introduce the 'regulatory committees with scrutiny' process that gives the European Parliament greater powers over the outcomes of comitology processes. However, this new procedure only applies to quasi-legislative implementing measures; for decisions of an administrative nature the old comitology procedures still apply.

[98] For an overview of how comitology processes operate see Sebastian Krapohl, 'Risk Regulation in the EU between Interests and Expertise: The Case of BSE', J. European Public Policy, 10(2) (2003), 189.

Given the constitution of comitology committees as expert bodies, the members often bring with them strong links to existing epistemic communities and their deliberations tend to be informed by shared professional norms. As highlighted in the previous chapter, commentators such as Christian Joerges and Jürgen Neyer have argued that the mix of government officials, experts and other stakeholders brought together by the relatively informal structure of the comitology process results in deliberative outcomes based on persuasion, argument and discursive processes.[99] However, invariably the lexicon of this dialogue is that of the technical expert, making it difficult for non-experts to participate, other than through the production of their own counter-expertise. Rather than a situation of experts providing decision-makers with advice 'on tap', which is then relied upon in the administration of a regime, the reality may well be one where the experts, and the scientific norms of discourse that they favour, end up 'on top'.[100]

Risk and its international regulation

Much of the scientific expertise deployed in contemporary international law is directed at the question of whether or not a health or environmental risk exists that requires action, whether by individual nation states or the international community as a whole. For example, the 'risk profiles' prepared by the POPs Committee discussed above form the basis for its decision as to whether a particular chemical 'is likely as a result of its long-range environmental transport to lead to significant adverse human health and/or environmental effects such that global action is warranted'.[101] Failure to produce credible scientific evidence of a risk is generally also fatal to the claims of any party seeking to assert the existence of adverse health or environmental effects in an international legal dispute.[102]

These examples are illustrative of the current penetration of scientific notions of risk, where 'risk' is seen as an objective, measurable entity combining the probability of an adverse event and the magnitude of its

[99] Christian Joerges and Jürgen Neyer, 'Transforming Strategic Interaction into Deliberative Problem-Solving: European Comitology in the Foodstuffs Sector', J. European Public Policy, 4(4) (1997), 609.

[100] Shapiro, '"Deliberative", "Independent" Technocracy v. Democratic Politics', 4.

[101] POPs Convention, Article 8(7)(a). See further, Chapter 6.

[102] See further, the SPS cases discussed in Chapter 5.

consequences.[103] However, risk connoting a threat of harm, assessed by way of expert evaluation, is a relatively recent phenomenon. Indeed, it is only a particular kind of society – what Ulrich Beck has termed the risk society – that conceives of risk in terms that require scientific definition.

Scientisation of risk

In the pre-industrial era, the term risk was virtually unknown, and if used, was as likely to signify the chance of something good as something bad. For societies living in these times, dangers in the form of famines, floods and earthquakes were aplenty and exacted a much higher toll on populations than they do today.[104] However, these events were viewed as a matter of fate or punishment, visited on humans by a wrathful God.[105]

The renewal of faith in human reason and the decline of religious orthodoxy and tradition that occurred in Europe during the sixteenth and seventeenth centuries saw the emergence of a distinct notion of risk based on the new mathematical theory of probability.[106] For the first time people could make decisions and forecast the future with the help of numbers. In the seventeenth century this notion of risk was applied mostly in a gambling context to predict the potential for losses and gains.[107] However, the idea was quickly taken up and developed into a powerful instrument for organising, interpreting and applying information; one which could be used to calculate life expectancies for annuities or to provide the foundation for a flourishing business in marine insurance.[108] The definition of risk in probabilistic terms thus allowed societies to bring the future into view in ways that had not previously been possible.

Today, the mathematical calculation of probabilities to determine risk underlies the tools of risk management and decision analysis used in a diverse range of fields from insurance to banking, to engineering

[103] John Adams, *Risk* (UCL Press, 1995), p. 8.

[104] Anthony Giddens, *The Consequences of Modernity* (Stanford University Press, 1990), pp. 106–10, noting that this is not necessarily so in less industrialised sectors of the globe.

[105] Deborah Lupton, *Risk* (London: Routledge, 1999), p. 5.

[106] *Ibid.*, 6.

[107] Mary Douglas, 'Risk as a Forensic Resource', *Daedalus*, 119(4) (1990), 1, 2.

[108] Peter Bernstein, *Against the Gods: The Remarkable Story of Risk* (New York: John Wiley, 1996), pp. 2–6.

and the management of health or environmental threats.[109] Concern with risk of all kinds reflects a society that is future-oriented and bent on controlling that future rather than leaving it in the hands of fate, faith or the vagaries of nature. Nevertheless, in the last few decades an important change has occurred in the understanding of the term risk that differentiates it sharply from its previously neutral meaning. In the contemporary lexicon, risk has come to mean hazard, and most usually hazards that are a product, not of fate or sin, but of human activities.[110] As Anthony Giddens puts it:

> At a certain point, somewhere over the past fifty years or so, we stopped worrying so much about what nature could do to us, and we started worrying more about what we have done to nature.[111]

In this setting, the risks that are frequently of greatest concern are those that result from industrialisation and the application of technologies that are a product of scientific discovery. These are generally risks which can result in widespread harm, are 'invisible' to the ordinary person and 'are localized in the sphere of physical and chemical formulas'.[112] Moreover, they are risks whose adverse effects may not become evident until far into the future, 'unleashed by the fathers [but] visited on the heads of their children, even to the *nth* generation'.[113]

Ulrich Beck sees the prominence of such risks in contemporary political debate as evidence of the emergence of the 'risk society' and argues that as industrialisation spreads across the globe, these risks also become globalised.[114] The imperceptibility of potential hazards and the need for complex models and formulae to understand the nature of risks in the risk society requires reference to scientific knowledge. Indeed, the term 'risk' may be applied to describe these new dangers of concern for the very reason that it has 'the aura of science' and hence 'the pretension of a possible precise calculation'.[115]

The growth and spread of the risk society has laid the foundation for today's extensive apparatus of risk-based decision-making, which has developed its own 'analytic infrastructure' of scientists, engineers,

[109] For different applications of risk management concepts see Alan Waring and A. Ian Glendon, *Managing Risk* (London: International Thomson Business Press, 1998).

[110] Lupton, *Risk*, pp. 8–9.

[111] Anthony Giddens, 'Risk and Responsibility', *Modern Law Review*, 62(1) (1999), 1, 3.

[112] Beck, *Risk Society*, p. 23.

[113] Douglas, 'Risk as a Forensic Resource', 5.

[114] Beck, *Risk Society*, p. 21.

[115] Douglas, 'Risk as a Forensic Resource', 4.

social scientists, decision-making theorists and specialised risk management agencies.[116] Risk analysis – covering the sub-disciplines of risk assessment, risk management and risk communication – has emerged as an area of formal study in its own right, with a specialised lexicon, an identifiable professional community and specialist journals. Contemporary risk assessors are most likely to be scientists, or those with scientific background, in fields such as toxicology, epidemiology, engineering, medicine, pharmacology and many others. These professionals engage in an exercise that seeks to select the priority risks for management from the myriad of potential technological hazards (a process which Mary Douglas and Aaron Wildavsky have described as 'the expert answer to the question of how much wealth should be sacrificed for how much health'.)[117]

In the majority of the research and literature on risk management, the mainstream view is of risk as being real, actual, objective and measurable.[118] Although this notion of risk differs from 'the more broadly defined risk of everyday English and everyday life', the technical perspective represents 'the prevailing international orthodoxy on the subject of risk'.[119] The measurement of probabilities and the collection of data on the magnitude of adverse events are central to the modern practice of risk analysis. Science supplies the necessary objectivity for this exercise by providing a knowledge base which can be used in the quantification (although sometimes only the estimation) and management of risk.

Development of risk regulation

In the industrialised world, risk regulation has become a central organising paradigm for modern society.[120] Governments manage political risk; businesses manage economic risks; public health authorities manage disease risks. Most sectors of society are thus seen as having to manage and contain risks inherent to their particular range of activities. Risk regulation has also entered the lexicon of administrative agencies and regulatory authorities as a means to aid decision-making when

116 Jaeger *et al.*, *Risk, Uncertainty, and Rational Action*, p. 19.

117 Douglas and Wildavsky, *Risk and Culture*, p. 67.

118 Ortwin Renn, 'Concepts of Risk: A Classification', in Sheldon Krimsky and Dominic Golding (eds.), *Social Theories of Risk* (Westport, CT: Praeger Publishers, 1992), p. 53.

119 Adams, *Risk*, p. 26.

120 Jaeger *et al.*, *Risk, Uncertainty, and Rational Action*, p. 16.

there is uncertainty about future effects and consequences. Systems for risk assessment and risk management are an integral part of the regulatory structures of governments, and are increasingly being transferred to the international arena with the designation of many risks as ones requiring global control or oversight.

Origins and models of risk regulation

As a technique for decision-making, risk regulation had its origins in the insurance and financial areas, and from there was taken up in mechanical, technological and engineering applications. The complex designs and technologies employed for the latter kind of projects require an understanding of statistical and probability calculation. This is especially the case where engineers are seeking to determine the manner in which constructions may interact with their surrounds, or in gauging how technologies will react in given circumstances, such as in the assessment of the potential for melt-down of a nuclear power plant reactor. In practice, though, the desire for quantitative probability calculations in engineering and technological settings is often attenuated by the reduction of the risk methodologies used to qualitative designations and categories such as 'high' or 'low' risk.

Models of risk assessment and risk management used in the health and environmental fields have been strongly influenced by practice in Western nations such as the USA and the United Kingdom (UK). In the USA risk assessment first became prominent as a means for identifying and evaluating health and environmental hazards in response to the need to fulfil the broad mandates of 1970s federal laws calling for the protection of public health and the environment.[121] At that time the mechanism was much maligned because it involved an uncomfortable mixture of science and policy, as well as dubious extrapolation from research findings to predict potential future harms. William D. Ruckelshaus, the director of the US federal Environmental Protection Agency during this period, candidly described risk assessment as:

> a kind of pretence; to avoid the paralysis of protective action that would result from waiting for 'definitive' data, we assume that we have greater knowledge than scientists actually possess and make decisions based on those assumptions.[122]

121 See, e.g., § 121r Clean Air Act, 42 U.S.C. §7401 et seq. (1970), § 5, 6 Toxic Substances Control Act, 15 U.S.C. §2601 et seq. (1976).

122 William Ruckelshaus, 'Risk, Science and Democracy', *Issues in Science and Technology*, 3(1) (1985), 26.

A turning point in the perception of risk assessment came with the release in 1983 of the US National Research Council's influential report, *Risk Assessment in the Federal Government: Managing the Process.*[123] In the same year, a parallel report was published by the UK's national academy of science, the Royal Society.[124] Both publications quickly became major works of reference on the topic of risk and risk assessment. In the USA the Red Book, as it became known, helped to systematise the various risk assessment procedures applied by US federal agencies by breaking the process down into a series of stages. These comprise:

1. Hazard identification – the determination of whether a particular substance is or is not causally linked to particular health or environmental effects;
2. Dose-response evaluation – the determination of the relation between the magnitude of exposure and the probability of occurrence of the health or environmental effects in question;
3. Exposure assessment – the determination of the extent of human or environmental exposure before or after the application of regulatory controls; and
4. Risk characterisation – the description of the nature and magnitude of risk, including any attendant uncertainty.[125]

Both the Red Book and the Royal Society report also insisted on the distinction between the scientific realm of risk assessment and the political realm of risk policy and management. For instance, the Royal Society report distinguished between objective risk (meaning that evaluated by experts) and perceived risk (entailing lay people's often quite different anticipation of future events). Similarly the Red Book stressed the importance of maintaining 'a clear conceptual distinction between assessment of risks and the consideration of risk management alternatives'.[126] The firm line drawn in both reports between scientifically assessed risk and politically informed risk management decision-making helped to give risk assessment a scientific imprimatur, thereby increasing the profile of the technique.

[123] Sheila Jasanoff, *Designs on Nature: Science and Democracy in Europe and the United States* (Princeton University Press, 2005), p. 265.

[124] Royal Society, *Risk Assessment: Report of a Royal Society Study Group* (London: Royal Society, 1983).

[125] National Research Council, *Risk Assessment in the Federal Government: Managing the Process* (Washington DC: National Academy Press, 1983), p. 3.

[126] *Ibid.*, pp. 19–20.

Risk assessment is now widely accepted as 'a principled approach to ordering knowledge and weighing alternatives' in many countries outside of the USA and UK.[127] The technique has also been enthusiastically adopted and deployed in the regulatory structures of regional organisations such as the EU. For example, the European Food Safety Agency, set up in 2002, prepares independent scientific risk assessments for the political institutions of the EU engaged in risk management. The Agency describes its function as the production of 'scientific opinions and advice to provide a sound foundation for European policies and legislation and to support the European Commission, European Parliament and EU member states in taking effective and timely risk management decisions'.[128]

Dominance of technical perspective on risk

The Red Book's boundary between the 'scientific basis' and the 'policy basis' of risk decision-making (or 'objective' and 'perceived' risk in the Royal Society's terms) is one that has come under increasing attack from social scientists, including those participating in reports issued by the two organisations.[129] The challenge mounted in the social scientific literature to the 'objectivity' of risk assessment is discussed further below. Such challenges notwithstanding, the distinction between objective, expert-assessed risk and subjective risk inaccurately perceived by non-experts retains a tenacious hold on the regulatory imagination and remains the predominant perspective in the risk management literature.[130]

The technical perspective regarding risk has proved resilient even in the face of uncertainties that place substantial obstacles in the way of quantifying many health and environmental risks. In the classic distinction between 'risk' and 'uncertainty', first introduced by Frank Knight in 1921, risk was defined as a 'measurable uncertainty', that is an event for which the probability is known or able to be measured even if the consequences are not, as opposed to uncertainties which are

127 Jasanoff, *Designs on Nature*, p. 267.

128 'About EFSA', European Food Safety Agency website: www.efsa.europa.eu.

129 Royal Society Study Group, *Risk: Analysis, Perception and Management* (London: The Royal Society, 1992); Paul Stern and Harvey Fineberg (eds.), *Understanding Risk: Informing Decisions in a Democratic Society* (Washington DC: National Academy Press, 1996). The former report was not adopted as representative of the Royal Society's views.

130 Adams, *Risk*, p. 10.

non-quantifiable.[131] However, true risk problems are a rarity in contemporary risk management.

This should place significant limits on the utility of technical risk assessment, but in practice the field has proved adept in reinventing itself to respond to prevalent uncertainties. One approach that has been pursued focuses on sharpening the analytical tools for modelling variability and characterising and expressing uncertainties.[132] On other occasions, where deficiencies in the available data are irremediable through further research, professional judgment or assumptions may be used to fill the gap.[133] A related development is the emergence of qualitative risk assessment alluded to above, which circumvents the problem of uncertainty by characterising risk in non-quantitative terms such as 'low', 'moderate' or 'high' risk.

Challenges posed by social scientific research to the objective/subjective risk distinction have proved more difficult to overcome, though in the end not insurmountable. One of the primary challenges in this regard has come from cognitive psychologists investigating the basis of lay and expert risk perceptions. Prominent researchers in this field, such as Paul Slovic, have concluded that lay people's risk assessments are not irrational but merely represent a different, equally valid, way of evaluating risk. Accordingly, lay people evaluate risk in light of a range of qualitative factors, such as the voluntariness and controllability of exposure to a hazard, the potential for catastrophic consequences and the degree of 'dread' associated with a particular risk.[134] In addition, some hazards may appear riskier to the ordinary person if instances of the hazard can be readily called to mind as, for example, when dramatised in the mass media.[135] Risk perception studies have also yielded the important insight that experts are as prone to such biases in their evaluation of risks as lay people.

What has been picked up in the risk management literature from this research, however, is the theme of the divergence of lay risk assessments from those made by experts, which is then overlain with the

[131] Frank H. Knight, *Risk, Uncertainty and Profit* (Boston, MA: Hart, 1921), p. 26.

[132] Andreas Klinke and Ortwinn Renn, 'A New Approach to Risk Evaluation and Management: Risk-Based, Precaution-Based, and Discourse-Based Strategies', *Risk Analysis*, 22(6) (2002), 1078–80.

[133] Mark A. Burgman, *Risks and Decisions for Conservation and Environmental Management* (New York: Cambridge University Press, 2005), pp. 75–81.

[134] See Paul Slovic, *The Perception of Risk* (London: Earthscan Publications Ltd, 2000).

[135] *Ibid*, pp. xxiii–xxviii.

assumption that lay evaluations misperceive risks.[136] New risk regulatory models devised as a result characterise the public's view of risks in terms of 'moral outrage', but counsel that this should nonetheless be taken into account alongside expert evaluations of hazard (calculated via 'objective measures' such as mortality rates).[137] Other approaches prevalent in the risk management literature aim to correct public 'misperceptions' through processes of risk communication, designed to help educate the public about the 'real risks'.[138] Neither approach questions the fundamental assumption of the technical risk perspective that expert processes of risk assessment yield the most rational and objective evaluations of risks to health and the environment. Indeed, if anything, they serve to reinforce the apparently superior rationality of expert risk assessment and the dangers of giving non-experts too great a role in the diagnosis of risk.

The technical risk perspective found in the professional risk literature finds a powerful echo in a number of policy and legal analyses of domestic risk regulation.[139] One of the most sophisticated advocates in this regard is Cass Sunstein. Sunstein has defended 'a highly technocratic approach to risk regulation' on the basis that governments will otherwise regulate on the basis of the public's irrational fears over risk.[140] Relying on the findings of cognitive psychological research demonstrating the intuitive basis of lay risk assessment and the potential for negative perceptions of risk to be amplified through the media, Sunstein argues that ordinary people are prone to 'misfearing'; that is, 'they fear things that are not dangerous, and they do not fear things that impose serious risks'.[141] According to Sunstein, when these fears

136 Something lamented by cognitive psychologists whose research has been co-opted by risk professionals: Baruch Fischhoff, 'Psychology and Public Policy: Tool or Toolmaker?', *American Psychologist*, 45(5) (1990), 647.

137 Peter M. Sandman, 'Risk Communication: Facing Public Outrage', *Management Communication Quarterly*, 2(2) (1988), 235.

138 Jaeger *et al.*, *Risk, Uncertainty, and Rational Action*, pp. 127–30.

139 E.g., Sunstein, *Risk and Reason*; Stephen Breyer, *Breaking the Vicious Circle: Toward Effective Risk Regulation* (Cambridge, MA: Harvard University Press, 1993); Howard Margolis, *Dealing with Risk: Why the Public and the Experts Disagree on Environmental Issues* (University of Chicago Press, 1996).

140 Sunstein, *Risk and Reason*, pp. viii, 294. Similar views are expressed by Sunstein in his more recent book: Cass Sunstein, *Laws of Fear: Beyond the Precautionary Principle* (Cambridge University Press, 2005).

141 Cass R. Sunstein, 'Misfearing: A Reply', *Harvard Law Review*, 119 (2006), 110, 110. As critics have pointed out, Sunstein's account fails to acknowledge the ways in which expert risk assessments may be shaped by similar biases: Dan M. Kahan *et al.*, 'Fear

are translated via democratically responsive institutions into regulatory measures, the result is 'public blunders'.[142] His recommended solution to this problem – where 'the public demand for regulation is likely to be distorted by unjustified fear' – is for a major role to be 'given to more insulated officials who are in a better position to judge whether risks are real'.[143]

Risk regulation in international law

As a mechanism for decision-making on health and environmental risks, expert risk assessment seems to hold increasing appeal for international policy-makers and lawyers, as much as for their domestic counterparts. In the international sphere, moreover, technical perspectives on risk evaluation – focused on the probability of harm and the magnitude of adverse consequences – offer a means of conceptualising health and environmental issues that ostensibly transcends cultural and political differences. Physical harm, quantified by measures such as individual deaths, is something which (almost) all social groups and cultures are able to agree is undesirable.[144]

The perception that risk assessment is an objective basis for decision-making has also contributed to the attractiveness of the approach from the perspective of states. As Sheila Jasanoff explains, this 'allows governing bodies to claim the cognitive high ground, a place from which they can be seen to be acting for the benefit of all without bowing to any particular interests or knowledge claims of the governed'.[145] The deployment of '[o]bjectivist and scientifically reductionist theories of risk' in international affairs holds out the hope of achieving basic agreement on hazards of concern and their seriousness; a prerequisite for programmes of greater harmonisation of national approaches to risk reduction.[146] Some have contended that even broader benefits may flow from the adoption of risk regulatory approaches at the international level. For example, Robert Howse

of Democracy: A Cultural Evaluation of Sunstein on Risk', Harvard L. Rev., 119 (2006), 1071, 1092–6.

142 Sunstein, 'Misfearing: A Reply', 111.

143 Sunstein, *Laws of Fear*, p. 126.

144 Renn, 'Concepts of Risk', p. 61, noting, however, that reducing harm to a single measure can vastly oversimplify multidimensional risks.

145 Jasanoff, *Designs on Nature*, pp. 264–5.

146 Aynsley Kellow, 'Accounting for Risk in Multilateral Negotiations', in David Robertson and Aynsley Kellow (eds.), *Globalization and the Environment: Risk Assessment and the WTO* (Cheltenham: Edward Elgar, 2001), p. 126.

(arguing against critiques of the WTO SPS regulations as undemocratic) sees the Agreement's requirements for scientific evidence and risk assessment 'as enhancing the quality of rational democratic deliberation about risk and its control'.[147]

At present, risk assessment is not yet as widespread a decision-making tool in international law as in national systems such as that of the USA. Nevertheless, those instruments which do include provisions on risk assessment would seem to have exercised a disproportionate effect on other areas of global regulatory activity concerned with health and environmental matters. Extending out from the sphere of international trade laws under the auspices of the WTO, over the course of the last few decades risk assessment has become a more and more prominent part of global institutions and treaty regimes in the health and environmental field.

The leading international instrument to adopt a risk assessment approach is the WTO SPS Agreement, which came into force at the beginning of 1995. The SPS Agreement explicitly requires WTO members to 'ensure that their sanitary or phytosanitary measures are based on an assessment, as appropriate to the circumstances, of the risks to human, animal or plant life or health, taking into account risk assessment techniques developed by the relevant international organizations'.[148] Similar provisions are found in many US-negotiated free trade agreements, commencing with the 1992 North American Free Trade Agreement,[149] negotiations for which ran in parallel with the WTO SPS negotiations. The emphasis on risk assessment in these treaties reflects a prevalent view in the trade community – particularly voiced by agricultural exporting countries in the SPS negotiations – that rigorous scientific and transparent evaluations of risk are a necessary antidote to neo-protectionism in the form of overly stringent regulatory requirements for the import of goods into a member country's territory.[150]

The institution of a requirement for risk assessment under the WTO SPS Agreement served to generate or augment cultures of

147 Robert Howse, 'Democracy, Science, and Free Trade: Risk Regulation on Trial at the World Trade Organization', Michigan L. Rev., 98 (2000), 2330.

148 SPS Agreement, Article 5.1.

149 North American Free Trade Agreement, 17 December 1992, 32 ILM 289, Articles 712, 715.

150 Ian Holland and Aynsley Kellow, 'Trade and Risk Management: Exploring the Issues', in David Robertson and Aynsley Kellow (eds.), *Globalization and the Environment: Risk Assessment and the WTO* (Cheltenham: Edward Elgar, 2001), p. 229.

risk assessment in the three international standard-setting bodies referenced by its provisions: the Codex Alimentarius Commission, the International Office for Epizootics and the International Plant Protection Convention. Of these bodies, the International Plant Protection Convention was the least equipped to meet the new demands imposed by the SPS Agreement, necessitating extensive revision of its provisions to introduce new requirements relating to 'pest risk analysis'.[151] The International Office for Epizootics was also lacking risk assessment guidelines for some areas, such as aquatic animal health, as a representative of the organisation acknowledged in the WTO *Salmon* dispute.[152]

In the case of Codex, it already had relatively well-developed (albeit much-contested) risk assessment procedures for establishing standards on residues and additives in food prior to 1995.[153] These processes, modelled on those in the regulatory systems of the USA and Europe, based standards for the acceptable daily intake of a particular residue or additive on technical risk assessments conducted by independent expert committees. Nevertheless, the added political saliency given to Codex's work by the SPS Agreement saw attempts to elaborate more detailed risk assessment guidelines. In July 2003 Codex adopted the 'Working Principles for Risk Analysis for Application in the Framework of the Codex Alimentarius', which are highly reminiscent of the US Red Book. Codex's 'Working Principles' thus advise that risk assessment 'should be based on all available scientific data' and embrace an expert evaluation of food safety risks divided into the four standard stages of 'hazard identification, hazard characterization, exposure assessment and risk characterization'.[154]

The effects of the adoption of risk assessment techniques in the SPS Agreement are also increasingly evident in MEAs concluded since 1995, negotiations for which often have one eye to the potential for clashes

[151] IPPC New Revised Text, approved by the FAO Conference at its 29th session, November 1997, arts. IV(2)(f), VII(2)(g), VIII(1)(c).

[152] *Australia – Measures Affecting Importation of Salmon*, Report of the Panel, WT/DS18/R, 12 June 1998 (*Salmon* Panel Report), Annex II, para. 47.

[153] See the discussion in Chapter 6. See also Terence Stewart and David Johanson, 'The SPS Agreement of the World Trade Organization and International Organizations: The Roles of the Codex Alimentarius Commission, the International Plant Protection Convention, and the International Office of Epizootics', Syracuse J. Int'l L. & Commerce, 26 (1998), 40–6.

[154] Report of the Codex Alimentarius Commission, 26th session, Rome, 30 June–7 July 2003, ALINORM 03/41, Appendix IV, [19], [20].

with the global trade regime.[155] In this regard, a prominent example of the adoption of risk regulatory requirements in an international environmental treaty is the Biosafety Protocol, concluded in 2000. The Biosafety Protocol explicitly focuses on 'risks to biological diversity, taking also into account risks to human health' flowing from the development, handling, transport, use, transfer and release of 'living modified organisms' (LMOs).[156] The decision-making procedure for national decisions on the first import of an LMO into a party's territory for the purpose of intentional environmental release must incorporate a risk assessment 'carried out in a scientifically sound manner' in accordance with the detailed specifications in Annex III of the Protocol. The purpose of the risk assessment is 'to identify and evaluate the possible adverse effects of living modified organisms on the conservation and sustainable use of biological diversity, taking also into account risks to human health'.[157] In addition, the Protocol requires parties to maintain appropriate risk management measures, 'based on risk assessment' and imposed only to the extent 'necessary' to prevent adverse effects on biodiversity, also taking into account human health risks.[158]

Even outside the sphere where MEAs may come into direct conflict with trade treaties such as the SPS Agreement, risk assessment has gained a significant foothold. Alongside the 2001 POPs Convention,[159] risk assessment is also a central element of listing processes for hazardous chemicals under the Rotterdam Convention on the Prior Informed Consent Procedure for Certain Hazardous Chemicals and Pesticides in International Trade (Rotterdam Convention).[160] The Chemical Review Committee established under the Rotterdam Convention is charged with reviewing national notifications of potentially hazardous chemicals to determine whether they are 'based on a risk evaluation'.[161] This review is a prerequisite for a determination by the treaty's decision-making body that chemicals should be made subject to the prior informed consent procedure.

Together with these prominent treaty examples, there is some evidence that risk assessment may even be replacing the previously

155 Robyn Eckersley, 'The Big Chill: The WTO and Multilateral Environmental Agreements', *Global Environmental Politics*, 4(2) (2004), 24.

156 Biosafety Protocol, Article 2.2.

157 *Ibid.*, Article 15.1.

158 *Ibid.*, Article 16.2.

159 POPs Convention, Article 8.

160 Rotterdam Convention, Annex 2.

161 Rotterdam Convention, Annex 2 (b)(iii).

favoured tool of environmental impact assessment (EIA) in the environmental field. For instance, the International Law Commission's 2001 draft articles on the Prevention of Trans-boundary Harm from Hazardous Activities uses a 'risk of causing significant trans-boundary harm' as its animating concept, defining this concept as 'the combined effect of the probability of occurrence of an accident and the magnitude of its injurious impact'.[162] The scope of coverage of the articles is potentially very broad as they were envisaged by the Commission to apply to '[a]ny activity which involves the risk of causing significant transboundary harm through the physical consequences'.[163] Before authorising any such activity, the draft articles call for states to base decision-making 'on an assessment of the possible transboundary harm caused by that activity, including any environmental impact assessment'.[164]

The Commission's commentary to the draft articles uses the concepts of risk assessment and EIA interchangeably, apparently seeing them as one and the same thing. However, the definition of risk employed – combining probability and consequences – bears the strong hallmarks of the technical risk perspective. By contrast, the practice of EIA has developed in a different direction, evaluating potential harms in a contextualised and often multidimensional fashion. This has encouraged the consideration of public views on environmental impact and the acceptance of participatory processes to a much greater degree in EIA than in standard processes of risk assessment.[165]

On the other hand, risk measured in terms of probability and magnitude of consequences tends to be distilled down to a single aggregate value (for instance a low or one-in-one-thousand risk), which encourages scientisation and 'the pretension of a possible precise calculation'.[166] The turn from EIA to risk assessment in international environmental law may thus strengthen the universalising, scientific discourse of risk assessment, resulting in an oversimplification of complex environmental problems and a reduction in the opportunities available for non-experts to contest the results of an evaluation.

162 *Draft Articles on Prevention of Transboundary Harm from Hazardous Activities*, Report of the International Law Commission on the Work of its Fifty-third Session, A/56/10, 2001, Article 2 (commentary [2]).

163 *Ibid.*, Article 1 (commentary, [2]).

164 *Ibid.*, Article 7.

165 See, generally, Sands, *Principles of International Environmental Law*, chapter 16.

166 Douglas, 'Risk as a Forensic Resource', 4.

Science, risk assessment and their discontents

Paralleling the advance of science-based risk assessment procedures in international law is a deepening malaise in the public reception of science and technology. Ironically such discontent has been most manifest in those societies which, in the past, were enthusiastic supporters of scientific progress.[167] In the closing decades of the twentieth century, the fallibility of science and technology was illustrated by a number of catastrophic events that still resonate in the public memory – Chernobyl, Bhopal and the discovery of the Antarctic ozone hole. On both sides of the Atlantic, there were also a series of incidents that shook public confidence in science in the world's centres of industrialisation and wealth: the Love Canal, Three Mile Island and the Challenger disasters in the USA; food and health scares, culminating in the mad cow disease crisis in the EU.

As a consequence, today people (particularly in Western societies) are much more aware of the potential hazards associated with even the most mundane of technologies. Moreover, as these risks become matters of political debate in industrialised countries, experts have been enlisted to support competing positions. Rancorous disputes between scientists have exposed science as a realm of plurality and disagreement – a 'supermarket for rationalizing political decisions'[168] – rather than the source of authoritative and stable consensus previously assumed.

Assumptions concerning the objectivity of science – as the earlier sections of the chapter highlighted – are deeply embedded within Western societies and the global culture of rationality they have spawned. Perhaps not surprisingly then, such assumptions were once not only prevalent in the public domain, but also within the community of sociologists of science. Since the 1960s, however, more critical social scientific literature has emerged that challenges the objectivity and neutrality of scientific knowledge and mechanisms such as risk assessment that rely on science for their authority. This literature is now well established, although it has often struggled for acceptance 'by an establishment dedicated to the promotion of orthodox, "hard", quantitative scientific activity'.[169] The view of science and risk proffered

[167] Yaron Ezrahi, *The Descent of Icarus: Science and the Transformation of Contemporary Democracy* (Cambridge, MA: Harvard University Press, 1990).

[168] Jaeger *et al.*, *Risk, Uncertainty, and Rational Action*, p. 219.

[169] Horlick-Jones and Sime, 'Living on the Border', 445–6.

by critical social scientific analysis is certainly not always palatable for those who yearn for a truly transcultural basis for risk regulation. Nonetheless, its insights have a new prominence and urgency for global legal structures that seem poised to place their legitimacy claims in the hands of science and expert risk evaluation.

Challenges to scientific objectivity

Up until the 1960s social scientists seemed as convinced of the objectivity of scientific knowledge as the natural and physical scientists themselves. There was a long-standing belief among social scientists at this time that the content of science was determined by nature and that the objectivity of its knowledge claims was assured by the rational and empirical processes of the scientific method. This positivist view of science was radically challenged in 1962 with the publication of Thomas Kuhn's seminal essay on 'The Structure of Scientific Revolutions'.[170] Kuhn's essay itself had a revolutionary impact on the scientific community because it questioned the dominant narrative in the history and philosophy of science that saw the discipline as a rational and progressive endeavour.[171]

The constructivist critique of science

Kuhn argued that scientific research is conducted within an overarching set of theories, methods and commitments that he labelled paradigms. These paradigms, he said, guide what problems are deemed acceptable for investigation and which results are treated as new scientific knowledge.[172] According to Kuhn, most of the time scientists are engaged in 'normal science': 'puzzle-solving' and theoretical 'mopping up' within the preset boundaries of a relevant paradigm.[173] Eventually sufficient, and sufficiently severe, 'anomalies' (the inability to rationalise findings within the constraints of a governing paradigm) may build up, precipitating a 'revolution' and acceptance of a new paradigm.[174] However, in Kuhn's view the transition to a new paradigm is not a linear or cumulative process. Rather, he argued, 'it is a reconstruction of the field from new fundamentals, a reconstruction that changes

[170] Thomas S. Kuhn, *The Structure of Scientific Revolutions*, 3rd edn, (University of Chicago Press, 1996).

[171] Kuhn's essay is by no means the only source of a relativist trend in the contemporary philosophy of science, although it has been one of the most influential.

[172] Kuhn, *The Structure of Scientific Revolutions*, p. 37.

[173] *Ibid.*, pp. 24, 36. [174] *Ibid.*, pp. 82–5.

some of the field's most elementary theoretical generalisations as well as many of its paradigm methods and applications.'[175]

The radical nature of Kuhn's theory lay in his contention that scientists' ways of seeing the natural world depended on the particular paradigm accepted for the purposes of normal science at any time.[176] Indeed, Kuhn's assertion that the accepted paradigm determines what problems are considered worth solving hinted that the choice between paradigms was made, not on the basis of objective criteria, but rather upon ones external to science. At its heart, this theory seems to advance a relativist understanding of science where 'anything goes',[177] although such a conclusion was denied by Kuhn himself.

Following in Kuhn's footsteps, a new breed of constructivist researchers in the field of sociology of science have been far less hesitant, taking his work as a departure point for developing theories of science that see its knowledge claims as a matter of social construction.[178] These researchers have produced numerous studies which show how inscriptions such as graphs in scientific papers are underlain by sophisticated instruments and practices that dictate certain ways of representing nature, that claims of replication of studies (which form the basis of the reliability of the scientific method) are rarely fulfilled in practice, and that the acceptance of papers in scientific journals is determined by the social and cognitive interests of reviewers rather than by the content of the paper itself.[179] The most radical of the social constructivist theories argue that nature is not defined by science, but rather that the social behaviour of scientists determines what is accepted as the laws of nature. Hence, since no group or world-view is more logical or rational than another, it is contended that scientific knowledge has no special claim to authority in its statements about the natural world.[180]

The social constructivist position presents a fundamental challenge to the authority, and indeed the utility, of science and, not surprisingly, is strongly rejected by many scientists. A common position is

[175] *Ibid.*, p. 85. [176] *Ibid.*, p. 111.

[177] Paul Feyerabend, *Against Method*, 3rd edn, (London: Verso, 1993), pp. 14–19.

[178] The approach of such authors to the study of science is introduced in Barry Barnes, David Bloor and John Henry, *Scientific Knowledge: A Sociological Analysis* (London: Athlone, 1996).

[179] The insights yielded by sociological studies of study are usefully summarised for a legal audience in Sheila Jasanoff, 'What Judges Should Know About the Sociology of Science', *Judicature*, 77(2) (1993), 77.

[180] Cole, *Making Science*, p. 11.

that even if some socio-cultural influences can be identified in science, 'it is surely the case that the demand for a fit with nature gives to scientific objectivity a hard edge which is ... lacking in other areas of human activity, such as the arts and literature, where acceptability is defined entirely in terms of human response.'[181] During the 1990s debate between positivists and constructivists over whether science represents truths about nature or simply a series of socially constructed ideas produced the infamous science wars.[182] While no resolution has been reached, for the time being a ceasefire seems to have been negotiated, with each camp agreeing to disagree (violently) with the other's position.

Critical realism and the co-production of knowledge

One of the most important contributions made by the constructivist account of science has been in alerting us to the ways that 'facts and values become linked in our understanding of the world'.[183] Positivist approaches to science and knowledge production fail to capture this critical value dimension of scientific activity. However, constructivist theories within the 'strong programme' in the sociology of knowledge go much further. They share in common the 'persuasion that "truth" is just a term honorifically attached to those items of belief that have managed to prevail – by whatever strategic or rhetorical means – in [the] contest for the high ground of scientific "knowledge" and "progress"'.[184] This leads to the conclusion that there are no truth claims in science, only relative truths, and no definitive way of counting any belief false.

For many – the present author included – this argument is counterintuitive, suggesting as it does that old theories, such as the miasmatic ('bad air') theory of disease prevalent in medieval times, stand on an equivalent footing with the widely accepted scientific beliefs of today, such as the germ theory of disease. Sociologists who oppose the 'strong programme' position have thus often sought a middle ground between a narrowly positivist conception of science and constructivist theories. For example, Stephen Cole puts forward a

[181] O'Hear, *Introduction to the Philosophy of Science*, pp. 214–15.

[182] Keith Ashman and Philip Baringer (eds.), *After the Science Wars* (London: Routledge, 2000).

[183] Horlick-Jones and Sime, 'Living on the Border', 448.

[184] Christopher Norris, 'Truth, Science, and the Growth of Knowledge', *New Left Review*, 210 (1995), 110.

'realist-constructivist' position which concedes that scientific knowledge 'is socially constructed both in the laboratory and in the wider community' but argues that 'this construction is influenced or constrained to a greater or lesser extent by input from the empirical world'.[185] Cole offers a pragmatic justification for his realist-constructivist view of science, arguing that the radical constructivist understanding of scientific knowledge will never win over the wider community. Instead he sees the potential for the nuanced realist-constructivist position to encourage the community to adopt a more sophisticated and realistic view of science.[186]

Key to Cole's approach, and that of other 'critical realists' such as Roy Bhaskar, is the distinction drawn between different types of scientific knowledge. Cole uses the terminology of the 'core' versus the 'research frontier' in science.[187] The former is said to 'consis[t] of a small set of theories, analytic techniques and facts which represent the given at any particular point in time' because they attract a high degree of consensus in the scientific community based on the belief that they are both 'true' and 'important'.[188] For instance, the germ theory mentioned above, which posits that microorganisms are the cause of many diseases, is now a cornerstone of modern medicine and clinical microbiology, and the basis for important innovations such as antibiotics and hygienic practices. By contrast, Cole identifies the research frontier as the site of production of new knowledge where considerable disagreement exists among scientists about what observations are true.[189] This is the kind of science often at issue in risk regulation, where scientists must extrapolate from more established theories in an attempt to predict future patterns of harm.

Other sociologists have taken an acceptance of socio-cultural influences in scientific research and used it to argue for 'the emergence of a new kind of science'.[190] In this new mode scientific knowledge

[185] Cole, *Making Science*, p. x.

[186] *Ibid.*, p. 238.

[187] Bhaskar uses different terms distinguishing between 'intransitive' and 'transitive' objects of knowledge. The former do not depend on human activity. The other is a social product: Roy Bhaskar, 'Philosophy and Scientific Realism', in Margaret Archer *et al.* (eds.), *Critical Realism: Essential Readings* (London: Routledge, 1998), p. 16.

[188] Cole, *Making Science*, p. 15. [189] *Ibid.*, pp. 15–16.

[190] See Michael Gibbons *et al.*, *The New Production of Knowledge: the Dynamics of Science and Research in Contemporary Societies* (London: Sage, 1994); Helga Nowotny, Peter Scott and Michael Gibbons, *Re-Thinking Science: Knowledge and the Public in an Age of Uncertainty* (Cambridge: Polity Press, 2001).

production is intended to transcend conventional disciplinary boundaries generating 'contextualized, or context-sensitive, science'.[191] Along similar lines is the idea of the co-production of knowledge put forward by the diverse collection of researchers in the emerging field of science and technology studies.[192] These researchers employ the concept of co-production as 'shorthand for the proposition that the ways in which we know and represent the world (both nature and society) are inseparable from the ways in which we choose to live in it'.[193] Accordingly, in the co-production framework, science is 'neither ... constituted by interests alone nor ... an unmediated reflection of nature'. Rather, notions of co-production presume 'that knowledge and its material embodiments are products of social work and, at the same time, constitutive of forms of social life'.[194]

What follows from such approaches is a dynamic understanding of the relationship between society and science where the social plays an important role alongside the scientific in shaping what comes to be regarded as knowledge of the natural world. We are only just beginning to come to grips with the implications of these modes of scientific knowledge production for governance systems, such as those operating in the field of global risk regulation. They suggest that in global risk governance, law and the legal processes involved may play a part shaping the very processes of scientific knowledge production that are drawn on to supply the basis for risk decision-making. A further implication is that the authority of science in global risk governance will depend not only on its content, but also its production and dissemination within normatively authoritative institutions and processes.[195]

Uncertainties in science

Even if we accept that science, to some degree, describes real phenomena, another challenge posed to the authority of science lies in the increasing recognition of the problem of uncertainty in many areas of scientific research, especially those concerned with health and the

[191] Nowotny *et al.*, *Re-Thinking Science*, p. vii.

[192] Science and technology studies bring together scholars in the fields of history, philosophy, sociology, politics, law, economics and anthropology.

[193] Sheila Jasanoff (ed.), *States of Knowledge: The Co-Production of Science and Social Order* (London: Routledge, 2004), p. 2.

[194] *Ibid.*, p. 274.

[195] Clark Miller, 'Climate Science and the Making of a Global Political Order', in Sheila Jasanoff (ed.), *States of Knowledge: The Co-Production of Science and Social Order* (London: Routledge, 2004), p. 65.

environment. At issue in this case is the reliability or credibility of scientific knowledge. Concern over the potential for scientific uncertainty adversely to affect decision-making underlies the emergence of the precautionary principle in international law (discussed further in the next chapter), as well as efforts by risk professionals to design better techniques for 'managing' uncertainties or 'factoring' them into risk assessment. Indeed, accommodating uncertainty in science, rather than reducing it to zero, is all that can realistically be achieved. The open-ended, empirical scientific method tolerates an inherent level of uncertainty, and indeed celebrates this as providing opportunities for ongoing inquiry (not to mention research funding!).

There is a common regulatory perception that the residual uncertainty underlying the scientific method must be accepted if all technological development is not to grind to a halt. Accordingly, in the technical risk perspective the task of risk management is seen as one of reducing risk to some acceptable (but not zero) level that is then certified as safe.[196] Global risk governance bodies, such as the WTO Appellate Body, have implicitly endorsed this approach, advising that 'theoretical uncertainty is not the kind of risk which ... is to be assessed' in SPS risk assessments.[197] This suggests that residual risks are not the proper subject of risk management measures under the SPS Agreement, even though if such risks later materialise, the consequences for affected populations could be very serious.

Spectrum of uncertainty in science

Assumptions that the residual uncertainty of the scientific method is the only type of uncertainty of any significance for risk regulatory purposes reflects an overly narrow perception of the challenges uncertainty poses for science. Rather than a mere artefact of the scientific method, uncertainty is increasingly conceived as a core issue in science, particularly in those areas where scientists are operating at the research frontier.

Not only is uncertainty 'inescapable' in scientific research and risk regulation,[198] it also takes a variety of forms. A number of approaches have been put forward in an effort to understand the spectrum of

[196] Chris Whipple, 'Inconsistent Values in Risk Management', in Sheldon Krimsky and Dominic Golding (eds.), *Social Theories of Risk* (Westport, CT: Praeger Publishers, 1992), p. 346.

[197] *Hormones*, [186]. See further Chapter 5.

[198] Adams, *Risk*, p. 26.

uncertainties present in science. One of the most helpful from an interdisciplinary law/science perspective is that developed in the work of Vern R. Walker. Walker has identified five common categories of uncertainty that affect scientific knowledge and science-based judgments about risk. He describes these as follows:

1. Conceptual uncertainty. This is the potential for error created by using particular variables to describe and study the world (as when one theoretical paradigm for study is selected or preferred over another).
2. Measurement uncertainty. This may result from systematic error (because the particular method used does not in fact measure what it purports to measure) or random error (which introduces unreliability into measurements because of faults in the measurement instruments or variability in the competence of those conducting the measurements).
3. Sampling uncertainty. This arises because scientists are often measuring highly variable phenomena, such as fluctuations in species' numbers in a particular ecosystem. Hence a single sample or set of samples may not be representative of actual patterns in the wider environment.
4. Modelling uncertainty. This is associated with using different parameters or formulae in computer modelling, a prevalent technique in areas such as climate change research. Depending on the parameters chosen, different predictions will be produced with differing levels of precision and accuracy.
5. Causal uncertainty. This describes the potential for error about the existence, direction or strength of the causal relationship postulated, which poses the problem that other theories of causation may be closed off prematurely.[199]

In part, the tendency to underestimate the extent of scientific uncertainty in risk assessment may stem from the use of the term 'risk' itself which, as discussed above, has a historical lineage in ideas of calculable probability. Where the notion of risk is employed, it implies that assessors 'know the odds' of a particular adverse event or have the means to measure them. However, as John Adams points out, while there are some problems where the odds are known or able to be accurately estimated with further research, these are 'trivial' in comparison with problems of uncertainty.[200]

[199] Vern Walker, 'The Siren Songs of Science: Toward a Taxonomy of Scientific Uncertainty for Decisionmakers', Connecticut Law Review, 23 (1991), 567. These categories are also summarised in Vern Walker, 'The Myth of Science as a "Neutral Arbiter" for Triggering Precautions', B.C. Int'l & Comp. L. Rev., 26 (2003), 204–11.

[200] Adams, *Risk*, p. 26.

The latter category encompasses instances of 'true uncertainty', where the probability of occurrence of adverse events is unknown, as well as situations of ignorance and indeterminacy. A condition of scientific ignorance exists in cases where both the parameters of the system and the odds of adverse events are unknown (for instance, because there has been little research into the problem). Indeterminacy, on the other hand, describes a situation where causal chains and networks are open so 'we can't know what we need to know'.[201] Indeterminacy might arise because the system under study is so complex that it can only be examined by making a series of assumptions for which scientists have no way of knowing whether they reflect the true position or not.

Ignorance and indeterminacy present difficult problems for science. Although several techniques are used by scientists to confront such issues – such as computer modelling – they do not provide a guarantee against the occurrence of surprises. In addition, ignorance and indeterminacy are often relegated to the 'too hard basket' in scientific research in favour of focusing on more tractable forms of uncertainty that can be addressed through statistical methods and other uncertainty management techniques. Brian Wynne thus argues that scientific knowledge proceeds 'by exogenizing some significant uncertainties, which thus become invisible to it'.[202]

In the realm of research science, putting intractable uncertainties to one side may be a pragmatic approach but 'becomes a problem when (as is usual) scientific knowledge is misunderstood and is institutionalised in policy making as if this condition did not pervade all competent scientific knowledge'.[203] Indeed, there are often significant pressures on decision-makers, as well as the scientists who advise them, to make science appear more certain than it is to support unambiguous decisions that will be defensible in political and legal terms. As one advising expert quipped in a WTO SPS case:

> You know the story of the two-handed scientists: on the one hand and on the other. Lawyers are often looking for one-handed scientists.[204]

201 Brian Wynne, 'Uncertainty and Environmental Learning: Reconceiving Science and Policy in the Preventative Paradigm', *Global Environmental Change*, 2(2) (1992), 111.

202 Brian Wynne, 'Science and Social Responsibility', in Jake Ansell and Frank Wharton (eds.), *Risk: Analysis, Assessment and Management* (Chichester: John Wiley & Sons Ltd, 1992), p. 141.

203 *Ibid.*, p. 142.

204 *EC – Measures Concerning Meat and Meat Products*, Report of the Panel, WT/DS26/R/USA, 18 August 1997, Annex, [800] (per Dr Ritter).

Rather than seeking the (false) comfort of certainty, however, Wynne contends that '[m]ature use of science in public policy would focus ... on the conditions under which it is valid, and whether those conditions prevail in the situation of interest'.[205]

New ways of doing science

Recognition of the prevalence of uncertainty in scientific risk research has also prompted calls in the social science literature for new ways of undertaking such research. The transdisciplinary movement in science is one response to this challenge, which focuses on 'developing a broad-based scientific and cultural approach capable of facilitating long-term dialogue between specialists informed by the new worldview of complexity in science'.[206] Transdisciplinary research takes as its focus complex and interdependent problems that exhibit characteristics of non-linearity, uncertainty and high political stakes in decision-making, with problems in the environmental field being a paradigmatic example.[207] When dealing with such complex problems it is argued that scientific knowledge alone is insufficient. Consequently, transdisciplinary research seeks to engage multiple stakeholders from the beginning in formulating a problem, contributing their heterogeneous skills and expertise to the task of problem-solving.[208]

One of the most well-developed ideas for a new scientific methodology in the transdisciplinary tradition is the notion of 'post-normal science' developed by Silvio Funtowicz and Jerome Ravetz. The starting point for their approach is a distinction between three different types of scenarios, defined in terms of the interaction of two variables: 'systems uncertainties' and 'decision stakes'.[209] Where both uncertainties and stakes are low, they characterise the resultant situation as one of puzzle-solving 'normal science' in the Kuhnian sense. In cases where one of either uncertainties or stakes is significant, decision-making is

[205] Wynne, 'Science and Social Responsibility', p. 142.

[206] Julie Thompson Klein, 'Prospects for Transdisciplinarity', *Futures*, 36 (2004), 515, 516. Transdisciplinarity can be regarded as a complement to the emergence of co-production models of knowledge development in the social sciences.

[207] See Lawrence and Després, 'Futures of Transdisciplinarity', 399–400 for an overview of the nature of this research approach and its differences from multidisciplinary and interdisciplinary work.

[208] Klein, 'Prospects for Transdisciplinarity', 517.

[209] Silvio Funtowicz and Jerome Ravetz, 'Three Types of Risk Assessment and the Emergence of Post-Normal Science', in Sheldon Krimsky and Dominic Golding (eds.), *Social Theories of Risk* (Westport, CT: Praeger Publishers, 1992), p. 253.

said to be in the arena of 'professional consultancy', which involves the exercise of professional skill and judgment.[210] The final situation, in which both uncertainties and decision stakes are high, represents the arena of post-normal science.

When in the 'wild' area of post-normal science, Funtowicz and Ravetz contend that all (scientific and professional experts included) are 'amateurs' because the questions at stake are essentially 'trans-scientific' (that is, they can be asked of, but not answered by, science).[211] The authors' prescription for a 'quality assessment' of scientific materials in such circumstances is to make use of an 'extended peer community' that will use 'extended facts', including anecdotal and community knowledge.[212] In effect, they see the problems of post-normal science as requiring a 'democratisation' of science itself, not 'out of some generalized wish for the greatest possible extension of democracy in society' but rather because 'an extension of peer communities, with the corresponding extension of facts, is necessary for the effectiveness of this new sort of science in meeting the great challenges of our age'.[213]

A looming challenge for practices of post-normal or 'democratised' science, as well as transdisciplinary research more generally, is how conventional scientific research can be combined with other forms of knowledge in risk decision-making processes. Institutional factors, such as the scope of relevant legal rules and the existence of appropriate structures to facilitate broader engagement and participation, will play an important part in determining the extent to which such new forms of science can be operationalised in global risk governance. This a topic to which we return in Chapter 7.

Contingencies in risk assessment

In the 1980s a parallel debate to that in the sociology of science over the objectivity and reliability of scientific knowledge began to emerge in the literature on risk.[214] Constructivist notions of risk were developed across a range of social scientific disciplines, united by the emphasis they placed on the role of social processes in the identification and

[210] *Ibid.*

[211] *Ibid.*, pp. 253–4. The notion of trans-science draws on the work of Alvin Weinberg, 'Science and Trans-Science', *Minerva*, 10(2) (1972), 209.

[212] Funtowicz and Ravetz, 'Three Types of Risk Assessment', p. 254. See also the concept of 'border work' developed by Horlick-Jones and Sime, 'Living on the Border', 445.

[213] Funtowicz and Ravetz, 'Three Types of Risk Assessment', p. 273.

[214] Royal Society Study Group, *Risk: Analysis, Perception and Management*, pp. 111–14.

assessment of risk. These perspectives see value judgments as inherent to the process of risk assessment. Nonetheless, constructivist understandings of risk embrace a wide spectrum of positions. At one end are those that treat 'risk' as entirely a product of historically, socially and politically contingent perspectives. At the other end of the spectrum are perspectives – such as those developed in the work of cognitive psychologists – that recognise risk to be an objective hazard, albeit one mediated through social and cultural processes.[215] The latter categories of constructivist thinking tend to be more compatible with the technical risk perspective. As discussed above, this has resulted in the assimilation of some of their findings into technical risk approaches, although this has occurred very much on the terms of the latter. Other, more radical constructivist theories remain essentially ignored by mainstream risk practice.

Expert versus lay framing of risks

One branch of constructivist risk research that takes very seriously the insight that social context matters when it comes to issues of risk assessment is work that looks at how different framings of hazards by experts and the public can influence their views about risk. A leading scholar in this field is Brian Wynne. Wynne has argued that expert definitions of risk incorporate a series of unarticulated assumptions about relevant actors, behaviours and control processes, which frequently misrepresent the intrinsically open-ended, indeterminate nature of social-situational factors.[216] His research suggests that, by contrast, members of the public, in their framing of controversial risk problems, tend to place more emphasis on the perceived trustworthiness of institutions charged with the assessment and management of risk.

Wynne has illustrated these arguments through a series of case studies of lay–expert interactions with respect to the risks of technological hazards. One such illuminating case study concerned the public controversy over risks posed by use of the herbicide 2,4,5-T in the UK. The UK Scientific Pesticides Advisory Committee set up to assess the risks of the herbicide based its evaluation on the research literature and repeatedly dismissed labour union claims of health damage as imaginary. Wynne details that it later emerged that a critical

[215] Horlick-Jones and Sime, 'Living on the Border', 447.

[216] For an overview of Wynne's approach see Brian Wynne, 'Risk and Social Learning: Reification to Engagement', in Sheldon Krimsky and Dominic Golding (eds.), *Social Theories of Risk* (Westport, CT: Praeger Publishers, 1992), p. 275.

assumption underlying the experts' assessment of safety was that the ideal, laboratory conditions used in toxicological studies of the herbicide would be replicated in the outside world. In other words, central to the experts' risk evaluation were a series of assumptions that extended beyond the available scientific data. These included that:

> the manufacturing process conditions for pesticides never varied sufficiently to produce dioxin and other toxic contaminants, ... drums of herbicide always arrived at the point of use with full instructions intact and intelligible [and] [i]n spite of the inconvenience, farmers and other users would comply with the stated conditions, such as correct solvents, proper spray nozzles, pressure valves, and other equipment, appropriate weather conditions, and full protective gear.[217]

The point of Wynne's studies, and that of other social science research that draws on these insights,[218] is not that either lay or expert participants in risk debates are better informed, but rather that they bring different knowledges to the table. Those different perspectives place emphasis on different hazards and different uncertainties that then affect subsequent attempts to assess the magnitude, seriousness and distribution of likely harms.[219]

Moreover, as case studies such as Wynne's analysis of the herbicide 2,4,5-T risk assessment process suggest, it may be lay citizens, as opposed to experts conditioned by their training to approach scientific knowledge in a particular way, who are better 'at making room for the unknown along with the known'.[220] The implication is that technical risk assessments, relying solely on scientific knowledge and dominated by expert framings of the problem at hand, may present only a partial picture of the risks and uncertainties of significance in any case.

Cultural theory and risk assessment

A more radical challenge to technical, expert-driven processes of risk assessment was posed by constructivist notions of risk that arose in the area of cultural theory. These notions of risk originated in the discipline of anthropology and emerged, initially, as a critique of the

217 *Ibid.*, p. 285.

218 E.g., Les Levidow, 'Risk: Analysis, Perception and Management', *New Scientist*, 136(1841) (1992), 44; Les Levidow and Susan Carr, 'How Biotechnology Regulation Sets a Risk/Ethics Boundary', *Agriculture and Human Values*, 14 (1997), 29.

219 David Winickoff *et al.*, 'Adjudicating the GM Food Wars: Science, Risk, and Democracy in World Trade Law', Yale J. Int'l L., 30 (2005), 97.

220 Jasanoff, *Designs on Nature*, p. 254.

theories of individual risk perception put forward in the work of cognitive psychologists. Cognitive psychology studies are generally conducted by asking individual respondents to give judgments of risk for a range of hazards, either in terms of the degree of similarity or dissimilarity between the hazards or their perceived characteristics.

By contrast, cultural theorists argue that it is artificial to assume perceivers of risk are isolated individuals. In their view, attitudes to risk are shaped by the social structures in which individuals are embedded and the 'cultural bias' they favour. A leading cultural theorist, Mary Douglas, together with her colleague Aaron Wildavsky, contended in a 1982 essay on 'Risk and Culture' that people - both lay and expert - select their risks of concern to conform with (and defend) a specific way of life or culture.[221] Douglas and Wildavsky identified a number of different cultural biases in society - labelled hierarchists, individualists and egalitarians - which they argued exhibit differing levels of group solidarity, differing confidence in the utility of regulatory approaches and different tolerances for risk. The designation of these categories has since come in for significant criticism as it is not clear that Douglas and Wildavsky's categories are the best, or only, explanations for cultural variation in risk perceptions.[222] Nonetheless, the central insight of cultural risk theory remains a powerful one: that cultural world-views are a vital component of the social mechanisms that determine risk perceptions. Accordingly, 'whatever objective dangers exist in the world social organisations will emphasise those that reinforce the moral, political or religious order that holds the group together.'[223] The insight that this offers for international risk regulation is that a failure to acknowledge risk assessment as a value-laden process risks universalising implicit socio-cultural judgments 'about the appropriateness of particular social roles, power relationships, public attitudes, and regulatory styles'.[224]

A considerable body of empirical research now provides support for the theory that cultural world-views exercise significant influence

[221] Douglas & Wildavsky, *Risk and Culture*, p. 9.

[222] Pat Caplan (ed.), *Risk Revisited* (London: Pluto Press, 2000), pp. 12–13.

[223] Steve Rayner, 'Cultural Theory and Risk Analysis', in Sheldon Krimsky and Dominic Golding (eds.), *Social Theories of Risk* (Westport, CT: Praeger Publishers, 1992), p. 87.

[224] Winickoff *et al.*, 'Adjudicating the GM Food Wars', 106.

over peoples' perception of environmental and health risks.[225] This research provides a strong counterpoint to the technical risk perspective advanced by authors such as Sunstein, which draws a sharp distinction between lay and expert risk assessment. It also highlights that experts, as much as members of the general public, 'are inclined to form attitudes toward risk that best express their cultural vision'.[226] Consequently, while experts may have 'a more accurate sense of the magnitude of various risks', they have 'no special competence to identify what vision of society … the law should endorse'.[227]

Nevertheless, cultural perspectives or risk can also lead to the conclusion that there is no defensible basis for risk regulatory policy and the law to distinguish between different risk perceptions if they simply reflect alternative cultural world-views. As Dan Kahan and his co-authors put it:

> If risk disputes are really disputes over the good life, then the challenge that risk regulation poses for democracy is less how to reconcile public sensibilities with science than how to accommodate diverse visions of the good within a popular system of regulation.[228]

One response to this dilemma looks to strategies of risk communication 'that enable citizens to accept new information, and ultimately change their minds, without experiencing a threat to their cultural identities'.[229] This approach suggests that deliberative democratic processes may 'help persons of diverse cultural worldviews converge on empirically sound beliefs about risk'.[230]

Alternatively, as Sunstein points out, it may be that it is only a subset of risk issues which trigger cultural conflicts and strongly divergent risk perspectives in the domestic context (though translated to the international sphere these are the risks that invariably give rise to legal disputes). Sunstein describes such risks as '"hot" risks' that separate people in moral and political terms.[231] Where risk disputes of this kind *are* in issue, Sunstein concedes there is the need for normative issues to be debated and explored in their own right, rather than being subsumed within expert risk assessment.[232]

225 See, e.g., the studies detailed in Kahan *et al.*, 'Fear of Democracy', 1068–88.

226 *Ibid.*, 1094. 227 *Ibid.*, 1105–6. 228 *Ibid.*, 1073.

229 Dan M. Kahan and Paul Slovic, 'Cultural Evaluations of Risk: "Values" or "Blunders"', *Harvard Law Review Forum*, 119 (2006), 171.

230 *Ibid.* 231 Sunstein, 'Misfearing: A Reply', 117. 232 *Ibid.*, 1123.

Conclusion

This chapter has considered both the reasons why science and expert risk assessment enjoy their current pre-eminence in international law, and also the challenges to scientific objectivity and neutrality which may undermine the special status of technical expertise as a 'universal legitimator' of global risk decision-making.[233] Two discourses, one of the authority of science and technical risk assessment, and the other of their inherent uncertainty and contingency, exist side by side in contemporary international debates, although sometimes it appears as if they are conversations taking place in parallel, but unconnected, universes.

Despite a greater awareness in the non-scientific community regarding the limitations of scientific knowledge, it seems that 'science still has a fatal attraction for policy-makers'.[234] In part, this may reflect the extent to which epistemic communities of scientists and experts have become entrenched in national bureaucracies, global governance networks and relevant international institutions, perpetuating ideas that the decisions taken are matters that can be settled on purely technical grounds. Alternatively, it may be that, although well aware of the questionable neutrality of expertise in many cases, government delegates in a global context adhere to the 'fiction' of objective science in order to establish a common set of ground rules 'helpful in overcoming politically constituted preferences'.[235]

The dominant positivist and technical perspectives on science and risk have been the subject of important challenges coming from social scientific research. Constructivist perspectives in this literature highlight that science and risk are not monolithic, objective entities, and moreover that expert processes of risk assessment are subject to significant uncertainties and contingencies. Such views have encouraged the development of new ways of doing science and undertaking risk assessment that are capable of extending beyond the inputs of technical expertise. Indeed, the institution of more innovative risk assessment techniques may be necessary in order to cope with today's

[233] Litfin, *Ozone Discourses*, p. 51.

[234] David Collingridge and Colin Reeve, *Science Speaks to Power: The Role of Experts in Policy Making* (London: Frances Pinter, 1986), p. ix.

[235] Joerges and Neyer, 'Transforming Strategic Interaction into Deliberative Problem-Solving', 619.

complex risk concerns, which encompass 'a multimedia, multisource and multiagent approach'.[236]

Yet the danger remains that the wider adoption of these approaches may tend to make risk disputes more difficult to resolve by dispelling the authority of science and exposing risk perception as an area where a diverse range of culturally influenced perspectives pertain. After all, the appeal of science and expert processes of risk assessment for global governance and international law has largely been that they organise information in a manner that is useful for the task of reaching an authoritative *decision* (rather than simply airing all perspectives) on a risk question. The legitimating force of these discourses is still powerful in terms of their capacity to produce social consensus around issues of health and environmental risk. Understandably there is thus a reluctance to undermine the utility of risk governance structures in international law by encouraging (endless) argument over issues of scientific evidence and risk evaluation.

Clearly, international law will need to find workable mechanisms to bring together more technically oriented approaches with broader views of science and risk. Already there are some useful ideas emerging at the theoretical level in this regard. For instance, transdisciplinary approaches, and analysis from the field of science and technology studies, suggest ways in which science can transform itself into a body of knowledge producing 'serviceable truths' for risk decision-making, which embrace non-scientific perspectives.[237] Equally, combining the insights of cultural risk theory with technical risk perspectives advanced by authors such as Cass Sunstein may yield a differentiated approach to risk assessment that treats 'hot risks' differently from ones where differing culturally based risk perceptions are susceptible to change through well-structured processes of risk communication. At the domestic level in some industrialised countries there have been moves to deploy such approaches within broader deliberative democratic frameworks. Whether international law has the institutional capacity to institute similar reforms is a topic to which we return in Chapter 7.

In the next chapter, however, we turn to consider how positivist and technical perspectives of science and risk, on the one hand, and new

[236] Oren Perez, *Ecological Sensitivity and Global Legal Pluralism: Rethinking the Trade and Environment Conflict* (Portland: Hart Publishing, 2004), p. 120.

[237] Jasanoff, *The Fifth Branch*, p. 250.

approaches to science and risk regulation, on the other hand, have been mobilised in global risk governance. The vehicles for this process have been two competing risk regulatory paradigms, which can be described as 'sound science' and 'precaution'. Each paradigm represents a distinctively different mixing of science, uncertainty concerns and value considerations in risk decision-making, which proponents of each approach have sought to 'internationalize'.[238] After a period of uneasy coexistence in the language of global politics and within international legal regimes, these two paradigms are increasingly being brought into conflict in global risk disputes and governance systems. Some see in these clashes the potential for the triumph of 'a singular conception of sound science' in furtherance of the goal of globally harmonised regulation of health and environmental risk.[239]

[238] Aarti Gupta, 'Governing Trade in Genetically Modified Organisms: The Cartagena Protocol on Biosafety', *Environment*, 42(4) (2000), 26.

[239] Winickoff *et al.*, 'Adjudicating the GM Food Wars', 106.

4 Competing risk regulatory paradigms: sound science and the precautionary principle

Introduction

In the previous chapter we saw how science and processes of expert risk assessment have become increasingly central to the workings of contemporary international law concerned with the regulation of risk. Requirements for science-based decision-making are now found in many treaties in the health and environmental field, from the SPS Agreement to international regimes dealing with hazardous chemicals, marine resources, industrial pollution and threats to biodiversity (such as those potentially posed by biotechnology and GMOs). In treaty language, a number of different formulae have been used to specify a requirement for scientific involvement in decision-making; for example, that regulations must consider 'the best scientific evidence available',[1] 'not [be] maintained without sufficient scientific evidence',[2] take into account 'relevant scientific and technical considerations',[3] or be based on risk assessment.[4]

As Chapter 3 highlighted, however, scientific evidence – even the best available – often goes hand-in-hand with problems of uncertainty, and

[1] For formulations of this kind see Convention on the Conservation of Migratory Species of Wild Animals, Bonn, 23 June 1979, 1651 UNTS 333, in force 1 November 1983, Article III; Convention on the Conservation of Antarctic Marine Living Resources, Canberra, 5 May 1980, 1329 UNTS 48, in force 7 April 1982, Article IX(1)(f); Protocol on Environmental Protection to the Antarctic Treaty, Madrid, 4 October 1991, 30 ILM 1455 in force 14 January 1998, Article 10(1); United Nations Framework Convention on Climate Change, Rio De Janeiro, 9 May 1992, 1771 UNTS 164 in force 24 March 1994, Article 4(2)(c).

[2] Agreement on the Application of Sanitary and Phytosanitary Measures, Geneva, 15 April 1994, 1867 UNTS 493, in force 1 January 1995 (SPS Agreement), Article 2.2.

[3] For formulations of this kind see Vienna Convention for the Protection of the Ozone Layer, Vienna, 22 March 1985, 1513 UNTS 293 in force 22 September 1988, Article 2(4); Basel Convention on the Control of Transboundary Movements of Hazardous Wastes and their Disposal, Basel, 23 March 1989, 1673 UNTS 57, in force 5 May 1992, Article 17(1).

[4] See the discussion of the SPS Agreement in Chapter 5.

the conclusions of expert risk evaluations may be dependent upon the way in which assessments are framed in light of differing sensitivities to uncertainty, the influence of socio-cultural values and institutional assumptions. Thus an important question in evaluating requirements for science-based decision-making in international law concerns the breadth of notions of science and risk intended by such provisions. Is what is contemplated restricted (as some SPS dispute settlement panels have suggested) to evidence 'gathered through scientific methods' with the result that only 'a complete, self-contained, scientific evaluation' will be considered an adequate risk assessment?[5] Or is there flexibility, where issues of scientific uncertainty arise, to embrace post-normal visions of science and broader forms of risk assessment that blend scientific findings, anecdotal information and value concerns?

In international policy and legal arenas, these questions are increasingly framed in terms of two competing paradigms for the assessment and management of health and environmental risk, which for convenience, rather than more precise articulation, can be designated 'sound science' and the 'precautionary principle'.[6] The latter paradigm based on the precautionary principle responds to the critiques of positivist science and technical perspectives on risk assessment discussed in the previous chapter. In general terms, it places emphasis on the need for risk regulation despite scientific uncertainty in order to avoid the possibility of serious health or environmental damage. While the precautionary principle has antecedents in international law, as well as in a number of domestic legal systems,[7] as a basis for

[5] Japan – Measures Affecting the Importation of Apples, Report of the Panel, WT/DS245/R, 15 July 2003, [8.92]; European Communities – Measures Affecting the Approval and Marketing of Biotech Products, Report of the Panel, WT/DS/291/R; WT/DS292/R; WT/DS293/R, 29 September 2006 (GMO case), [7.3188].

[6] Rosie Cooney and Andrew T. F. Lang, 'Taking Uncertainty Seriously: Adaptive Governance and International Trade', European J. Int'l Law, 18(3) (2007), 540–2 point out that a risk regulatory approach that is responsive to scientific uncertainty can encompass more than simply adherence to precautionary principle. It might also extend to requirements for adaptive governance and processes for policy learning.

[7] The origins of the precautionary principle are often said to lie in Germany: see Sonja Boehmer-Christiansen, 'The Precautionary Principle in Germany – Enabling Government', in Timothy O'Riordan and James Cameron (eds.), *Interpreting the Precautionary Principle* (London: Earthscan Publications Ltd, 1994), p. 31. However, other countries, including the USA, also have a long history of reliance on precautionary approaches to regulation: Jonathon B. Weiner, 'Whose Precaution After All? A Comment on the Comparison and Evolution of Risk Regulatory Systems', *Duke Journal of Comparative and International Law*, 13 (2003), 207.

risk regulation it has been most vigorously pursued in recent years by the EU.[8]

The paradigm of sound science, on the other hand, has been championed in North America, most particularly by the USA. The notion of sound science is not inherently in conflict with precautionary approaches to science and risk assessment.[9] Nevertheless, the associated risk regulatory paradigm has come to stand for reliance in risk decision-making on 'hard science':[10] scientific studies that are verifiable, reproducible and certified through rigorous processes of expert peer review. Indeed, in the hands of the former US administration of President George W. Bush, some allege sound science took on a distinctly anti-regulatory meaning, with implications for international risk regulation in fields such as biotechnology, chemicals regulation and climate change.[11]

This chapter explores the competing risk regulatory paradigms of sound science and the precautionary principle, the different ways in which they approach questions of the appropriateness of regulatory action to address human health or environmental risks, and how they have been translated into the international legal arena. To gain an understanding of the development of each paradigm it is necessary to delve into the histories of their use within the US and EU systems, respectively. Consequently, the chapter considers the US risk regulatory system and its reliance on sound science, as well as EU risk regulation and the role of the precautionary principle in that setting, particularly as the basis for Community-wide food safety and health measures. What emerges from this analysis is not that one system is science-based and the other 'anti-scientific',[12] but rather that both systems rely

[8] Alberto Alemanno, 'The Shaping of European Risk Regulation by Community Courts', (2008) Jean Monnet Working Paper 18/08, available at www.jeanmonnetprogram.org/papers/08/081801.html.

[9] Daniel McGarvey and Brett Marshall, 'Making Sense of Scientists and "Sound Science": Truth and Consequences for Endangered Species in the Klamath Basin and Beyond', *Ecology Law Quarterly*, 32 (2005), 73; Cooney and Lang, 'Taking Uncertainty Seriously', 540.

[10] Philippe Sands, *Principles of International Environmental Law*, 2nd edn (Cambridge University Press, 2003), p. 7.

[11] See, particularly, Holly Doremus, 'Science Plays Defense: Natural Resource Management in the Bush Administration', *Ecology Law Quarterly*, 32 (2005), 249; Thomas O. McGarity, 'Beyond the Dirty Dozen: The Bush Administration's Cautious Approach to Listing New Persistent Organic Pollutants and the Future of the Stockholm Convention', *William and Mary Environmental Law and Policy Review*, 28 (2003), 1.

[12] The latter criticism has been levelled at the precautionary principle: see, e.g., Editorial, 'Fear of the Future', *The Wall Street Journal* (New York), 10 February 2000,

on (often much the same) science that is used in different ways in risk regulation. Key to the divergence between the two regulatory paradigms is the differing conclusions regulators attach to considerations of uncertainty, cost–benefit calculations and social values in reaching decisions about levels of acceptable risk.

Via the kinds of transnational governance mechanisms discussed in Chapter 2, as well as in policy-making and dispute settlement fora under international risk regulatory treaties, the competing risk regulatory approaches advanced in the USA and Europe have achieved global prominence.[13] Given that much international risk regulatory activity is concerned with products which are the subject of international trade, the WTO and its dispute settlement system have been the site of some of the most bruising encounters between the paradigms of sound science and the precautionary principle. The *GMO* case, for example, saw a WTO panel presented with competing understandings of biotechnology risk informed, respectively, by the need for 'a basis in science'[14] and a 'prudent and precautionary approach'.[15]

International law and lawyers have responded in different ways to conflicts of this kind. One prominent approach sees such regulatory divergence as part of a broader phenomenon of fragmentation in international law, whereby differing interpretations of the requirements and scope of international legal norms are developed within separate, specialised regimes.[16] In the field of international trade law, such thinking

A18; Ron Brunton, 'The Precautionary Principle: The Greatest Risk of All', *IPA Environmental Backgrounder*, 20 (1994), 1. It is vigorously refuted by other authors who argue that 'applying the precautionary principle requires the application of the most rigorous science criteria with a view to characterizing uncertainties, filling gaps in knowledge and furthering research': Nicolas de Sadeleer, 'The Precautionary Principle in European Community Health and Environmental Law: Sword or Shield for the Nordic Countries?', in Nicolas de Sadeleer (ed.), *Implementing the Precautionary Principle: Approaches from the Nordic Countries, EU and USA* (London: Earthscan, 2007), p. 29.

13 David Vogel, 'Risk Regulation in Europe and the United States', in Han Somsen (ed.), *Yearbook of European Environmental Law* (Oxford University Press, 2004) vol. 3, available at http://faculty.haas.berkeley.edu/vogel/uk%20oct.pdf.

14 *EC – Measures Affecting the Approval and Marketing of Biotech Products*, WTO Docs WT/DS291, WT/DS292, WT/DS293 (2004) (First Submission of the US), [86].

15 *EC – Measures Affecting the Approval and Marketing of Biotech Products*, WTO Docs WT/DS291, WT/DS292, WT/DS293 (2004) [1] (First Written Submission by the EC), [12].

16 Pierre-Marie Dupuy, 'The Danger of Fragmentation or Unification of the International Legal System and the International Court of Justice', N.Y.U. J. Int'l L. Politics, 31 (1999), 797.

has given rise to a well-developed 'trade and' (or 'linkage')[17] literature that deals with the question of whether, and to what extent, the WTO regime should take account of non-trade norms, values or decision-making approaches. Conceptualised in this way, conflicts between risk regulatory paradigms at the global level might either be irresolvable – because such conflicts are simply 'a legal reproduction of collisions between the diverse rationalities within global society'[18] – or susceptible to case-by-case resolution employing the 'tool-box' of general and customary international legal rules of treaty interpretation.[19]

However, other international legal commentators view this debate as illusory, potentially even misleading; some because they believe it overlooks the potential for global governance institutions to evolve to accommodate new perspectives, and others because they argue it bypasses fundamental questions about the purpose of leading international legal regimes, such as that of the WTO. It is not necessary for the purpose of this book to seek to resolve these questions. Rather the final sections of the chapter position competition between different risk regulatory paradigms at the global level within the relevant, broader debates in international law and the international legal literature in order to illustrate how such perspectives may impact perceptions of the relationship between the paradigms and their capacity to influence global risk regulation.

The sound science risk regulatory paradigm

The rhetoric of sound science has long been part of the political arsenal of governments, invoked in support of official policy positions on various health and environmental issues.[20] In the last few decades, calls for risk regulation to be 'based on science rather than prejudice' have also become regular refrains in international debates and negotiations

[17] For a discussion of this concept see José E. Alvarez, 'Symposium: The Boundaries of the WTO – Foreword', Am. J. Int'l L., 96 (2002), 2.

[18] Andreas Fischer-Lescano and Gunther Teubner, 'Regime-Collisions: The Vain Search for Legal Unity in the Fragmentation of Global Law', *Michigan Journal of International Law*, 25 (2004), 1017.

[19] Martti Koskenniemi, 'Fragmentation of International Law: Difficulties Arising from the Diversification and Expansion of International Law', (A/CN.4/L.682, *International Law Commission*, 2006), 210.

[20] Robin Grove-White, 'Afterword: On "Sound Science", the Environment, and Political Authority', *Environmental Values*, 8(2) (1999), 279–80.

dealing with issues of health and environmental risk. While this trend has been most prominent in the areas of food safety, biotechnology and chemical pollution control, similar calls are beginning to emerge in fora concerned with the more traditional environmental concerns of nature conservation and resource management.[21] In addition, some global bodies draw on concepts of sound science in an attempt to reinforce the legitimacy of (contested) risk decision-making processes. Codex, for example, endorses a general policy that its food standards, guidelines and recommendations should be based on the principle of sound scientific analysis and evidence.[22]

Many of the most strident invocations of sound science in recent times have come from advocates of minimal regulation for new technologies.[23] Sound science in such manifestations is used as a bulwark against what is seen to be unnecessary risk regulation, hindering scientific research and technological progress. The legitimacy-enhancing power of science is drawn on in a negative way to question the need for domestic or international action to address a particular risk in the absence of sound scientific proof of possible harm.

Examples of sound science being invoked in this way have been prevalent in the biotechnology field. For instance, Henry Miller and Gregory Conko argue that national and international introduction of 'more restrictive and burdensome regulatory regimes [for biotechnological products] that fly in the face of scientific consensus' will 'diminish the overall potential application of gene splicing to agriculture and food production, and ... delay or deny the benefits of the "gene revolution" to the poorest and neediest parts of the world'.[24] Similar arguments often underlie critiques of the Biosafety Protocol, notwithstanding its requirements for national risk assessments to be 'carried out in a

[21] For an example within the Western and Central Pacific Fisheries Commission see Transform Aqorau, 'Challenges for Fisheries Management in the Pacific', Paper presented at the ANZSIL Annual Meeting 29 June – 1 July, Wellington, (2006).

[22] Codex Alimentarius Commission, 'Statements of Principle Concerning the Role of Science in the Codex Decision-making Process and the Extent to which Other Factors are Taken into Account', Decision of the 21st Session of the Commission, 1995, reproduced in *Procedural Manual* (Geneva: Codex Alimentarius Commission, 2004), Appendix, 159 (Principle 1). See Chapter 6 for further details of Codex processes.

[23] Henk van den Belt and Bart Gremmen, 'Between Precautionary Principle and "Sound Science": Distributing the Burdens of Proof', *Journal of Agricultural and Environmental Ethics*, 15 (2002), 112.

[24] Henry Miller and Gregory Conko, 'The Science of Biotechnology Meets the Politics of Global Regulation', *Issues in Science and Technology*, 17(1) (2000), 48. Miller is a former director of the US Food and Drug Administration's Office of Biotechnology;

scientifically sound manner'.[25] In this context, the concern expressed is that the precautionary elements of the Protocol (discussed further below) give licence to parties to ignore sound science, hence enhancing risks of trade protectionism.[26]

Sound science and US risk regulation

While calls for sound science are now regularly heard on both sides of the Atlantic, as well as in international fora, the concept is one that in its evolution and regulatory implementation has peculiarly North American antecedents. In this setting, 'sound science' has come to mean that regulation is affirmatively supported by evidence that satisfies the stringent burden of proof applied in research science of a 95 per cent confidence level.[27] The efficacy of a sound science rationale for risk policy and regulation in the USA is thought to reflect the unparalleled strength of the American public's faith in science and its capacity to drive technological progress.

A short history of US health and environmental risk regulation

Tracing the origins of the sound science paradigm requires a short history of US risk regulation in the health and environmental field. The early 1970s marked the beginnings of US legislative regulation of risks to human health and the environment. This was a period of intense activity at the federal level as administrative agencies sought to elaborate a wide range of standards under broad new Congressional statutes dealing with health and environmental issues.[28] Laws enacted during this time included environmentally focused measures, such as the Clean Air Act (CAA),[29] the Clean Water Act[30] and the Endangered Species Act

Conko is a research fellow with the pro-industry Washington-based think-tank, the Competitive Enterprise Institute. The two co-authored a book in 2004 on *The Frankenfood Myth: How Protest and Politics Threaten the Biotech Revolution.*

25 Cartagena Protocol on Biosafety to the Convention on Biological Diversity, Cartagena, 29 January 2000, in force 11 September 2003, 2226 UNTS 208, Article 15, Annex III.

26 Deborah Katz, 'The Mismatch between the Biosafety Protocol and the Precautionary Principle', Georgetown Int'l Envtl L. Rev., 13 (2001), 966–7.

27 Doremus, 'Science Plays Defense', 263. A 95% confidence level requires that the results of a study strongly support an inference of harm, with only a 5% probability that the recorded observations are due to chance alone.

28 Sidney Shapiro and Robert Glicksman, *Risk Regulation at Risk: Restoring a Pragmatic Approach* (Stanford University Press, 2003), pp. 31–3.

29 Clean Air Act of 1970, 42 U.S.C. §§ 7401–767q (2009).

30 Clean Water Act of 1977, 33 U.S.C. §§ 1251–387 (2009). This legislation adopts a precautionary goal of zero emissions.

(ESA),[31] as well as health and safety statutes such as the Occupational Safety and Health Act (OSHA)[32] and the Toxic Substances Control Act (TSCA).[33] A feature of this suite of legislation was its acknowledgement of the problems created by insufficient scientific data in setting risk management standards. For instance, the CAA required the promulgation of national ambient air quality standards 'requisite to protect the public health' while 'allowing an adequate margin of safety'.[34] Indeed, the health and environmental laws of the 1970s have been described as introducing 'a much-needed precautionary turn' in US environmental regulation, displacing previous common law and statutory liability regimes that focused on remedying or compensating for damage once it had occurred.[35]

The Congressional laws themselves followed on from a wave of environmental activism that swept the USA in the 1960s, inspired by the work of, and often led by, natural scientists such as Aldo Leopold and Rachel Carson. Using 'their scientific expertise to weave effective stories about the disastrous consequences of careless treatment of the environment', these scientists successfully advocated for strong conservation measures in areas such as species protection.[36] Although the conservation-oriented world-view of such experts was influential in their assessments of the implications of the limited data available in many cases, their training and experience nonetheless led them to present their perspective as science-based. Holly Doremus has reflected that this emphasis on science as dictating conservation outcomes may simply have been a consequence of the invisibility of their value-frames to the scientists concerned, or it may have been the result of a conscious decision to conceal value elements in order to harness the political advantages of sound science rhetoric.[37]

In the 1980s the stringent precautionary protections introduced by US federal health and environmental agencies – sometimes on the basis of minimal scientific evidence indicating harm – began to

[31] Endangered Species Act of 1973, 16 U.S.C. §§ 1531–44 (2009).

[32] Occupational Safety and Health Act of 1970, 29 U.S.C. §§ 651–78 (2009).

[33] Toxic Substances Control Act of 1976, 15 U.S.C. §§ 2601–92 (2009).

[34] Clean Air Act of 1970 §109, 42 U.S.C. § 7409. For hazardous pollutants the standard is an ample margin of safety: s 112.

[35] Sheila Jasanoff, 'Between Risk and Precaution – Reassessing the Future of GM Crops', *Journal of Risk Research*, 3(3) (2000), 278.

[36] Doremus, 'Science Plays Defense', 259.

[37] *Ibid.*, 260–1.

earn a reputation for 'too much' caution both at home and abroad.[38] Allegations of 'regulatory overkill' surfaced, with claims that agencies were locked in a 'vicious circle' of overreaction to public risk concerns, generating further pressure for regulation based on questionable science.[39] Such concerns resonated with the Republican Reagan administration, elected to office in 1981 on a strong deregulatory platform. The new administration quickly introduced reforms, such as the President's Executive Order 12291, to constrain the regulatory discretion of federal agencies.[40] The principal tool employed for this purpose was cost–benefit analysis effected via a requirement for the review of significant new regulatory actions to ensure their potential benefits to society outweighed potential costs. Under President Clinton's 1993 executive order, cost–benefit review of regulatory measures was retained and supplemented with a requirement for agencies to submit information on anticipated costs and benefits quantified 'to the extent feasible'.[41] Cost–benefit analysis of this kind marries well with regulation based on 'sound science' since the latter also strives towards quantifiable measures of harm and benefit.

Challenges in the US courts to risk regulatory measures

In parallel with such developments at the regulatory level, risk-producing industries (such as those involved in the manufacture of pharmaceuticals, hazardous chemicals and energy) began to realise that the agencies' regulations targeting them were vulnerable to science-based attacks.[42] Legislative provisions permitting judicial review of agency action under the major federal health and environmental statutes lent themselves to industry-based challenges before the courts. For example, under OSHA (covering toxic substances in the workplace) and TSCA

[38] Jasanoff, 'Between Risk and Precaution', 278.

[39] Stephen Breyer, *Breaking the Vicious Circle: Toward Effective Risk Regulation* (Cambridge, MA: Harvard University Press, 1993). For a more recent critique along these lines see Cass Sunstein, *Risk and Reason: Safety, Law, and the Environment* (Cambridge University Press, 2002).

[40] Exec. Order No. 12,291, 3 CFR 127 (1981), reprinted as 5 USC § 691 note (2009). President Clinton's 1993 executive order revoked that of President Reagan but retained cost–benefit review of regulatory measures: Exec. Order No. 12,866, 58 Fed. Reg. 51735 (Sept. 30 1993) 5 USC § 601 note (2009).

[41] Exec. Order No. 12,866, 58 Fed. Reg. 51735 (Sept. 30 1993) 5 USC § 601 note (2009), § 3(f) and 6(a)(3)(B).

[42] Thomas McGarity, 'Our Science is Sound Science and Their Science is Junk Science: Science-Based Strategies for Avoiding Accountability and Responsibility for Risk-Producing Products and Activities', Uni. Kansas L. Rev., 52 (2004), 904–5.

(dealing with chemicals that present unreasonable health or environmental risks), the relevant standard for judicial review requires a showing by the agency concerned of 'substantial evidence on the record as a whole'.[43] In the DC Circuit Court of Appeals, which has jurisdiction with respect to judicial review challenges brought pursuant to the CAA, the relevant standard is the avoidance of action that is 'arbitrary or capricious'.[44]

During the 1970s US courts had generally been highly deferential to agencies' assessment of science where challenges were brought to risk regulatory measures. Reviewing a new stringent OSHA standard for exposure to asbestos in 1974, the DC Court of Appeals hence deferred to an agency determination that a more protective level was required. It ruled:

> Some of the questions involved in the promulgation of these standards are on the frontiers of scientific knowledge, and consequently ... insufficient data is presently available to make a fully informed factual determination.[45]

In the context of the CAA, the DC Circuit Court of Appeals suggested an even more flexible standard for the review of regulatory measures. The Court determined that the CAA permitted the Environmental Protection Agency to act 'before the threatened harm occurs' and refused to demand 'rigorous step-by-step proof of cause and effect' as a prerequisite to regulation given the precautionary thrust of the legislation and the uncertain effects of exposure to potentially toxic fuel additives.[46] Accordingly, the agency was authorised to 'apply [its] expertise to draw conclusions from suspected, but not completely substantiated, relationships between facts, from trends among facts, from theoretical projections from imperfect data, from probative preliminary data not yet certifiable as "fact", and the like'.[47] In a later decision considering the CAA in *Lead Industries Association, Inc* v. *Environmental Protection Agency* the Court once again affirmed 'Congress directed the administrator to err on the side of caution' in making decisions about ambient air standards, and specifically mandated allowance of an adequate margin of safety in order to provide

[43] Occupational Safety and Health Act of 1970, 29 USC §660 (2009); Toxic Substances Control Act of 1976, 15 USC §2618 (2009).

[44] Administrative Procedure Act, 5 USC §706(2)(A) (2009).

[45] *Industrial Union Department, AFL-CIO* v. *Hodgson* 499 F.2D 467 (DC Cir 1974), 474. See also *The Society of Plastics Industry Inc* v. *OSHA* 509 F.2d 1301 (2nd Cir 1975).

[46] *Ethyl Corp* v. *EPA* 541 F.2D 1 (D.C.Cir 1976) 13, 28.

[47] *Ibid.*

some protection against harmful effects not yet uncovered by scientific research.[48]

However, the judicial trend to endorse precautionary regulation in the USA came to an abrupt halt with the Supreme Court's decision in 1980 striking down a new OSHA standard for benzene exposure. The basis of the Supreme Court's ruling in *Industrial Union Department* v. *American Petroleum Institute* was that the regulatory agency seeking to introduce the standard had not demonstrated a 'significant risk'.[49] This decision had the effect of increasing 'the legal pressure on agencies to perform detailed, science-intensive risk assessments in support of regulations'.[50] The risk assessment requirement was developed further by the courts in later decisions. For instance, in *Natural Resource Defense Council, Inc* v. *Environmental Protection Agency*, Judge Bork, writing for a three-judge panel of the DC Circuit Court of Appeals, read into the CAA's provisions regarding the regulation of hazardous air pollutants a requirement for the agency to make an initial determination of what pollutant level is 'safe'.[51] As Nicholas Ashford suggests, this can be interpreted as the insertion of a 'de minimis' risk requirement into the statute as a basis for risk regulatory measures.[52]

Judicial imposition of risk assessment requirements on federal US agencies regulating in the health and environmental field combined with a trend for courts to take a progressively harder look at the science underpinning agency decisions. The 'hard look' doctrine of judicial review, and its hardening over time, encouraged agencies to pay more and more attention to assembling a detailed evidentiary record to support their measures, including attempts to quantify any remaining uncertainties surrounding risks.[53] The resulting situation

48 647 F.2d 1130 (DC Cir. 1980), 1136.

49 647 F.2d 1130 (DC Cir. 1980). The Court gave some guidance as to what 'significant risk' might be, indicating it would lie somewhere in the realm of a risk to health of 10^{-3} to 10^{-9}.

50 Carl F. Cranor, 'Asymmetric Information, the Precautionary Principle, and Burdens of Proof', in Carolyn Raffensperger and Joel A. Tickner (eds.), *Protecting Public Health and the Environment: Implementing the Precautionary Principle* (Washington DC: Island Press, 1999), p. 92.

51 824 F.2d 1146 (DC Cir. 1987).

52 Nicholas A. Ashford, 'The Legacy of the Precautionary Principle in US Law: The Rise of Cost-Benefit Analysis and Risk Assessment as Undermining Factors in Health, Safety and Environmental Protection', in Nicolas de Sadeleer (ed.), *Implementing the Precautionary Principle: Approaches from the Nordic Countries, the EU and USA* (London: Earthscan, 2007), p. 353.

53 Joel A. Tickner and Sara Wright, 'The Precautionary Principle and Democratizing Expertise: a US Perspective', *Science and Public Policy*, 30(3) (2003), 213.

has been described as an 'ossification' of the regulatory process or, more colloquially, 'paralysis by analysis'.[54] It has led to other, probably unintended, perverse effects on the risk regulatory process. Donald Hornstein notes, for instance, that risk assessments born out of the agencies' need to survive the hard look doctrine of judicial review became a means by which government officials could avoid the value questions raised by risk management by claiming that those problems were being solved through the application of 'sound science'.[55] Some authors, such as Wendy Wagner, are even more scathing, describing agencies' actions to 'exaggerate the contributions made by science in setting toxic standards in order to avoid accountability for the underlying policy decisions' as a 'science charade'.[56]

'Sound science' versus 'junk science'

Ultimately, US risk regulatory agencies' attempts to shield themselves from scrutiny by amassing information and undertaking detailed risk assessments have not proved successful. Given the scientific uncertainties that abound in most risk assessment exercises, it is questionable whether greater emphasis on information collection and evaluation makes decision-making either more efficient or more effective as a precise assessment of risk (or cost–benefit) generally remains out of reach.[57] Rather than buttressing the scientific credibility of regulatory decision-making processes, agencies' attempts to put forward sufficient information and analysis to convince courts that their actions are evidence-based have provided a ready avenue for industry to deconstruct, and thereby call into question, the soundness of the agencies' science.

In taking this line, industries attacking US federal health and environmental regulation have often reworked the well-rehearsed arguments of tobacco companies in toxic tort and other litigation designed to deny the risks associated with cigarette smoke.[58] This strategy in

[54] Thomas O. McGarity, 'Some Thoughts on "Deossifying" the Rulemaking Process', Duke L.J., 41 (1992), 1385.

[55] Donald T. Hornstein, 'Reclaiming Environmental Law: A Normative Critique of Comparative Risk Assessment', Colum. L. Rev., 92 (1992), 565–7.

[56] Wendy Wagner, 'The Science Charade in Toxic Risk Regulation', Columbia Law Review, 95 (1995), 1617.

[57] Elizabeth Fisher, 'Drowning by Numbers: Standard Setting in Risk Regulation and the Pursuit of Accountable Public Administration', *Oxford Journal of Legal Studies*, 20(1) (2000), 128.

[58] McGarity, 'Our Science is Sound Science'.

turn draws on academic analysis of scientific information that seeks to draw a hard line between 'good' or 'sound' science, and so-called 'junk science'. One of the best known works in this genre is Peter Huber's book on 'junk science' in the courtroom. Huber calls for enhanced judicial powers to screen out 'junk science', which he describes as 'a hodgepodge of biased data, spurious inference, and logical legerdemain, patched together by researchers whose enthusiasm for discovery and diagnosis far outstrips their skill'.[59] For industries seeking to resist the introduction of more stringent risk regulatory measures, attacking agency science as unsound or junk science has proved highly effective.

Despite many members of the scientific community in the USA rejecting the 'junk science' concept,[60] it found some favour in 1993 in the eyes of the Supreme Court. In its well-known *Daubert* decision, the Court ruled that trial judges were required to assess the admissibility of expert evidence by reference to its scientific validity, measured against the methods and procedures of conventional positivist science.[61] Since the Supreme Court's decision, various proposals have appeared in the US legal literature advocating a similar 'Daubertization' of regulatory science so that only evidence produced via rigorous processes of testing and certified by impartial peer reviewers may be relied upon by agencies establishing risk measures.[62] While such efforts have not persuaded Congress to legislate generally applicable science-based controls on risk regulatory activity, a number of 'sound science' reforms have been introduced in the form of riders to federal appropriation bills.[63] These have provided a vehicle for industry and others to gain greater access to the science

[59] Peter Huber, *Galileo's Revenge: Junk Science in the Courtroom* (New York: Basic Books, 1991), p. 3.

[60] See, e.g., Brief of Amici Curiae of Physicians, Scientists and Historians of Science in Support of Petitioners, at 2, *Daubert* v. *Merrell Dow Pharmaceuticals, Inc.*113 S.Ct. 2786 (1993).

[61] *Daubert* v. *Merrell Dow Pharmaceuticals*, Inc.113 S.Ct. 2786 (1993).

[62] Alan Raul and Julie Zampa Dwyer, '"Regulatory Daubert": A Proposal to Enhance Judicial Review of Agency Science by Incorporating Daubert Principles into Administrative Law', *Law and Contemporary Problems*, 66 (2003), 7. For a critique of such proposals see Thomas O. McGarity, 'On the Prospect of "Daubertizing" Judicial Review of Risk Assessment', *Law and Contemporary Problems*, 66 (2003), 155.

[63] See particularly the Information (Data) Quality Act 2000, §515 of the Treasury, Postal Service and General Government Appropriations Act for Fiscal Year 2001, enacted on 21 December 2000, Pub L. No. 106–554, 114 Stat. 2763A-153 to 2763A-154.

underlying government regulation and so to mount further 'sound science' challenges in the courts.[64]

Sound science and the Bush administration's anti-regulatory agenda

According to prominent American science policy commentators, such as Sheila Jasanoff, the trends in US health and environmental risk regulation over the past forty years have brought about a 'swing back toward a technocratic model of governance in the United States'. Characteristic of this shift has been a retreat from early precautionary approaches to regulation and 'a rise … in official discourses of "risk assessment", "sound science" and "evidence-based decisionmaking"'.[65]

For many US scientists and commentators, the period of President George W. Bush's administration marked a new, disturbing phase in the use (and abuse) of science in policy and regulatory decision-making.[66] Early on in his first term of office, President Bush announced that his administration was 'going to make decisions based upon sound science, not some environmental fad or what may sound good'.[67] On the domestic front, however, many instances and reports began to emerge suggesting 'sound science' was drawn on primarily as a way of discrediting expert evidence (even that prepared by federal regulatory agencies) that suggested the need for health or environmental protection.[68] Such efforts were directed, for example, against proposals for the listing of a seabird that favoured old-growth forest as a nesting habitat, an environmental impact study carried out in relation to proposed oil drilling in the Arctic National Wildlife Refuge, and the implementation of a stricter standard for arsenic levels in drinking water.

[64] Wendy Wagner, 'The "Bad Science" Fiction: Reclaiming the Debate over the Role of Science in Public Health and Environmental Regulation', *Law and Contemporary Problems*, 66 (2003), 63.

[65] Sheila Jasanoff, '(No?) Accounting for Expertise', *Science and Public Policy*, 30(3) (2003), 158.

[66] Patrick Parenteau, 'Anything Industry Wants: Environmental Policy under Bush II', Duke Environmental Law and Policy Forum, 14 (2004), 363. See also Union of Concerned Scientists, 'Scientific Integrity in Policy Making 7/04', available at www.ucsusa.org/scientific_integrity/abuses_of_science/scientific-integrity-in-1.html, alleging an 'unprecedented' level of political interference in regulatory science by the Bush administration.

[67] Office of the Press Secretary, 'Remarks by the President to the Environmental Youth Award Winners', White House, 24 April 2001, available at http://georgewbush-whitehouse.archives.gov/news/releases/2001/04/20010424-1.html.

[68] For an overview of controversies in this regard see Holly Doremus, 'Scientific and Political Integrity in Environmental Policy', *Texas Law Review*, 86 (2008), 1603–19.

In addition, attempts were made by members of the Bush executive government, such as the Secretary for the Interior (responsible for the domestic management of natural resources), to introduce sound science requirements into central pieces of environmental legislation such as the ESA. Legislative amendments pushed by the Secretary for the Interior would have expanded the ESA's obligation to make listing determinations 'solely on the basis of the best scientific and commercial data available' to require greater weight to be given to empirical, field-tested or peer-reviewed data.[69] Such requirements could potentially have been used to implement a very high scientific evidence threshold for the listing of species despite well-known difficulties in compiling comprehensive data on rare or endangered populations.[70]

The Bush administration also sought to transfer its understanding of sound science to the international sphere in key health and environmental areas such as the regulation of GMOs, chemicals regulation and the production of greenhouse gas (GHG) emissions. In the field of biotechnology, its efforts built on a longer standing US policy (also shared by the Clinton administration) that was designed 'to assign a weak burden of evidence for safety and a strong burden for evidence of risk', so facilitating the commercial approval of GMO products and the further development of the biotechnology sector.[71] This approach is also reflected in the domestic US regulatory framework for GMOs, which regulates these products on the 'basis of risks defined in straightforwardly biophysical terms' rather than in light of the broader social, ethical and environmental questions raised by processes of genetic modification.[72]

In the area of chemicals regulation, the Bush administration initially signalled its support for, and signed, the Stockholm Convention on Persistent Organic Pollutants (POPs Convention) concluded in 2001.[73] However, the domestic implementing legislation later introduced by the administration sought to undermine the Convention's

69 S. 911, the Endangered Species Recovery Act of 2001, 17 May 2001, Senators Gordon Smith (R-OR) and Max Baucus (D-MT).

70 James Hein, 'The "Sound Science" Amendment to the Endangered Species Act: Why It Fails to Resolve the Klamath Basin Conflict', B.C. Envtl. Aff. L. Rev., 32 (2005), 207.

71 Les Levidow and Susan Carr, 'Sound Science or Ideology?', *Forum for Applied Research and Public Policy*, Fall (2000), 44.

72 Jasanoff, 'Between Risk and Precaution', 277. Accordingly, GMOs only attract regulation by US agencies if they pose risks substantially different in kind from conventional crops.

73 Convention on Persistent Organic Pollutants, Stockholm, 23 May 2001, (2001) 40 ILM 532, in force 17 May 2004 (POPs Convention).

processes pertaining to the listing of new POPs by the treaty's expert assessment body, the Persistent Organic Pollutants Review Committee.[74] As the previous chapter highlighted, this Committee employs a science-based risk assessment process in recommending whether additional chemicals are appropriate candidates for international listing. However, in considering the Committee's recommendations and deciding on the listing of a new chemical, the Convention's Conference of the Parties is instructed to act 'in a precautionary manner'.[75] It was this requirement to which the Bush administration took strong exception.

Its proposed legislation thus did not provide for the automatic listing of new POPs approved under the Convention. This meant that in order for additional chemicals to be regulated under US law, case-by-case decisions by the relevant regulatory agency (the Environmental Protection Agency (EPA)) would be needed, followed by legislative amendment.[76] Moreover, the administration, via the Office of Management and Budget (OMB), sought to impose cost–benefit and 'data quality' requirements on the process. Their effect would have been to subject any decision under domestic law to add a chemical to the POPs list to a cost–benefit criterion, as well as to require the EPA to evaluate independently the scientific merits of the data used by Convention bodies in risk assessment and decision-making. Thomas McGarity suggests that OMB's ultimate goal in insisting on these additional procedures was to ensure that the chemical and pesticide industries would have ample opportunity to derail any attempt to augment the POPs list applicable under US law.[77]

An even starker example of the Bush administration's use of 'sound science' to advance an anti-regulatory international agenda was provided by its decision in 2001 to reject the Kyoto Protocol to the global climate change convention. Since the conclusion of the Kyoto Protocol in 1997, US administrations of both political persuasions

[74] POPs and PIC Implementation Act of 2002, S. 2507, 107th Cong. (2002).

[75] POPs Convention, Article 8(9).

[76] The relevant legislation affected is the Federal Insecticide, Fungicide, and Rodenticide Act, 7. U.S.C. § 136 f., which regulates the production, sale and use of pesticides; and TSCA, 15 U.S.C. § 2601 f., which regulates industrial chemicals. McGarity, 'Beyond the Dirty Dozen', 32 notes that as both statutes have rarely been amended over the years, such a requirement would amount to 'a prescription for failure'.

[77] McGarity, 'Beyond the Dirty Dozen', 17.

have voiced opposition to its lack of binding restraints on the GHG emissions of developing countries. However, President Bush took the further step of seeking to undermine the scientific credentials of the treaty. In combination with the American fossil fuel industry, the Bush administration 'worked hard to present the climate change science as highly uncertain particularly with respect to the extent to which anthropogenic emissions are driving global warming'.[78] This activity took various forms, including editing of US agency reports on global warming to emphasise uncertainty, restricting the ability of government scientists to communicate information on climate change to Congress and the media, and sowing doubt about whether global warming would be a harmful thing for the environment. By emphasising the remaining uncertainties, the administration played into the 'sound science' rhetoric used by opponents of climate change regulation who argued 'that truly "scientific" decision making requires near-certain proof of responsibility and efficacy before drastic policy changes are implemented'; proof that can never be provided in time in the face of a rapidly evolving environmental problem.[79]

Backed by the administration's understanding of sound science, President Bush criticised the Kyoto Protocol in press releases as 'fatally flawed' and argued that its targets for the reduction of greenhouse gases were 'arbitrary and not based on science'. In place of acceptance of the Kyoto Protocol regime, President Bush vowed that the USA would work within the United Nations and elsewhere to develop 'an effective and science-based response to the issue of global warming'.[80] In the meantime, scientific understanding and consensus surrounding the severity of likely climate change impacts continued to grow with the issue of two assessment reports (in 2001 and 2007) by the international scientific body dealing with climate change issues, the Inter-governmental Panel on Climate Change (IPCC). Nevertheless, the Bush administration continued to stress 'the incomplete state of scientific knowledge

[78] Holly Doremus, 'Lots of Science, Not Much Law: Why Knowledge Has Not (Yet) Been Power Over Greenhouse Gas Emissions', in William H. Rodgers, Jr., Jeni Barcelos, Anna T. Moritz and Michael Robinson-Dorn (eds.), *Climate Change: A Reader* (Durham, NC: Carolina Academic Press, forthcoming 2010) (copy on file with the author).

[79] *Ibid.*

[80] Office of the Press Secretary, 'President Bush Discusses Climate Change', White House, 11 June 2001, available at http://georgewbush-whitehouse.archives.gov/news/releases/2001/06/20010611-2.html.

of the causes of, and solutions to, global climate change'.[81] This stance permitted the administration to acknowledge the seriousness of, and even the human contribution to, the global problem of climate change, while at the same time delaying regulatory action by insisting that it should be put in place only as '(sound) science justifies'.[82]

With the swearing-in of the Obama administration in 2009, it appears that the US is charting a new course with respect to the use of science in domestic and international policy. In his Inaugural Address, the new president stated that his administration would 'restore science to its rightful place'.[83] President Obama has also pledged that addressing climate change will be a 'leading priority' for the USA and has appointed a well-respected scientist, Stephen Chu, as his Energy Secretary.[84]

Yet, while such statements of the US administration have pleased many scientists – disheartened after eight years of seeing science, at times, politicised, underfunded, or ignored – they do not necessarily signal a significant move away from the sound science regulatory paradigm. As the history of US health and environmental regulation shows, this paradigm has deep roots in American regulatory culture. Moreover, it is one that resonates well with other aspects of that culture, which include the powerful influence of industry lobbyists in political processes, a reluctance to impose costly regulation on business that might impede innovation without strong evidence of harm, and a litigious society that has long looked to the courts to impose penalties on those who cause health or environmental damage as a disincentive to abusive practices.[85] Consequently adherence to sound science or evidence-based decision-making in health and environmental regulation amounts to more than just an exercise in cynical political spin. As Mark Schapiro has argued, 'at the core' of the environmental path followed by the USA is a particular understanding of *risk*, which dictates

[81] Reed McManus, 'The Science of Stalling', Sierra Watch, September (2001), 17, quoting from a 13 March letter from Bush to Republican senators Chuck Hagel, Jesse Helms, Larry Craig and Pat Roberts.

[82] Sean D. Murphy, 'Bush Administration Proposal for Reducing Greenhouse Gases', American J. Int'l L., 96 (2002), 488.

[83] 'Barack Obama's Inaugural Address', *The New York Times* (online), 20 January 2009, available at www.nytimes.com/2009/01/20/us/politics/20text-obama.html.

[84] Barack Obama, 'Announcement of Energy and Environment Team', Press Conference, 15 December 2008, Chicago.

[85] See generally, Mark Schapiro, *Exposed: The Toxic Chemistry of Everyday Products and What's at Stake for American Power* (White River Junction, VT: Chelsea Green Publishing, 2007).

decisions about the evidence that is necessary to prove it and the level which is sufficient to prompt government action.[86]

Risk regulation and the precautionary principle

In many ways the emergence of the precautionary principle can be seen as 'a sobering rejoinder to the overzealous promotion of "sound science" in public policy'.[87] In the USA, for example, the embrace of precautionary regulation in the 1970s followed recognition of the limitations of science as a foundation for proactive environmental decision-making. Methodologies that insist upon a strong scientific basis for environmental standards proved to be insufficiently protective in the face of complex, highly variable natural ecosystems.[88] Waiting for conclusive scientific evidence of harm before taking regulatory action has thus often been a recipe for permitting ongoing environmental degradation. In addition, sound science practices have, on occasion, been blindsided by ignorance leading to the occurrence of harmful 'surprises', such as the Antarctic ozone hole created by synthetic chemicals that had previously been considered benign.[89] Overall such instances have motivated changes in health and environmental regulation to embrace a more cautious approach.

In the 1980s these various currents in regulatory thinking began to coalesce at the global level. An early instance was the World Charter for Nature, adopted by the United Nations General Assembly, which advised that 'where potential adverse effects are not fully understood' activities likely to pose a significant risk to nature 'should not proceed'.[90] This was followed by more explicit calls for a 'precautionary

[86] *Ibid.*, p. 12.

[87] Elizabeth Fisher, Judith Jones and René von Schomberg (eds.), *Implementing the Precautionary Principle: Perspectives and Prospects* (Cheltenham: Edward Elgar, 2006), p. 3.

[88] Charmian Barton, 'The Status of the Precautionary Principle in Australia: its emergence in legislation and as a common law doctrine', Harv. Envtl. L. Rev., 22 (1998), 511–14 critiquing the science-based assimilative capacity approach to standard-setting.

[89] European Environment Agency, *Late Lessons from Early Warnings: the Precautionary Principle 1896–2000* (Luxembourg: European Union, 2001), pp. 3–4.

[90] World Charter for Nature, General Assembly 48th plenary meeting, 28 October 1982, A/RES/37/711(b). The USA was the only member of the General Assembly to vote against adoption of the resolution.

approach' in international environmental instruments.[91] For example, in the 1987 Ministerial Declaration of the Second North Sea Conference the delegates from the various European countries concerned accepted that:

> in order to protect the North Sea from possibly damaging effects of the most dangerous substances, a precautionary approach is necessary which may require action to control inputs of such substances even before a causal link has been established by absolutely clear scientific evidence.[92]

In the same year states concluded the Montreal Protocol regulating the global production and consumption of chemicals, such as chlorofluorocarbons, despite the scientific uncertainty surrounding the link between such substances and depletion of the Earth's protective ozone layer. The Protocol's preamble documents the parties' determination 'to protect the ozone layer by taking precautionary measures'.[93]

However, it was not until the United Nations Conference on Environment and Development (UNCED) held in 1992 that the precautionary principle achieved broad recognition as one of the central elements of international environmental law and policy. The Rio Declaration resulting from UNCED includes Principle 15, which provides:

> In order to protect the environment, the precautionary approach shall be widely applied by States according to their capabilities. Where there are threats of serious or irreversible damage, lack of full scientific certainty shall not be used as a reason for postponing cost-effective measures to prevent environmental degradation.

The principle was also included in the two major multilateral environmental treaties concluded at UNCED – the Convention on Biological Diversity[94] and the United Nations Framework Convention on Climate Change.[95]

[91] The majority of international scholars do not see a meaningful distinction between a precautionary approach and the precautionary principle: see further Jacqueline Peel, 'Precaution: A Matter of Principle, Approach or Process?', Melb. J. Intl L., 5(2) (2004), 483.

[92] Ministerial Declaration, Second International Conference on the Protection of the North Sea, London, 24–25 November 1987, [VII]. See also [XV(ii)], [XVI(1)].

[93] Montreal Protocol on Substances that Deplete the Ozone Layer, Montreal, 16 September 1987, 1522 UNTS 3 in force 1 January 1989, preamble.

[94] Convention on Biological Diversity, Rio De Janeiro, 5 June 1992, 1760 UNTS 79 in force 29 December 1993, preamble.

[95] UNFCCC, Article 3.

Since the early 1990s the precautionary principle has been incorporated into a number of international environmental treaties and legal instruments,[96] including the POPs Convention[97] and the Biosafety Protocol. The latter treaty is commonly cited as one of the most extensive international implementations of the precautionary principle, with both the preamble and the 'objective' recalling 'the precautionary approach contained in Principle 15 of the Rio Declaration on Environment and Development'.[98] The provisions of the Protocol dealing with decision-making on imports of 'living modified organisms' also implicitly reference the precautionary principle by providing that decisions shall not be prevented by:

> Lack of scientific certainty due to insufficient relevant scientific information and knowledge regarding the extent of the potential adverse effects of a living modified organism on the conservation and sustainable use of biological diversity in the Party of import, taking also into account risks to human health.[99]

In addition to its prominence in international environmental treaties, the precautionary principle has expanded into a range of other areas of international law. This has been facilitated by its linkage to the broad notion of sustainable development,[100] which animates environmental practices such as impact assessment in international financial institutions,[101] and also is a declared 'objective' of the WTO.[102] A pertinent example of the penetration of precautionary concepts into non-environmental spheres is Article 5.7 of the SPS Agreement. This provision, which allows WTO members to adopt provisional SPS measures '[i]n cases where relevant scientific evidence is insufficient', provided they also 'seek to obtain the additional information necessary for a more objective assessment of risk and review the sanitary or phytosanitary measure accordingly within a reasonable period of time', is often cited as a (weak) version of the precautionary principle.

96 See generally Arie Trouwborst, *Evolution and Status of the Precautionary Principle in International Law* (The Hague: Kluwer International, 2002).

97 POPs Convention, Articles 1, 8(9).

98 Biosafety Protocol, preamble, Article 1.

99 *Ibid*, Articles 10(6), 11(8).

100 This concept calls for the integration of environmental concerns into all aspects of economic and development processes at the international level: Sands, *Principles of International Environmental Law*, p. 263.

101 E.g., World Bank Operational Policy 4.01 and Bank Procedures 4.01 (1999).

102 Marrakesh Agreement Establishing the World Trade Organisation, Marrakesh, 15 April 1994, 1867 UNTS 3, in force 1 January 1995 preamble, recital 1.

The WTO Appellate Body has endorsed this view, ruling that 'the precautionary principle has been incorporated and given a specific meaning in Article 5.7 of the SPS Agreement'.[103]

The widespread adoption of the precautionary principle in international law – particularly in areas of health and environmental regulation – reflects the appeal of caution as a prudent regulatory response in the face of scientific uncertainty and possibilities of serious health or environmental damage.[104] Further, the emphasis on uncertainties in science inherent to the precautionary principle provides a strong counterpoint to the global proliferation of scientific risk assessment processes described in the previous chapter. International requirements for risk assessment and science-based decision-making now sit alongside one another, often within the same treaty. As Philippe Sands has commented, this leaves the 'tension' between the two regulatory approaches unresolved at the level of international legislation, meaning that it will generally fall to international adjudicators to determine on a case-by-case basis the degree of scientific evidence and certainty necessary to justify restrictions on health or environmental grounds.[105]

EU risk regulation and the role of the precautionary principle

Over the last few decades the precautionary principle has become an established element not only of international environmental law, but also the domestic environmental law of a number of countries, such as Australia, Canada, India and most European nations.[106] However, it is in the EU that the principle has achieved its greatest prominence as the foundation of Community health and environmental risk regulation.[107] Internationally, the EU has also become a strong proponent of precautionary regulation in a range of health and environmental areas.[108]

[103] *EC – Measures Concerning Meat and Meat Products*, Report of the WTO Appellate Body, WT/DS26/AB/R & WT/DS48/AB/R, 16 January 1998, (*Hormones*), [120].

[104] Even strident critics of precaution, such as Cass Sunstein, acknowledge 'that a lack of decisive evidence of harm should not be grounds for refusing to regulate': Cass R. Sunstein, 'Beyond the Precautionary Principle', *University of Pennsylvania Law Review*, 151 (2003), 1012.

[105] Sands, *Principles of International Environmental Law*, p. 7.

[106] Fisher *et al.*, *Implementing the Precautionary Principle*.

[107] José Luís da Cruz Vilaça, 'The Precautionary Principle in EC Law', *European Public Law*, 10(2) (2004), 369.

[108] Most notably climate change, biotechnology and chemicals regulation. By contrast, it was the USA throughout the 1970s and 1980s that pushed most strongly for precautionary international environmental agreements for endangered species protection and the regulation of ozone-depleting substances.

For long-time observers of transatlantic regulatory trends, the EU's current embrace of the precautionary principle is more than a little ironic. David Vogel notes that from the 1960s through to the mid-1980s it was US health and environmental standards that were generally regarded as more stringent than those of Europe.[109] By contrast, the European standard-setting process in the health and environmental area was highly technocratic, reflecting an assumption that citizens' interests and needs could be adequately taken care of by experts and national regulatory agencies alone.[110] Since 1990, however, commentators such as Vogel detect a change in the 'transatlantic direction of regulatory emulation', with European and American regulatory policies effectively having 'traded places' in terms of their relative stringency, comprehensiveness and innovativeness.[111] Relevant examples of this trend include European regulation of GMOs (discussed further below), bans introduced on chemicals thought to pose health risks, such as phthalate softeners in children's toys, and European initiatives in the area of recycling.[112]

Vogel identifies three interrelated factors as contributing to the shift to more risk-averse regulatory politics and policies in Europe. First is the series of regulatory failures and 'crises' that took place in Europe in the late 1980s (in respect of nuclear and chemical accidents) and the latter half of the 1990s (involving food safety and health protection).[113] The most prominent of these was the crisis over mad cow disease and its transmission to humans through the consumption of contaminated beef, which severely undermined public trust in EU food safety regulations and the scientific expertise on which they were based. Second, the ensuing political environment in Europe, characterised by a heightened public sense of vulnerability to, and anxiety about, health and environmental risks, led to broader and stronger support for more stringent risk regulatory standards in the EU.[114] Finally, treaty reforms beginning in the late 1980s produced a growth in the regulatory competence

[109] David Vogel, 'The Hare and the Tortoise Revisited: The New Politics of Consumer and Environmental Regulation in Europe', British J. Pol. Sci., 33 (2003), 557.

[110] Bruna de Marchi, 'Public Participation and Risk Governance', *Science and Public Policy*, 30(3) (2003), 173.

[111] Vogel, 'The Hare and the Tortoise Revisited', 558, 565. See also, Ragner E. Löfstedt and David Vogel, 'The Changing Character of Regulation: A Comparison of Europe and the United States', *Risk Analysis*, 21(3) (2001), 399, suggesting an apparent 'flip-flop' of the two regulatory systems.

[112] See also the various instances discussed in Schapiro, *Exposed: The Toxic Chemistry of Everyday Products and What's at Stake for American Power.*

[113] Vogel, 'The Hare and the Tortoise Revisited', 568–71.

[114] *Ibid.*, 571–3.

of the EU. Concomitantly, the importance of EU risk measures in governing the health and environmental safety of products distributed throughout the single market area became increasingly obvious.[115] The consequence, Vogel contends, is that 'protecting the health and safety of Europeans as well as the European environment has become critical to the EU's legitimacy and its claim to represent the broader interests and concerns of Europeans'.[116]

Within this constellation of influences, Vogel identifies '[t]he emergence of the precautionary principle as a guide to regulatory decision-making' as 'an important dimension of the new European approach to risk regulation'.[117] The precautionary principle stresses responsiveness to scientific uncertainty rather than the need for conclusive evidence of potential harms before taking regulatory action. It has accordingly provided an important rallying point for growing perceptions in Europe regarding the insufficiency of science as a guide to risk policy, and decreasing public confidence in the benefits of technological innovation.[118]

Incorporation of the precautionary principle in Community law and policy

One reason that the precautionary principle has achieved such salience in Europe is as a result of its incorporation in EU treaty law in association with the EU's burgeoning mandate and infrastructure for risk regulation. While not originally envisioned by the founding treaties of the European Communities, contemporary European commentators remark that 'today the most important and widespread form of EU regulation in the internal market is concerned with the government of risk', particularly risks to health, the environment and consumer safety.[119]

Although the precautionary principle was a long-standing element of the environmental law of some European countries, its formal inclusion as part of EU law did not occur until 1993 with the entry into

[115] *Ibid.*, 573–5.

[116] *Ibid.*, 575. See also Theofanis Christoforou, 'The Regulation of Genetically Modified Organisms in the European Union: The Interplay of Science, Law and Politics', *Common Market Law Review*, 41 (2004), 689.

[117] Vogel, 'The Hare and the Tortoise Revisited', 566.

[118] *Ibid.*, 567.

[119] Alemanno, 'The Shaping of European Risk Regulation by Community Courts', 4.

force of the Treaty on European Union.[120] Subsequent amendment of the Treaty Establishing the European Community (EC Treaty) by the Treaty of Amsterdam[121] saw the precautionary principle become part of Article 174(2) of the EC Treaty. This article provides:

> Community policy on the environment shall aim at a high level of protection taking into account the diversity of situations in the various regions of the Community. It shall be based on the precautionary principle and on the principles that preventive action should be taken, that environmental damage should as a priority be rectified at source and that the polluter should pay.

While Article 174(2) suggests that the operation of the precautionary principle is confined to the area of Community environmental policy, it has been interpreted as having broader application.[122] In particular, various EC Treaty provisions seeking 'a high level' of health, environmental and consumer protection,[123] coupled with requirements for the integration of environmental concerns into a wide range of Community policies and activities,[124] are read together to constitute the precautionary principle as a central element of EU law and risk regulation.

Precautionary health and environmental legislation in the EU

Prompted by the public reaction to health and safety crises, such as that over mad cow disease, the Community institutions have relied on the relevant EC Treaty provisions to implement a range of highly precautionary regulations.[125] Despite its environmental origins in the EC Treaty, the precautionary principle has broken free from its roots, being most prominent in EU laws relating to food safety. For example,

120 Treaty on European Union, Official Journal C 191, 29 July 1992 (entered into force 1 November 1993). This treaty changed the name of the former European Economic Community to simply the 'European Community' and added new provisions to the Community's constitutive treaty document. These new provisions included Article 130r concerning the role of the precautionary principle in Community environmental policy.

121 Treaty of Amsterdam Amending the Treaty on European Union, the Treaties Establishing the European Communities and Related Acts, Official Journal C 340, 10 November 1997 (entered into force 1 May 1999).

122 Case T-74/00 *Artegodan* v. *Commission* [2002] II-ECR 4945, [183].

123 Treaty Establishing the European Communities, Rome, 25 March 1957, [1992] OJ C 224, 6, in force 1 January 1058 (EC Treaty) Articles 2, 3(p), 152(1) and 153(1) (as amended).

124 EC Treaty, Articles 6 and 153(2).

125 Nicolas de Sadeleer, 'The Precautionary Principle in EC Health and Environmental Law', *European Law Journal*, 12(2) (2006), 139.

Article 7 of the EC food safety regulation, entitled 'Precautionary principle', provides:

> In specific circumstances where, following an assessment of available information, the possibility of harmful effects on health is identified but scientific uncertainty persists, provisional risk management measures necessary to ensure the high level of health protection chosen in the Community may be adopted, pending further scientific information for a more comprehensive risk assessment.[126]

One of the few EU *environmental* risk instruments to be explicitly based on the precautionary principle is Directive 2001/18/EC on the deliberate release into the environment of GMOs (the GMO directive).[127] This directive is part of a complex of legislation introduced by the Community political institutions that requires the approval of GMO crops and associated food products prior to their Community-wide release.[128] Under the scheme of the GMO directive, which has evolved significantly since its initial introduction in 1990, applications for approval proceed through a multi-layered process of member state and Community level risk assessment and decision-making, informed by scientific assessments and political considerations brought to the process by national authorities and various EC-level expert

[126] Regulation (EC) No 178/2002 of the European Parliament and of the Council of 28 January 2002 laying down the general principles and requirements of food law, establishing the European Food Safety Authority and laying down procedures in matters of food safety, Official Journal L 31, 1 February 2002, Article 7(1).

[127] Council Directive 2001/18/EC, [2001] OJ L 106/1 (replacing Council Directive 90/220/EEC, [1990] OJ L 117/15).

[128] See Regulation 1829/2003 of the European Parliament and the Council of 22 September 2003 on genetically modified food and feed, Official Journal L 268, 18 October 2003; Regulation 1830/2003 of the European Parliament and the Council of 22 September 2003 concerning the traceability and labelling of genetically modified organisms and the traceability of food and feed products produced from genetically modified organisms, Official Journal L 268, 18 October 2003; Commission Regulation 65/2004 of 14 January 2004 establishing a system for the development and assignment of unique identifiers for genetically modified organisms, Official Journal L 10/5, 16 January 2004; Commission Regulation 641/2004 of 6 April 2004 on detailed rules for the implementation of Regulation (EC) No 1829/2003, Official Journal L102/14, 7 April 2004. For detailed discussion of this legislative regime see Theofanis Christoforou, 'Genetically Modified Organisms in European Union Law', in Nicolas de Sadeleer (ed.), *Implementing the Precautionary Principle: Approaches from the Nordic Countries, the EU and USA* (London: Earthscan, 2007), p. 197; Gregory C. Shaffer and Mark A. Pollack, 'The EU Regulatory System for GMOs', in Michelle Everson and Ellen Vos (eds.), *Uncertain Risks Regulated* (Abingdon: Routledge-Cavendish, 2009), p. 269.

committees. While environmental risk assessment 'carried out in a scientifically sound and transparent manner based on available scientific and technical data' is a fundamental principle underpinning the GMO Directive, the general principles governing risk assessment, the Directive's objective and general obligations sections also contain express references to the precautionary principle.[129] Moreover, the recital of the GMO Directive declares that the precautionary principle 'must be taken into account when implementing it'.[130]

Another 'precautionary' feature of the GMO Directive is its 'safeguard clause' in Article 23.[131] This provision permits a member state, on a provisional basis, to restrict or prohibit the use and/or sale of a particular GMO, which has received approval under the EC Directive, as or in a product on the member state's territory. A member state may only take safeguard action, if it has 'detailed grounds' for considering that the GMO 'constitutes a risk to human health or the environment' on the basis of 'new or additional scientific knowledge' affecting the original risk assessment. In addition, the member state concerned must notify the Commission and other member states of its safeguard action and the reasons for it, including its review of the environmental risk assessment for the GMO and any new or additional information on which its decision is based. This notification is intended to prompt a Community-level review and decision on the matter within a defined period (though this has not always been the case in practice).

The explicitly precautionary basis of the GMO Directive has brought it into conflict in the trading arena with the sound science risk paradigm that animates the limited regulation of GMOs across the Atlantic. Such differences gave rise to the *GMO* dispute between the EC, the USA, Canada and Argentina that was determined by a WTO panel in 2006. This dispute and the approach taken by the panel to the differing risk regulatory paradigms advanced by the parties are discussed in detail in the following chapter.

Commission's Communication on the precautionary principle

Though the treaty and legislative instruments of the EU confirm the prominent status of the precautionary principle in EU law, they do not

[129] GMO Directive, Annex II, B, Article 1, Article 4(1).

[130] *Ibid.*, Recital 8.

[131] The ECJ has described this clause as an expression of the precautionary principle: see Case C-236/01, *Monsanto Agricoltura Italia* v. *Presidenz del Consiglio dei Ministri* [2003] ECR I-8105.

provide a definition of the principle's content, nor any instructions as to how it should be implemented in risk regulation. This is an issue of no little moment as there is no common definition of the principle in either domestic or international environmental law, and a marked variation in opinion as to what its operationalisation requires. Indeed, some commentators view 'inconsistent' formulations as a key problem faced by the precautionary principle,[132] leading to the potential for arbitrary application by decision-makers.[133] Others fear that the precautionary principle will be deployed in a highly risk-averse fashion as the basis of regulations that lack any scientific or other rational basis.[134]

Such criticisms of the precautionary principle have gained particular momentum in the sphere of international trade where the precautionary regulations introduced by the EU have often been viewed as a form of protectionism in disguise.[135] In response the EC has variously sought to argue (unsuccessfully so far) that the precautionary principle is 'a general customary rule of international law', 'a general principle of law' or 'a general principle of international law',[136] which modifies the effect of scientific evidence and risk assessment requirements under WTO law.[137] These arguments have been refuted by the USA on the basis that 'invocation of a "precautionary principle" cannot create a risk assessment where there is none, nor can a "principle" create "sufficient scientific evidence" where there is none'.[138]

In response to pressures (both internal and external) for clarification of the EU's understanding and approach to implementing the precautionary principle, the European Commission in 2000 issued a

[132] Christopher D. Stone, 'Is there a Precautionary Principle?', *Environmental Law Reporter*, 31 (2001), 10790.

[133] Gary Marchant and Kenneth Mossman, *Arbitrary and Capricious: The Precautionary Principle in the European Union Courts* (Washington DC: AEI Press, 2004).

[134] See, e.g., Giandomenico Majone, 'What Price Safety? The Precautionary Principle and its Policy Implications', J. Common Market Studies, 40 (2002), 89.

[135] Kent Jones, 'The WTO Core Agreement, Non-trade Issues and Institutional Integrity', World Trade Review, 1(3) (2002), 264.

[136] *Hormones*, [121]; *GMO*, [7.67].

[137] E.g. in *Hormones*, [16], the EC put forward precaution as a principle that applies 'not only in the management of a risk, but also in the assessment thereof' (a proposition at odds with the Commission's later characterisation of the precautionary principle as a tool of risk management in its Communication discussed below). It also asserted that the SPS Agreement provisions requiring risk assessment 'do not prescribe a particular type of risk assessment' and hence 'do not prevent Members from being cautious when setting health standards in the face of conflicting scientific information and uncertainty'.

[138] *Ibid.*, [43].

'Communication on the Precautionary Principle'.[139] Although the Communication falls within the category of soft law guidelines, there is scope for the Community courts to ascertain whether a particular disputed measure 'is consistent with the guidelines that the institutions have laid down for themselves by adopting and publishing such communications'.[140] Central to the Commission's Communication is the proposition that the precautionary principle is a risk management tool applied as part of a risk analysis framework.[141]

The Communication's starting point in discussing the implementation of the precautionary principle is that the principle presupposes 'a scientific evaluation of the risk', even though 'insufficiency of the data, their inconclusive or imprecise nature' may make it 'impossible to determine with sufficient certainty the risk in question'.[142] Despite the imprecision of the risk assessment exercise in conditions of uncertainty, the Commission seeks to instil rigour into the process, advising that the 'scientific evaluation' should be 'as complete as possible, and where possible, identifying at each stage the degree of scientific uncertainty'.[143] This evaluation then forms the basis of a 'political decision', which determines the need for action in light of the risk level that is considered 'acceptable' to the society upon which the risk will be imposed.[144] In addition, risk management measures based on the precautionary principle must adhere to a number of other general principles of EU law.[145] These include requirements that precautionary measures:

- are proportionate to the desired level of health, environmental or consumer protection (without aiming at achieving 'zero risk');[146]

[139] Natalie McNelis, 'EU Communication on the Precautionary Principle', *Journal of International Economic Law*, 3(3) (2000), 546.

[140] Case T-13/99, *Pfizer Animal Health SA* v. *Council of the European Union* [2002] ECR II-3305, [119]. The Court of First Instance did not evaluate consistency with the guidelines in this case as the Commission's Communication had been issued after the adoption of the contested Community measure.

[141] De Sadeleer, 'The Precautionary Principle in European Community Health and Environmental Law', p. 12.

[142] European Commission, 'Communication from the Commission on the Precautionary Principle', (European Union, 2000), 15.

[143] *Ibid.*, 16. [144] *Ibid.*, 15.

[145] These constraints on precautionary measures have been seen as a welcome step in converting the precautionary principle into an 'operational decision rule': Mark Geistfeld, 'Reconciling Cost-Benefit Analysis with the Principle that Safety Matters More Than Money', N.Y.U. Law Rev., 76 (2001), 114.

[146] European Commission, 'Communication from the Commission on the Precautionary Principle', 17.

- do not discriminate so as to treat comparable risk situations differently and are consistent with previously adopted, similar measures;[147]
- are subject to an examination of their costs and benefits (though this is not equivalent to an economic cost–benefit analysis and may include non-economic considerations such as health protection);[148] and
- take account of, and are updated in light of, new scientific developments.[149]

By casting the precautionary principle as a risk management tool, the Communication may have been seeking to protect the scientific integrity of the risk assessment process. However, it is arguable that too much was conceded to the dictates of science-based decision-making.[150] Given the difficulty of quarantining risk assessment and risk management in conditions of scientific uncertainty (see Chapter 3), restricting application of the precautionary principle to the political stage of decision-making may simply serve to privilege expert ways of evaluating uncertainties and reduce transparency with regard to the role played by value concerns in risk assessment.

Perhaps for this reason, a resolution on the precautionary principle adopted by the EC Council in December 2000 departed from the Commission's Communication in two important respects. First, the resolution watered down the requirement for a comprehensive scientific evaluation prior to invoking the precautionary principle, noting that 'owing to insufficient data and the nature or urgency of the risk, it may not always be possible to complete every stage [of risk assessment] systematically'.[151] Second, it affirmed the need for functional separation between risk assessors and risk management decision-makers, but provided that risk assessment should be undertaken in a 'multidisciplinary, independent and transparent manner [ensuring] that all views are heard' and should allow for 'ongoing exchange' between risk assessors and decision-makers.[152]

[147] *Ibid.*, 18. [148] *Ibid.*, 18–19. [149] *Ibid.*, 20.

[150] Susan Carr, 'Ethical and Value-based Aspects of the European Commission's Precautionary Principle', *Journal of Agricultural and Environmental Ethics*, 15(1) (2002), 31.

[151] Council Resolution on the Precautionary Principle, Nice Summit, 7–10 December 2000, [8].

[152] *Ibid.*, [9], [11].

Precautionary jurisprudence of Community courts

To a large extent it has been the courts of the Community, rather than its political institutions, that have done the most to give shape to EU risk regulation and the role of the precautionary principle in that context. Indeed, the Commission's Communication relied to a large extent on understanding of the precautionary principle developed in previous Community case law. The jurisprudential definition of the precautionary principle in the EU goes as follows:

> where there is uncertainty as to the existence or extent of risks to human health, protective measures may be taken without having to wait until the reality and seriousness of those risks become fully apparent.[153]

This definition emphasises scientific uncertainty as the essence of the principle. In what is now an extensive body of case law, Community courts – the European Court of Justice (ECJ) and the Court of First Instance (CFI) – as well as the European Free Trade Association (EFTA) Court,[154] have elaborated more detailed requirements regarding the proper invocation of the precautionary principle and its relationship to other elements of the risk regulatory process. There are even some indications in the case law that the precautionary principle may be relied upon to compel Community action despite scientific uncertainty over threatened harm. For instance, in its *Artegodan* judgment, the CFI described the precautionary principle as:

> requiring the competent authorities to take appropriate measures to prevent potential risks to public health, safety and the environment, by giving precedence to the requirements related to the protection of those interests over economic interests.[155]

[153] Case C-236/01, *Monsanto Agricoltura Italia* [2003] ECR I-8105, [111]. See also Case C-180/96, *United Kingdom* v. *Commission* [1998] ECR I-2265, [99]; Case C-157/96 *National Farmers' Union and Others* [1998] ECR I-2211, [63]; Case E-3/00 *EFTA Surveillance Authority* v. *Norway* [2001] EFTA Ct. Rep. 73, [31].

[154] This Court corresponds to the ECJ of the European Communities in matters relating to the European Economic Agreement which unites the EU member states as well as Iceland, Liechtenstein, and Norway into an internal market governed by the same basic rules.

[155] Joined Cases T-74/00, 76/00, 83/00, 84/00, 85/00, 132/00, 137/00 and 141/00, *Artegodan GmbH* v. *Commission* [2002] ECR II-4945, [184]. The CFI's decision was appealed to the ECJ but the latter did not provide any further guidance on the status of the precautionary principle: Case C-39/03 *Artegdon & Ors* v. *Commission* [2003] ECR I-7783.

Within the confines of this book it is not possible to provide a comprehensive analysis of judicial interpretation of the precautionary principle in EU law. Moreover, for the purposes of elaborating the risk regulatory paradigm that has developed on the foundations of the precautionary principle more critical is an examination of the central principles that can be distilled from the case law and the nuances of their application across the wide range of risk disputes that have confronted European courts.

In this respect, the context of decision-making has proven an important factor in shaping the stringency of the interpretation of the precautionary principle advanced in particular judgments. One keen observer of the evolving European case law on the precautionary principle – Nicolas de Sadeleer – notes that the jurisprudence has arisen mostly out of food safety disputes where scientific knowledge regarding the risks concerned is relatively advanced, rather than 'genuine environmental cases' in which issues of uncertainty and ecosystem variability tend to be more prominent.[156] De Sadeleer observes that stricter interpretations of the precautionary principle have often been favoured by the courts in disputes over precautionary national food safety regulations that depart from European harmonised standards. By contrast, a more permissive understanding of the precautionary principle is characteristic of environmental cases involving the interpretation of relevant EU regulations (for example on waste management and nature conservation).[157] For instance, in a recent case concerning species protection the ECJ ruled:

> Where it proves impossible to determine with certainty the existence or extent of the risk envisaged because of the insufficiency, inconclusiveness or imprecision of the results of the studies conducted, but the likelihood of real harm to human or animal health or to the environment persists should the risk materialise, the precautionary principle justifies the adoption of restrictive measures.[158]

Another general trend that can be discerned in the case law relates to the intensity of review exercised by European courts in examining risk regulatory measures. Here a distinction can be observed between

[156] De Sadeleer, 'The Precautionary Principle in European Community Health and Environmental Law', p. 11.

[157] For discussion of precautionary principle case law in this context see Nicolas de Sadeleer, 'The Precautionary Principle as a Device for Greater Environmental Protection: Lessons from EC Courts', *Review of European Community and International Environmental Law*, 18(1) (2009), 3.

[158] Case C-219/07 Nationale Raad van Dierenkwekers en Liefhebbers VZW and Andibel VZW v. Belgische Staat [2008] ECR I-4475, [38].

disputes brought by private business or national interests in relation to a precautionary Community risk regulatory measure, and enforcement actions brought by the Commission against member states.[159]

Precautionary principle as a 'sword': Commission enforcement actions

In cases involving action by the Commission against member states,[160] courts must weigh the national public interest in achieving a high(er) level of health protection against the Community-wide public interest in ensuring the free movement of goods, a principle of EU law enshrined in the EC Treaty.[161] De Sadeleer comments that when faced with disputes of this kind, 'the ECJ appears to apply more strictly the principle of precaution to the extent that [Member State] measures could jeopardize the functioning of the internal market'.[162]

A relevant example is the case of *European Communities* v. *Denmark* involving Commission proceedings in respect of a Danish measure applicable to enriched foodstuffs.[163] The Danish authorities sought to prevent, on a precautionary basis, the marketing in Denmark of vitamin-enriched foods lawfully produced in other member states given the potential health risk posed by nutrient additives in those foods. The ECJ found that the Danish authorities bore the burden of undertaking a risk assessment on the basis of the latest scientific data in order to found their claim of a real risk to public health.[164] That

159 De Sadeleer, 'The Precautionary Principle as a Device for Greater Environmental Protection', 6.

160 A similar pattern can be detected in 'insufficient precaution challenges' brought by non-governmental organisations against Community actions as in Case C-6/99 *Association Greenpeace France* v. *Ministere de l'Agriculture et de la Peche* [2000] ECR I-1651 and Case C-132/03, *Ministerio della Salute* v. *Codacons and Federconsumeratori* [2005] ECR I-4167, which were both dismissed by the ECJ.

161 See EC Treaty, Articles 28–29. Article 30 of the EC Treaty, building on Case C-120/78 *Cassis de Dijon* [1979] ECR 649, permits departures from the principle of free movement of goods for member states to introduce 'prohibitions or restrictions on imports, exports or goods in transit justified on grounds of public morality, public policy or public security; the protection of health and life of humans, animals or plants; the protection of national treasures possessing artistic, historic or archaeological value; or the protection of industrial and commercial property' provided the measures do not constitute 'a means of arbitrary discrimination or a disguised restriction on trade between Member States'.

162 De Sadeleer, 'The Precautionary Principle as a Device for Greater Environmental Protection', 6.

163 Case C-192/01 *European Communities* v. *Denmark* [2003] ECR I-9693.

164 *Ibid.*, [48].

risk assessment, the ECJ acknowledged, 'could reveal that scientific uncertainty persists as regards the existence or extent of real risks to human health' permitting the Member State concerned to take precautionary measures 'without having to wait until the reality and seriousness of those risks are fully demonstrated'.[165] However, the ECJ stressed:

> A proper application of the precautionary principle presupposes, in the first place, the identification of the potentially negative consequences for health of the proposed addition of nutrients, and, secondly, a comprehensive assessment of the risk to health based on the most reliable scientific data available and the most recent results of international research.[166]

Moreover, the risk assessment undertaken by the national authorities could not be based on 'purely hypothetical considerations'.[167]

A slightly more lenient approach appeared to be adopted by the ECJ in a more recent case involving Denmark, this time over standards restricting the use of sulphites, nitrites and nitrates as food additives.[168] Relying on the procedures for derogation from harmonised Community requirements under Article 95(4)–(7) of the EC Treaty, Denmark had submitted a request to the Commission to maintain its pre-existing national standards for such additives that were stricter than a new Community standard on the issue. This request was rejected by the Commission, a decision subsequently appealed by Denmark to the ECJ. The ECJ partly upheld the Danish appeal ruling that:

> the applicant Member State may, in order to justify maintaining such derogating national provisions, put forward the fact that its assessment of the risk to public health is different from that made by the Community legislature in the harmonisation measure. In the light of the uncertainty inherent in assessing the public health risks posed by, *inter alia*, the use of food additives, divergent assessments of those risks can legitimately be made, without necessarily being based on new or different scientific evidence.[169]

Veerle Heyvaert comments that this decision would seem to have 'opened the door for differentiation in market regulation between the member states on the basis of precautionary considerations, even in

[165] *Ibid.*, [49]. [166] *Ibid.*, [51].

[167] *Ibid.*, [49]. This requirement has been reiterated by the CFI in recent case law: Case T-229/04, *Sweden* v. *Commission* [2007] ECR II-2437, [161].

[168] Case C-3/00 *Denmark* v. *Commission* [2003] ECR I-2643.

[169] *Ibid.*, [63].

areas subject to harmonised Community standards'.[170] Ultimately, however, she does not see the case as permitting extensive use of the precautionary principle to 'unravel' harmonised risk regulation. She notes that the dispute pertained to the narrow situation of a Member State's request to maintain *pre-existing* national regulations, rather than broader instances where a Member State submits a request for the introduction of *new* measures derogating from Community standards. Indeed, in the latter scenario, constraints on divergent national regulations – such as the requirement in Article 95(5) that the national measures concerned be 'based on new scientific evidence relating to the protection of the environment' that are introduced 'on grounds of a problem specific to that Member State arising after the adoption of the harmonisation measure' – have been construed relatively strictly by the courts.

An example is the case of *Austria* v. *Commission* involving an Austrian province's ban on GMOs in order to protect nature as well as organic farming interests. The Austrian province's ban was intended as a derogation from the harmonised notification, assessment and authorisation regime established by the GMO Directive (discussed above). The province relied on a report entitled 'GMO-free agricultural areas: Design and analysis of scenarios and implementing measures' as the basis for precautionary measures prohibiting the cultivation of seed and planting material composed of, or containing, GMOs and the breeding and release of transgenic animals. However, the European Food Safety Authority (EFSA), following a referral from the Commission, concluded that the report did not contain any new scientific evidence that could justify banning GMOs in the Land Oberösterreich. In particular, the EFSA took the view that no scientific evidence had been submitted proving the existence of unusual or unique ecosystems that required separate risk assessments from those conducted for Austria as a whole or in other similar areas of Europe. The Commission subsequently decided to disallow the Austrian derogation, a decision upheld by the CFI and, on appeal, the ECJ. Implicitly rejecting Austria's arguments for application of the precautionary principle, the ECJ ruled that the CFI did not appear 'to have erred in law by stating that EFSA's findings concerning the absence of scientific evidence demonstrating the existence of a specific problem had been taken into consideration by the Commission'.[171]

[170] Veerle Heyvaert, 'Facing the Consequences of the Precautionary Principle in European Community Law', European L. Rev., 31(2) (2006), 193.

[171] Joined cases C-439/05P and C-454/05P, *Land Oberösterreich and Republic of Austria* v. *Commission of the European Communities* [2007] ECR I-7441, [64].

Precautionary principle as a shield: challenges to EC regulatory measures

In Commission enforcement actions brought against member states the precautionary principle has largely been deployed as a sword in order to contest Community measures or to justify a departure from the principle of free movement of goods. By contrast, in cases involving private business or Member State challenges to Community risk regulation, the precautionary principle is generally invoked as a shield by the Community institutions to support stringent health and environmental measures in light of the Treaty objectives of achieving a high level of health, environmental and consumer protection.[172] In such cases the courts have tended to afford Community institutions a wide margin of discretion regarding the required evidentiary basis for precautionary regulation. This is effected by the adoption of a more lenient standard of judicial review, such that in situations where the Community institutions are required to evaluate 'highly complex scientific and technical facts', the judicature:

> must confine itself to ascertaining whether the exercise by the institutions of their discretion in that regard is vitiated by a manifest error or a misuse of powers or whether the institutions clearly exceeded the bounds of their discretion.[173]

Accordingly, 'under the precautionary principle the Community institutions are entitled, in the interests of human health to adopt, on the basis of as yet incomplete scientific knowledge, protective measures which may seriously harm legally protected positions, and they enjoy a broad discretion in that regard.'[174]

A striking illustration of this approach is provided by some of the ECJ's earliest precautionary decisions in cases fought over Community

[172] De Sadeleer, 'The Precautionary Principle in European Community Health and Environmental Law', p. 36. Heyvaert, 'Facing the Consequences of the Precautionary Principle in European Community Law', 185 describes the two categories of case law differently as 'insufficient precaution challenges' and 'excessive precaution challenges'.

[173] Case T-13/99, *Pfizer Animal Health SA* v. *Council of the European Union* [2002] ECR II-3305 (*Pfizer*), [169]. Judicial deference to discretion in decision-making in the context of scientific uncertainty has a long tradition in EU law. See, e.g., Case C-331/88, *R* v. *Minister of Agriculture, Fisheries and Food and Secretary of State for Health ex parte Fedesa* [1990] ECR I-4023.

[174] *Pfizer*, [170].

measures to safeguard against the spread of mad cow disease. At the height of the mad cow disease scare in Europe, the Commission decided to introduce emergency measures designed to prevent the transmission of disease by the export of meat from countries with infected cattle. The Commission's decision was strongly opposed by affected farmers and some member states who objected that there was little scientific evidence, available at the time, linking mad cow disease to human health effects, such as the neurological disorder Creutzfeldt-Jakob disease. However, the ECJ found that the Commission 'did not clearly exceed the bounds of its discretion' or act in manner that was 'manifestly inappropriate' in introducing the emergency measures, given the existence of 'risks regarded as a serious hazard to public health' and the 'great uncertainty' surrounding those risks at the time the decision was adopted. Reflecting its acknowledgment of the need for free market considerations to be reconciled with the goal of achieving a high level of health protection at the Community level, the ECJ sought support for its decision in the EC Treaty provisions relating to the objectives and underlying principles of Community environmental policy.[175]

Where Community risk regulatory measures based on the precautionary principle are at stake, the nature of judicial review exercised by the courts is ostensibly procedural, rather than substantive. Rather than conducting a detailed inquiry into the adequacy of the science or risk assessments on which Community institutions' measures rely, the courts thus place a heavy emphasis upon the process underlying the adoption of risk regulations. This approach enables use of the precautionary principle as a protective shield by Community institutions wishing to act in the face of scientific uncertainty to protect health or the environment. However, as Joanne Scott notes, '[t]he protection offered is contingent not absolute.'[176] In several cases European courts have 'talked the language of deference' while in fact subjecting institutions' decisions to detailed scrutiny and rigorous scientific risk assessment requirements.[177] Evident in the case law is hence the courts'

175 *Pfizer*, [170].Case C-180/96, *United Kingdom* v. *Commission* [1998] ECR I-2265, [69]–[70], [97], [100]; Case C-157/96 *National Farmers' Union and Others* [1998] ECR I-2211, [61], [64].

176 Joanne Scott, 'The Precautionary Principle before the European Courts', in Richard Macrory, Ian Havercroft and Ray Purdy (eds.), *Principles of European Environmental Law* (Groningen: Europa Law Publishing, 2004), p. 60.

177 *Ibid.*, p. 62. In this respect the trend in the EU case law may be echoing the evolution of 'hard look' judicial review of science-based regulatory measures in the USA.

simultaneous struggle to validate the broad discretion of Community institutions taking precautionary action while also guaranteeing the substantive quality of such decision-making.[178]

This trend is well illustrated by decisions of the CFI in cases such as *Pfizer*, *Alpharma* and *Solvay*.[179] All three cases concerned a precautionary Council Regulation that revoked previous authorisations for certain antibiotics to be used as growth promoters in animal feed. In each case the parties also agreed that there was scientific uncertainty surrounding the potentially harmful effects of the antibiotics at the time of the Council Regulation. Consistently with previous Community jurisprudence, the CFI found that, applying the precautionary principle in circumstances of scientific uncertainty, the Community institutions were entitled to take protective measures without having to wait until the reality or seriousness of risks to human health became fully apparent.[180] Moreover, in such situations a risk assessment could not be required 'to provide the Community institutions with conclusive scientific evidence of the reality of the risk and the seriousness of the potential adverse effects were that risk to become a reality'.[181] Nonetheless, the impossibility of carrying out a full scientific risk assessment did not prevent the competent authority from taking preventative measures if such 'appear essential given the level of risk to human health which the authority has deemed unacceptable for society', a decision reserved to the institution concerned.[182]

This latitude was matched by the CFI's concern that the adoption of Community precautionary measures should abide by procedural guidelines designed to safeguard against arbitrary decision-making.[183] The CFI articulated a number of conditions regarding the adoption of valid precautionary measures, including that:

1. Action is only justified where there is a risk which, albeit not fully demonstrated, is more than 'purely hypothetical' (that is, it is not simply founded on mere conjecture which has not been scientifically verified);[184]

[178] Heyvaert, 'Facing the Consequences of the Precautionary Principle in European Community Law', 185.

[179] Case T-70/99 *Alpharma Inc.* v. *Council of the European Union* [2002] ECR II-3495 (Alpharma); Case T-392/02, *Solvay Pharmaceuticals BV* v. *Council* [2003] ECR II-4555 (Solvay).

[180] *Pfizer*, [139], [140].

[181] *Pfizer*, [142]; *Alpharma*, [155].

[182] *Pfizer*, [151], [153], [160].

[183] Wybe Th. Douma, 'Fleshing Out the Precautionary Principle by the Court of First Instance', *Journal of Environmental Law*, 15 (2003), 405.

[184] *Pfizer*, [143]; *Solvay*, [129].

2. Decisions are 'adequately backed up by the scientific data available at the time when the measure was taken';[185]
3. Even if no full risk assessment is available, action must still be based on 'as thorough a scientific risk assessment as possible' and taken only when 'sufficient scientific indications' support it;[186]
4. Risk assessment should be entrusted to experts whose role is to supply the political institutions with scientific advice, based on the principles of 'excellence, independence and transparency', which will provide the competent public authority with 'sufficiently reliable and cogent information to allow it to understand the ramifications of the scientific question raised and decide upon a policy in full knowledge of the facts';[187]
5. Community institutions are not bound by the opinion of the Community's scientific committees, but if they opt to disregard these opinions they must provide specific reasons at a scientific level at least commensurate with that of the opinion in question.[188]

As Scott has observed, the preconditions for precautionary action specified by the CFI in these cases appear to be more than an insistence that 'the political institution in question clothe its decision in the garb of science'. Rather there is an evident concern 'that those with responsibility for providing the scientific evidence be appropriately qualified and independent, and that they operate in an environment which is transparent and hence susceptible to critique and dissent'.[189] She sees this as constituting the precautionary principle in Community jurisprudence as a 'weak' version of the principle that 'liberates' Community institutions to take discretionary measures in the face of scientific uncertainty rather than mandating stringent precautionary action.[190]

On the other hand, the to-and-fro visible in a number of cases decided by the Community courts – where there is a vacillation between confirming the authority of institutions to take action in conditions of scientific uncertainty and insisting on a sound scientific basis for decision-making – may simply be a reflection of institutional insecurities that still pertain in the EU system. Heyvaert comments that the jurisprudence 'betrays a keen awareness of the high level of

185 *Pfizer*, [144], [157], [159], [162]; *Alpharma*, [157], [175].
186 *Pfizer*, [165].
187 *Pfizer*, [144], [157], [159], [162]; *Alpharma*, [157], [172], [175].
188 *Pfizer*, [199], [213]. However, this requirement does not appear to be stringently applied as the decision in *Alpharma* indicates.
189 Joanne Scott, 'International Trade and Environmental Governance: Relating Rules (and Standards) in the EU and the WTO', *European Journal of International Law*, 15 (2004), 320.
190 Scott, 'The Precautionary Principle before the European Courts', p. 62.

discretionary prowess that the precautionary principle bestows on EC decision-makers, and perhaps a concern on the part of the courts that these institutions do not have sufficient credibility to carry off this level of discretion'.[191] Thus, even though the CFI in *Pfizer* implied the superior 'democratic legitimacy' of Community institutions compared with their advisory scientific committees,[192] this was accompanied by reassurances that the institutions would nonetheless base their decisions on expert advice and assessment. Such qualms over institutional legitimacy are also characteristic of many international risk regulatory bodies, as was discussed in Chapter 2. As we have seen, the tendency in such circumstances is to fall back to the familiar territory of expertise-based legitimacy, although this sits at odds with the acknowledgment inherent in precautionary regulation of the desirability of action despite scientific uncertainty.

The precautionary principle and sound science: conflict or convergence?

As has often been pointed out, there is nothing inherently contradictory between a requirement for decision-making with a sound scientific basis and application of the precautionary principle. The core concern of the latter is with scientific uncertainty, but the precautionary principle as it has developed in international and EU law is rarely divorced from the trappings of scientific risk assessment. Given the limitations of the scientific method based on empirical observation to offer absolute proof of any given hypothesis, conventional science is also tolerant of inherent uncertainty and 'good' scientists are forthright about the uncertainties and areas of ignorance or indeterminacy that underlie their research claims.[193] For example, the global climate change scientific body, the IPCC, has adopted methodologies for its assessment reports that distinguish between a 'robust finding' (one that holds under a variety of approaches, methods, models and assumptions) and 'key uncertainties' (those that, if reduced, may lead to new and robust findings). Its reports also supply information on 'assigned

[191] Heyvaert, 'Facing the Consequences of the Precautionary Principle in European Community Law', 203.

[192] *Pfizer*, [201].

[193] Consequently, 'an epistemologically humble precautionary approach is arguably more scientific than the traditional narrow risk approach called "sound science"': Les Levidow, 'Precautionary Uncertainty: Regulating GM Crops in Europe', *Social Studies of Science*, 31(6) (2001), 848.

confidence levels' that represent the scientific authors' 'collective judgment in the validity of a conclusion based on observational evidence, modelling results, and theory that they have examined'.[194] This mode of expressing uncertainties was introduced in order to make 'inevitably partly subjective estimations more transparent and understandable for policymakers'.[195]

However, while scientific methodologies can be employed in order to disclose uncertainties and value judgments in a transparent fashion, they also lend themselves to misuse for political ends. Holly Doremus notes that the Bush administration's use of sound science to 'advance its anti-regulatory, anti-conservation agenda' was achievable within the boundaries of science:

> None of [the strategies] requires falsification or outright rejection of clear scientific data, yet together they allow the administration to undermine conservation efforts without openly acknowledging its disdain for conservation values.[196]

Equally some of the fears raised with respect to 'strong' versions of the precautionary principle – that mandate regulatory action in the face of any level of scientific uncertainty – reflect a concern that allegations of uncertainty might be used to conceal a fervently anti-technological agenda. Science deployed as 'just a tool ... turned to a variety of ends' is thus susceptible to abuse at the extremes of both the sound science and precautionary approaches.[197]

This might seem to indicate that any distinction between the sound science and precautionary principle risk regulatory paradigms – other than at the margins – is more imagined than real. Such a perspective is reinforced by some of the literature on transatlantic regulatory trends that takes issue with claims that EU risk regulation in the health and environmental field is currently 'more precautionary' than that in the USA.[198] While there is more than a hint of hurt national pride in American refutations of the idea that the Europeans may be

194 IPCC, Climate Change 2001: Synthesis Report (Geneva: IPCC, 2001) 44.

195 Bernd Siebenhüner, 'Can Assessments Learn, and If So, How?', in Alexander Farrell and Jill Jäger (eds.), *Assessments of Regional and Global Environmental Risks: Designing Processes for the Effective Use of Science in Decisionmaking* (Washington DC: Resources for the Future, 2006), p. 174.

196 Doremus, 'Science Plays Defense', 266.

197 *Ibid.*, 258.

198 See particularly, Jonathan Wiener and Michael Rogers, 'Comparing Precaution in the United States and Europe', *Journal of Risk Research*, 5 (2002), 317; Jonathon

more precautionary when it comes to health and environmental protection, the literature does make the important point that in the last few decades there has been a fair degree of convergence in risk regulation across the two regions. An emphasis on high-quality scientific data and processes of risk assessment is thus a feature of both the US and EU systems. Indeed, the Community political institutions seem no less cognisant than their American counterparts of the strategic value of science-based evaluations as:

> Community law may, wherever it manages to promote science-based standards of validity, ensure its own authority without the usual entanglements in complex controversies over competencies, conflicting economic interests and highly sensitive issues of political accountability.[199]

Even David Vogel, who, as discussed above, has been one of the strongest proponents of the view that the USA and the EU have 'traded places' in terms of the stringency of their health and environmental risk policies, has remarked that 'somewhat paradoxically, European regulatory administration is also becoming more scientifically rigorous' with 'increased recognition of the need to strengthen the capacity of government agencies to conduct risk assessments and to improve the quality of scientific information available to decisionmakers'.[200] He sees an important factor underlying this development as being:

> an increase in judicial review of regulatory decisions at both the European and international levels. Just as American regulatory agencies engaged in more formal risk assessment in order to defend their decisions in federal court from challenges by both public interest groups and industry, so Europe's national authorities and the EU are undertaking similar steps in order to defend their decisions before the European Court of Justice (ECJ) and World Trade Organization dispute panels.[201]

More broadly, we might add that the convergence towards scientific risk assessment observable in the USA and the EU is entirely consistent

B. Wiener, 'Whose Precaution After All? A Comment on the Comparison and Evolution of Risk Regulatory Systems', *Duke Journal of Comparative and International Law*, 13 (2003), 207.

[199] Christian Joerges, 'Scientific Expertise in Social Regulation and the European Court of Justice: Legal Frameworks for Denationalized Governance Structures', in Christian Joerges, Karl-Heinz Ladeur and Ellen Vos (eds.), *Integrating Scientific Expertise into Regulatory Decision-Making* (Baden-Baden: Nomos Verlagsgesellschaft, 1997), pp. 297–8.

[200] Vogel, 'The Hare and the Tortoise Revisited', 567.

[201] *Ibid.*

with the privileged place science has been afforded in Western democracies (see Chapter 3).

As Jonathan Wiener points out, the reality is thus 'more complex' than simplistic claims of EU regulation being 'more precautionary' might suggest;[202] 'relative precaution appears to depend on the risk and the consequences of specific policies'.[203] Put another way, there are certain types of risks (and surrounding uncertainties) that appear to worry Europeans more than Americans (and vice versa). Contrary to Wiener's analysis, however, there are detectable patterns in these differing risk concerns that would seem to have been productive of divergent regulatory responses. In the case of the EU, the kinds of health and environmental risks that cause public concern – such as GMOs, hormone and other food additives, toxic chemicals, synthetic chemical additives in products targeted at children, climate change and marine pollution – are characterised by factors such as uncertainty over long-term effects, technological causes and public distrust in responsible regulatory authorities based on past failures. By contrast, the US risk regulatory system has paid less attention to these kinds of risk, emphasising instead issues of cost associated with their regulation, the potential for risk–risk trade-offs (for example where banning one kind of food additive may lead to a more harmful substitute being used), the danger of stifling technological innovation and the capacity for the courts to provide redress for any harms that do emerge over time.

Based on the experience of the US risk regulatory system, the momentum for increased regulatory stringency in Europe may not last indefinitely.[204] However, whatever the particular dynamics of US–EU regulatory trends in the health and environmental field, sound science and the precautionary principle are likely to remain signifiers of competing risk regulatory philosophies in the broader international sphere. There is some evidence of developing countries taking up the precautionary approach to risk regulation in areas such as hazardous waste control and the transboundary movement of GMOs. Moreover, US and EU advocacy of their alternative risk regulatory strategies, most particularly in trade disputes such as the *Hormones* and *GMO* cases in the WTO, has already left an indelible imprint on international law in the health and environmental field. The deep political divisions that exist on this issue are evident in general international

[202] Weiner, 'Whose Precaution After All?', 215–48.
[203] *Ibid.*, 225. [204] *Ibid.*, 580.

instruments such as the Plan of Implementation issued by the World Summit on Sustainable Development in 2002. This document represents an unwieldy compromise between the different positions advocated respectively by the USA (supported on some issues by Japan and Australia) and the EU (with Norway and Switzerland).[205] The end result is that states are simultaneously called upon to:

> Promote and improve science-based decision-making and reaffirm the precautionary approach as set out in principle 15 of the Rio Declaration on Environment and Development.[206]

International legal responses to paradigm conflict

Where sound science and precautionary risk regulation come into conflict in the international sphere, the disputes that arise are generally not about specific scientific evidence. Indeed, very often the domestic risk regulatory bodies concerned will have gathered and evaluated much the same scientific material prior to devising risk management measures. Instead their points of difference relate to questions about which evidence is relevant for assessment purposes (for example, should minority scientific views and concerns over areas of uncertainty and ignorance be taken into account?) and what weight should be placed on scientific evidence relative to other factors, such as economic considerations or public views. Whereas in the sound science paradigm the focus is on accumulating sufficient data to be satisfied of risk, in precautionary regulation questions of safety predominate, with science 'consulted less for the knowledge that it has to offer than for the doubts and concerns that it is in a position to raise'.[207]

Importantly, science alone is not capable of resolving either enquiry: at most scientific information can inform decisions about relative safety or risk.[208] Thus even though contemporary Western regulation of health and environmental risk draws (heavily) on science, it invariably retains a political element because of the questions of value

[205] IISD, 'Summary of the World Summit on Sustainable Development: 26 August – 4 September 2002', *Earth Negotiations Bulletin*, 22(51) (2002), 5.

[206] Plan of Implementation of the World Summit on Sustainable Development, A/CONF/199/20, (2002) [109f].

[207] De Sadeleer, 'The Precautionary Principle in EC Health and Environmental Law', 159.

[208] Vern Walker, 'The Myth of Science as a "Neutral Arbiter" for Triggering Precautions', B.C. Int'l & Comp. L. Rev., 26 (2003), 200.

implicated in judgments about harms of importance and risk levels deemed acceptable.[209] These hard decisions will be rendered even more difficult and contestable where there is a necessity to rely on assessments that, because of uncertainties, have been made on soft science.

For international law the dilemma posed when different risk regulatory paradigms come into conflict raises an important question as to what might be the appropriate legal response. If we take the view that international law is simply a decentralised collection of fora where states (and occasionally private actors) may air divergent views as to the priority of different regulatory modes or different regimes, international rules may seem to offer no coherent solution to paradigm conflicts other than that produced by the unequal institutional attributes and powers of separate treaty arrangements or institutions.[210] Outcomes on this basis might simply be dictated by the particular 'structural bias' of different regimes.[211] For instance, Robyn Eckersley laments the institutional weaknesses of, and lack of coordination between, multilateral environmental agreements compared with the strong, centralised system of WTO dispute settlement, fearing that the latter is having a 'chilling' effect on the development of the former.[212] John Applegate voices similar concerns, arguing that as a result of interactions with the trade regime, the precautionary principle in international law is being 'tamed' and assimilated within the 'risk-based paradigm', such that a precautionary response to uncertainty becomes 'at best a temporary substitute for real analysis'.[213]

However, at the level of international negotiations and legislation, a different approach is evident in respect of the conflict between the precautionary and sound science paradigms, particularly as it arises in the trade-and-environment sphere. One formulation that is becoming increasingly common declares the 'mutual supportiveness' of

[209] Jeffery Atik, 'Science and International Regulatory Convergence', Northwestern J. Int'l L. & Business, 17 (1996–1997), 737.

[210] Accordingly there may be little to prevent 'the most powerful regime from eventually devouring a less potent one': Dirk Pulkowski, 'Book Review: Pauwelyn, Joost. Conflict of Norms in Public International Law: How WTO Law Relates to Other Rules of International Law. Neumann, Jan. Die Koordination Des WTO-Rechts Mit Anderen Völkerrechtlichen Ordnungen: Konflickte Des Materiellen Rechts Und Konkurrenzen Der Streitbeilegung', European J. Int'l Law, 16 (2005), 153.

[211] Koskenniemi, 'Fragmentation of International Law', 208.

[212] Robyn Eckersley, 'The Big Chill: The WTO and Multilateral Environmental Agreements', *Global Environmental Politics*, 4(2) (2004), 24.

[213] John Applegate, 'The Taming of the Precautionary Principle', William & Mary Envtl L. & Policy Review, 27 (2002), 50.

different regimes, suggesting the possibility of coordinated implementation.[214] A pertinent example is the Biosafety Protocol that, following extensive negotiations on the question of the relationship between its precautionary risk assessment provisions and the more science-focused requirements of the SPS Agreement, sets forth the following in its preamble:

> *Recognizing* that trade and environment agreements should be mutually supportive with a view to achieving sustainable development,
>
> *Emphasizing* that this Protocol shall not be interpreted as implying a change in the rights and obligations of a Party under any existing international agreements,
>
> *Understanding* that the above recital is not intended to subordinate this Protocol to other international agreements …

There are many commentators who believe that the most that can be drawn out of such apparently contradictory recitals is an affirmation of the equal importance of all international agreements to which states might be party.[215] Some others, though, view calls for mutual supportiveness or the use of similar 'compromise formulas' as implying 'a willingness to acknowledge the existence of parallel and potentially conflicting treaty obligations', albeit 'fall[ing] short of indicating clearly what should be done in case conflicts emerge'.[216]

Other signs of a movement towards reconciliation of different regulatory approaches are found in judgments of the WTO Appellate Body. For example, in the *Shrimp/Turtle* dispute, which saw trading rights of Asian developing countries opposed to the environmental risk of turtle by-catch posed by shrimping practices, the Appellate Body's interpretation of relevant trade law provisions took into account the 'contemporary concerns of the community of nations about the protection

[214] E.g., Plan of Implementation, [91]; Ministerial Declaration, WT/MIN(01)/DEC/1 (14 November 2001), [6].

[215] E.g., Aarti Gupta, 'Governing Trade in Genetically Modified Organisms: The Cartagena Protocol on Biosafety', Environment, 42(4) (2000), 22; Ruth Mackenzie *et al.*, *An Explanatory Guide to the Cartagena Protocol on Biosafety*, IUCN Environmental Policy and Law Paper No. 46 (Gland: IUCN, 2003); Olivette Rivera-Torres, 'The Biosafety Protocol and the WTO', B.C. Int'l & Comp. L. Rev., 26 (2003), 263; Sabrina Safrin, 'Treaties in Collision? The Biosafety Protocol and the World Trade Organization Agreements', Am. J. Int'l L., 96 (2002), 606; Terence Stewart and David Johanson, 'A Nexus of Trade and the Environment: The Relationship between the Cartagena Protocol on Biosafety and the SPS Agreement of the World Trade Organization', Colorado J. Int'l Envt'l Law & Policy, 14 (2003), 1.

[216] Koskenniemi, 'Fragmentation of International Law', 118.

and conservation of the environment'.[217] This decision evidenced the Appellate Body's willingness to situate trade law in the broader normative environment of international law, albeit in a context where the trade agreement at issue recognised the scope for competing policy concerns (such as natural resource conservation) to override trading rights in appropriate cases.[218]

Beyond the trade-and-environment sphere there is also an increasing, although not substantial, number of cases that reflect scope for paradigm coexistence. In some instances this is manifested in 'a growing recognition as to some appropriate role for precautionary measures'.[219] In other cases, such as the decision of the *Iron Rhine Railway* arbitral tribunal, there has been an overt attempt to reconcile different legal and regulatory approaches with declarations that:

> Environmental law and the law on development stand not as alternatives but as mutually reinforcing, integral concepts, which require that where development may cause significant harm to the environment there is a duty to prevent, or at least mitigate, such harm.[220]

Of course, international law need not favour either the triumph of one paradigm over another or attempt reconciliation between them, but merely recognise the possibility for a pluralistic approach. Arguably such diversity is important in providing the necessary legal complexity to cope with a world that is itself 'irreducibly pluralistic'.[221] If international law were to respond in this manner in the field of risk regulation, it would seek to preserve possibilities for 'multiple regimes and multiple modes of thought'.[222]

217 *United States – Importation Prohibition of Certain Shrimp and Shrimp Products*, Report of the WTO Appellate Body, WT/DS58/AB/R, 12 October 1998 (*Shrimp/Turtle*), [129].

218 This case was decided under the GATT. For further discussion of the Appellate Body's approach see Chapter 6.

219 Sands, *Principles of International Environmental Law*, p. 8. Two notable decisions in this regard are those in the *Southern Blue-fin Tuna Cases* (*New Zealand* v. *Japan*; *Australia* v. *Japan*), Provisional Measures (27 August 1999) ITLOS Case Nos 3 and 4, [77] and *The MOX Plant Case*, (*Ireland* v. *United Kingdom*), Provisional Measures (3 December 2001), ITLOS Case No 10, [84].

220 *In the Arbitration Regarding the Iron Rhine Railway* (*Belgium* v. *Netherlands*), Award of the Permanent Court of Arbitration, The Hague, 24 May 2005, available from www.pca-cpa.org/showpage.asp?pag_id=363.

221 Koskenniemi, 'Fragmentation of International Law', 208.

222 Martti Koskenniemi, *Global Legal Pluralism: Multiple Regimes and Multiple Modes of Thought* (Erik Castrén Institute of International Law and Human Rights, University of Helsinki, 2005), available at www.helsinki.fi/eci/Publications/MKPluralism-Harvard-05d%5B1%5D.pdf.

The following sections of the chapter look more closely at the international legal literature that has developed around the question of how international law does or might respond to a divergence in regulatory approaches. Insights from three different bodies of literature are considered: (1) the literature on fragmentation in general international law; (2) the 'trade and' literature that focuses on fragmentation issues as they play out in the WTO context; and (3) perspectives that emphasise the possibilities for international institutional evolution to accommodate different regulatory approaches and values, even to embrace pluralism. As indicated in the introduction to the chapter, the purpose of this discussion is not to seek to resolve what have become much-debated questions in international law and the legal literature. Rather the intention is to illustrate how each approach would conceive the best response to the issues posed by conflicts between the different regulatory paradigms that have come to dominate international debate and governance structures concerning health and environmental risk.

Fragmentation in international law

The dominant discourse that has emerged in the international legal literature to describe and respond to phenomena of diversification and divergence in international law is that of fragmentation. In international law, fragmentation is seen as the flip side of globalisation, leading 'to the emergence of specialized and relatively autonomous spheres of social action and structure'.[223] For many international lawyers, including former and current justices of the International Court of Justice (ICJ), fragmentation in international law is seen in a negative light as something that may erode general international law, generate conflicting jurisprudence, encourage forum shopping in dispute settlement and result in a loss of legal security.[224] Underlying such concerns is the fear that if international law becomes too disjointed, complex or conflict-ridden this will undermine what has been achieved through

[223] Koskenniemi, 'Fragmentation of International Law', 11. See also Fischer-Lescano and Teubner, 'Regime-Collisions', 1017, 1045.

[224] E.g., Dupuy, 'The Danger of Fragmentation or Unification of the International Legal System and the International Court of Justice'; Georges Abi-Saab, 'Fragmentation or Unification: Some Concluding Remarks', N.Y.U. J. Int'l Law & Politics, 31 (1999), 919; Jonathan Charney, 'The Impact on the International Legal System of the Growth of International Courts and Tribunals', N.Y.U. J. Int'l Law & Politics, 31 (1999), 697; Gerhard Hafner, 'Pros and Cons Ensuing from Fragmentation of International Law', Michigan J. Int'l Law, 25 (2004), 849; John Jackson, 'Fragmentation or Unification Among International Institutions: The World Trade Organization',

expansion of the discipline and the development of global regimes following the Second World War.[225] For some this suggests the need for the constitutionalisation of international law,[226] or at the very least a more central role in dispute settlement for general international law institutions such as the ICJ.[227]

Others view fragmentation as merely a technical problem consequent upon the increase in international legal activity and the growth of global governance in a range of fields. Fragmentation in this perspective can be 'controlled by the use of technical streamlining and coordination', largely employing existing rules and legal structures.[228] This was the approach taken, for example, in the 2006 report of the International Law Commission on 'Difficulties arising from the Diversification and Expansion of International Law'.[229] According to the Commission, international law is not a 'random collection' of rules and principles but rather 'a legal system' with 'meaningful relationships' discernible between its norms'.[230] In applying international law the Commission recognises that two or more rules or principles may both be valid and applicable in respect of a situation. In such cases, the differing norms may be able to be applied in conjunction such that one norm assists in the interpretation of another. Indeed, the Commission declares that it

N.Y.U. J. Int'l Law & Politics, 31 (1999), 823; Pemmaraju Rao, 'Multiple International Judicial Forums: A Reflection of the Growing Strength of International Law or its Fragmentation?', Michigan J. Int'l Law, 25 (2004), 929; Tullio Treves, 'Conflicts between the International Tribunal for the Law of the Sea and the International Court of Justice', N.Y.U. J. Int'l Law & Politics, 31 (1999), 809.

225 A further political concern, reflected in works such as Philippe Sands, *Lawless World: America and the Making and Breaking of Global Rules* (London: Allen Lane, 2005), is that a lack of coherence in international law could encourage a practice of 'a la carte multilateralism' by powerful national executives.

226 E.g., Ernst-Ulrich Petersmann, 'Constitutionalism and International Adjudication: How to Constitutionalize the U.N. Dispute Settlement System?', N.Y.U. J. Int'l Law & Politics, 31 (1999), 753.

227 E.g., Dupuy, 'The Danger of Fragmentation or Unification of the International Legal System and the International Court of Justice'.

228 Koskenniemi, 'Fragmentation of International Law', 12.

229 International Law Commission, 'Conclusions of the work of the Study Group on the Fragmentation of International Law: Difficulties arising from the Diversification and Expansion of International Law', *Yearbook of the International Law Commission (ILC Report)*, 2(2) (2006). The General Assembly took note of the ILC report in A/RES/61/34 of 18 December 2006.

230 International Law Commission, 'Conclusions of the work of the Study Group on the Fragmentation of International Law', 2. See also Joost Pauwelyn, 'Bridging Fragmentation and Unity: International Law as a Universe of Inter-connected Islands', Michigan J. Int'l Law, 25 (2004), 903.

is 'a generally accepted principle that when several norms bear on a single issue they should, to the extent possible, be interpreted so as to give rise to a single set of compatible obligations'.[231]

However, the Commission also acknowledges the possibility of conflict between norms – where the two rules or principles at issue point to incompatible decisions, necessitating a choice between them. This choice, it insists, should be made on the basis of technical treaty interpretation rules set out in the Vienna Convention on the Law of Treaties (Vienna Convention).[232] These include:

- the *lex specialis* rule: that more specific rules and special regimes should be accorded priority over general rules, which are limited to a 'gap filling' role;[233]
- the *lex posteriori* rule: that later rules on the same subject supersede earlier law where the same treaty parties are concerned;[234] and
- the principle of 'systemic integration' based on Article 31(3)(c) of the Vienna Convention. This principle requires a treaty interpreter to take into account 'any relevant rules of international law applicable in the relations between the parties'.[235]

While the Commission's fragmentation report gives the impression that technical solutions are possible in all instances of normative conflict that arise in international law, the preparatory report produced by its study group led by Martti Koskenniemi is more forthright about potential limits to the accommodation between different norms or regulatory approaches that can be fashioned by employing technical tools. As the study group's report observed:

> Public international law does not contain rules in which a global society's problems are, as it were, already resolved. Developing these is a political task.[236]

In particular, situations involving genuine conflicts, animated by strong differences over the basic values at stake in dealing with a given issue,

[231] International Law Commission, 'Conclusions of the work of the Study Group on the Fragmentation of International Law', 2. The ILC describes this as 'the principle of harmonization': 14.

[232] Vienna Convention on the Law of Treaties, 23 May 1969, 1155 UNTS 331, in force 27 January 1980 (Vienna Convention).

[233] ILC Report, A/RES/61/34 of 18 December 2006, 2–6.

[234] *Ibid.*, 9–14. Subject to the operation of peremptory norms, such as rules of *jus cogens*.

[235] For discussion of this principle see Campbell McLachlan, 'The Principle of Systemic Integration and Article 31(3)(c) of the Vienna Convention', Int'l Comp. L.Q., 54 (2005), 279.

[236] Koskenniemi, 'Fragmentation of International Law', 247.

may offer limited possibilities for the application of 'a "coordinating" solution'.[237] Authors such as Christian Joerges and Jürgen Neyer have suggested that in such disputes, 'where the law ends' comity should intervene. International courts or dispute settlement bodies in this case would refrain from adjudication, although they might still play a role 'as fora or as instigators of fair and workable compromises'.[238]

Absent such restraint there is the likelihood that solutions to genuine normative conflicts will be resolved simply on the basis of the institutional strength of different dispute settlement regimes. In this regard, the WTO has become a central institution given the fact that it unites broad competence over trade-related issues with a compulsory, and highly effective, dispute settlement system. Indeed, under Article 23 of the organisation's Dispute Settlement Understanding (DSU), WTO Members are required to have recourse exclusively to the WTO dispute settlement organs in order to determine any dispute arising under the 'covered agreements', regardless of the other areas or norms of international law that might be implicated. From the perspective of contests between risk regulatory rules based on sound science and those based on the precautionary principle, many such disputes will be funnelled preferentially into the WTO system rather than being dealt with under environmental treaties that generally lack 'the coherence, reach, financial backing and organizational structure of the WTO'.[239]

'Trade and' literature on linkage of WTO law

The institutions and dispute settlement system of the WTO now play a prominent role in dealing with situations that involve not just trade issues, but also a range of other social areas.[240] It is notable, for instance, how many disputes over health and environmental risks have come before the WTO dispute settlement system in recent years, from cases involving food safety, to quarantine risks and more general health

[237] *Ibid.*, 210.

[238] Christian Joerges and Jürgen Neyer, 'Politics, Risk Management, World Trade Organisation Governance and the Limits of Legalisation', *Science and Public Policy*, 30(3) (2003), 224.

[239] Eckersley, 'The Big Chill', 24. It is notable that several of the environmental treaties that embody a precautionary approach – e.g. the Convention on Biological Diversity, the Biosafety Protocol and the Kyoto Protocol – are not ones to which proponents of sound science such as the USA are party.

[240] Joost Pauwelyn, 'Remarks on "The Jurisdiction of the World Trade Organization"', American Society International L. Proceedings, 98 (2004), 135.

and environmental concerns.[241] Many more questions about risks and appropriate regulatory structures for their assessment and management have been raised in WTO political bodies such as the Committee on Trade and Environment and the SPS Committee.[242] How the WTO responds to risk disputes is thus likely to be an important part of the overall international legal response to questions regarding science-based decision-making and the scope of permissible risk regulation.

At the theoretical level, the dominant approach to issues of interaction between the international trade regime and other systems or values found in international law is represented by the 'trade and' (or 'linkage') literature. This literature has formed a distinctive body of academic work in international trade law from the early 1990s onwards. Controversial dispute settlement decisions issued around this time, such as the *Tuna/Dolphin* cases (as well as the later *Shrimp/Turtle* cases), attracted academic attention to the question of the WTO's potential 'pro-trade bias'.[243] There have also been numerous calls for broadening of the WTO's regulatory agenda to areas such as environmental protection, based on a perception of the WTO as a powerful and effective institution for achieving various social ends.[244]

Like its generalist cousin in the literature on international fragmentation, a major focus of the 'trade and' scholarship is whether, and if so how, the WTO should accommodate norms and values from other systems. 'Linkage' in this context may take a variety of forms, from the conditioning of access to markets on the satisfaction of non-trade goals, to amendment of the WTO agreements to accommodate exceptions based on other social concerns, or modest options such as research and information exchange through the various WTO committees.[245] Linkage can also occur in more subtle ways as a result of arguments and decisions regarding the need for WTO dispute settlement organs to interpret relevant trade rules in light of other substantive rules of public international law.[246] As Joel Trachtman notes:

[241] There is also an increasing caseload examining health and environmental risk questions in trade-related and investment tribunals under free trade agreements.

[242] The SPS Committee's deliberations are discussed further in the next chapter.

[243] Sungjoon Cho, 'Linkage of Free Trade and Social Regulation: Moving Beyond the Entropic Dilemma', *Chicago Journal of International Law*, 5 (2005), 640.

[244] For one of the broadest such proposals, putting the case for a 'World Economic Organization' with expansive competence to address societal values, see Andrew Guzman, 'Global Governance and the WTO', Harv. J. Int'l L., 45 (2004), 303.

[245] Alvarez, 'Symposium: The Boundaries of the WTO – Foreword', 1.

[246] *Ibid.*, 2.

> The general issue raised by most linkage claims is whether trade rules and environmental, labor, human rights, or other nontrade rules *should* somehow be combined at the WTO in a different way than they now are.[247]

At stake is seen to be 'nothing less than inquiring into the values or policy objectives that ought to "trump the value of freer trade"'.[248]

For some international trade lawyers there is a simple response to the 'trade and' debate which is, in essence, that the WTO is a trade organisation exclusively concerned with free trade, and is not empowered to take on a broader mandate. Some international lawyers have also been averse to according the WTO a more expansive role in international governance.[249] Other commentators see substantial benefits for the WTO and broader international law from the trade regime incorporating 'non-trade' elements. Much of the 'trade and' literature is thus 'a plea for trade institutions to be more comprehending of other values'.[250]

Debate on the issue of WTO linkage with other areas or values of international law in the 'trade and' literature has been particularly vigorous with respect to the role of the organisation's dispute settlement bodies. For authors such as Trachtman this reflects a view that WTO dispute resolution 'is not simply a mechanism for neutral application of legislated rules but is itself a mechanism of legislation and of governance'.[251] Within this domain, the key questions that have arisen relate to whether WTO panels and the Appellate Body should rely upon other international treaties, such as those concerning environmental protection, to *interpret* provisions of trade law, as well as whether they may *apply* outside norms, such as the precautionary principle, within the WTO context to displace a competing principle of trade law.[252]

[247] Joel P. Trachtman, 'Institutional Linkage: Transcending "Trade and …"', Am. J. Int'l L., 96 (2002), 77.

[248] Alvarez, 'Symposium: The Boundaries of the WTO – Foreword', 4.

[249] See, e.g., Philip Alston, 'The Myopia of the Handmaidens: International Lawyers and Globalization', European J. Int'l Law, 3 (1997), 435; Daniel C. Esty, 'The World Trade Organization's Legitimacy Crisis', World Trade Review, 1(1) (2002), 17.

[250] Jeffery Atik, 'Uncorking International Trade, Filling the Cup of International Economic Law', American University International Law Review, 15 (2000), 1233.

[251] Joel P. Trachtman, 'The Domain of WTO Dispute Resolution', Harv. Int'l L.J., 40 (1999), 336.

[252] The distinction between interpretative and application functions has been most developed in the work of Joost Pauwelyn, *Conflict of Norms in Public International Law: How WTO Law Relates to Other Rules of International Law* (Cambridge University Press, 2003), pp, 478–86.

Central to the debate over the judicial competence of trade dispute resolution bodies as regards other international law is Article 3.2 of the WTO DSU, which provides:

> The dispute settlement system of the WTO is a central element in providing security and predictability to the multilateral trading system. The Members recognize that it serves to *preserve the rights and obligations of Members under the covered agreements*, and to clarify the existing provisions of those agreements *in accordance with customary rules of interpretation of public international law.*

Customary rules of interpretation in international law are generally taken to mean Articles 31 and 32 of the Vienna Convention. The Appellate Body has in fact identified, as 'a rule of customary or general international law', Article 31(1) which speaks of a treaty being 'interpreted in good faith in accordance with the ordinary meaning to be given to the terms of the treaty in their context and in the light of its object and purpose'.[253] In other cases the Appellate Body has discussed the process of treaty interpretation to be applied to the WTO covered agreements, referring to the Vienna Convention articles.[254] As yet unanswered is the question of whether other rules of international law, besides customary law principles of interpretation, can be taken into account in construing WTO law. Such an approach would seem to be allowed by Article (3)(1)(c) of the Vienna Convention, mentioned earlier, via its direction to consider 'any relevant rules of international law applicable in the relations between the parties'. There is also some suggestion in WTO case law that the reference in Article 3.2 of the DSU to customary rules 'of interpretation' does not limit the dispute settlement organs to mere reliance on Articles 31 and 32 of the Vienna Convention but rather that customary international law 'applies generally to the economic relations between the WTO Members'.[255]

However, a narrower interpretation of Article 3.2 of the DSU emphasises the importance of *preserving* the rights and obligations of

[253] United States – Standards for Reformulated and Conventional Gasoline, Report of the WTO Appellate Body, WT/DS2/AB/R, 29 April 1996, (Gasoline) p. 11.

[254] *European Communities – Customs classification of Certain Computer Equipment*, Appellate Body Report, WT/DS62/AB/R, WT/DS67/AB/R, WT/DS68/AB/R, 22 June 1998, [86]; *Shrimp/Turtle*, [114]; *India – Patent Protection for Pharmaceutical and Agricultural Chemical Products*, Report of the Appellate Body, WT/DS50/AB/R, 16 January 1998, [45]; *US – Hot Rolled Steel*, Report of the Appellate Body, WT/DS184/AB/R, 24 July 2001, [60].

[255] *Korea – Measures Affecting Government Procurement*, Panel Report, WT/DS163/R, 1 May 2000, [7.96] and fn. 753.

members under the WTO agreements and limiting the application of outside international law to the assistance offered by customary rules of *interpretation* as contained in the Vienna Convention. Adopting this perspective, Trachtman has argued that 'the mandate to WTO dispute settlement panels and the Appellate Body is to apply as substantive law only WTO law: the covered agreements'.[256] Taking this approach, the WTO dispute settlement bodies in a case involving the SPS Agreement, for example, would not be able to apply environmental treaty provisions or principles of international general law based on the precautionary principle to develop an understanding of 'risk assessment' in the SPS context. This was essentially the conclusion reached by a WTO panel in the *GMO* case when it was asked to rule on the relevance of the Biosafety Protocol to the interpretation of SPS risk assessment undertaken for GMOs. Construing Article 31(3)(c) of the Vienna Convention strictly, the Panel held that only agreements to which *all* the WTO Members were party could be taken into account.[257] This condition did not pertain in respect of the Biosafety Protocol, which despite being widely ratified, does not have identical membership to that of the WTO.[258]

A broader approach to the role of international law and other values within the WTO system could see the trading system being integrated into the mainstream of international law, rather than standing apart as a separate, self-contained regime. In a debate with Trachtman at the 2004 Annual Conference of the American Society of International Law, Joost Pauwelyn put the view that WTO dispute settlement bodies can and should refer to or apply rules or principles agreed to by countries outside the WTO.[259] Pauwelyn argued that the WTO judiciary may both apply general international law as a 'fallback' when faced with questions not regulated by the WTO and also apply other international law invoked in defence of a claim of WTO violation.[260] Accordingly, 'if two

[256] Joel Trachtman, 'Remarks on "The Jurisdiction of the World Trade Organization"', American Society International L. Proceedings, 98 (2004), 139.

[257] *GMO* case, [7.68].

[258] The Biosafety Protocol has received 157 ratifications and accessions though critically not from the complainants in the *GMO* case: the USA, Canada and Argentina. It seems that this requirement for congruent membership of the WTO and outside treaties is unlikely to be fulfilled in any situation of conflict between two multilateral regimes: Koskenniemi, 'Fragmentation of International Law', 200.

[259] Pauwelyn, 'Remarks on "The Jurisdiction of the World Trade Organization"', 135.

[260] *Ibid.*, 136.

WTO members have explicitly agreed elsewhere (perhaps under the Biosafety Protocol ...) that certain trade restrictions can or even must be imposed between them, should a WTO panel not then apply such agreement in defense of any subsequent claim of a WTO violation?'[261]

Pauwelyn draws support for his view that the WTO must be considered 'part of the territorial domain of international law' from various rulings of the Appellate Body.[262] For instance, in the *Reformulated Gasoline* dispute, the Appellate Body declared, in a case under the GATT, that global trade rules are 'not to be read in clinical isolation from public international law'.[263] In the *Shrimp/Turtle* case, also decided under the GATT, the Appellate Body illustrated how this interpretative philosophy might be put into practice by construing the phrase 'exhaustible natural resources' in Article XX(g) to extend beyond mineral or non-living resources to living biological resources such as the endangered species of marine turtles at issue in the case. The Appellate Body commented:

> The preamble of the WTO Agreement - which informs not only the GATT 1994, but also the other covered agreements - explicitly acknowledges 'the objective of sustainable development' ... From the perspective embodied in the preamble of the WTO Agreement, we note that the generic term 'natural resources' in Article XX(g) is not 'static' in its content or reference but is rather 'by definition, evolutionary'.[264]

Transcending the 'trade and' debate

Some commentators see in rulings such as the *Shrimp/Turtle* decision not merely a means for importing other forms and values of international law into the global trade regime but rather the genesis of an approach that might transcend the 'trade and' debate altogether. As several authors have noted, the 'trade and' approach 'presupposes a certain perspective on the issues';[265] one that assumes that WTO procedures are 'institutionally "programmed"' to prioritise trade concerns at the expense of others.[266] However, the Appellate Body's rulings in a number of contentious disputes give some reason to question such assumptions.

[261] *Ibid.*, 137.
[262] Pauwelyn, *Conflict of Norms in Public International Law*, p. xi.
[263] *Gasoline*, p. 11.
[264] *Shrimp/Turtle*, [129]–[130]. This ruling is discussed further in Chapter 6.
[265] Jeffrey L. Dunoff, 'Rethinking International Trade', University of Pennsylvania *Journal of International Economic Law*, 19 (1998), 383.
[266] Koskenniemi, 'Fragmentation of International Law', 208.

Citing decisions such as *Shrimp/Turtle*, *Hormones* and *Asbestos*, Robert Howse argues that the Appellate Body has demonstrated sensitivity to 'value pluralism',[267] using 'a variety of jurisprudential techniques to do justice to the delicate interrelationship of values and interests in such cases, some internal and some external to the trading "system"'.[268] Likewise, Sungjoon Cho finds 'evidence of the transformation of the telos of the global trading system' in the Appellate Body's refashioning of the role of the Article XX *chapeau*, which guides the application of national public policy measures that seek to derogate from the trade liberalisation requirements of the GATT.[269]

Cho also points to institutional changes within the WTO – such as the reference to 'sustainable development' in the preamble to the WTO Agreement and members' affirmation of a commitment to the 'mutual supportiveness' of open markets and adequate social regulation – as further evidence of the emergence of a 'new telos' within the global trading system. He foresees that this may lead to strengthening of the free trade/social regulation linkage, even while maintaining the WTO's identity and capacity as a trade organisation.[270] Robert Howse remarks that transformation at the institutional level is having some effects in terms of the personnel of the WTO Secretariat. Generational change is 'yielding some incremental advance towards the embrace of the political ethics of democracy', displacing the previously dominant trade 'insider network', which had more fixed views as to the purpose of the global trading regime.[271]

Indeed, an increasing number of commentators stress the need to reinvigorate debate as to the meaning of free trade and the purpose of the WTO. Andrew Lang criticises the 'trade and' scholarship for reinforcing and naturalising a particular meaning of 'free trade' that concomitantly serves to shut down discussion over the fundamental purposes of the trade regime and the problems that it addresses.[272] Taking a historical perspective, Lang and other authors have emphasised that the

[267] Robert Howse, *The WTO System: Law, Politics and Legitimacy* (London: Cameron May, 2007), p. 72.

[268] Robert Howse, 'From Politics to Technocracy – and Back Again: The Fate of the Multilateral Trading Regime', *American Journal of International Law*, 96 (2002), 109.

[269] Cho, 'Linkage of Free Trade and Social Regulation', 652–3.

[270] *Ibid.*, 646–8. [271] Howse, *The WTO System*, p. 73.

[272] Andrew T. F. Lang, 'Reflecting on "Linkage": Cognitive and Institutional Change in the International Trading System', *Modern Law Review*, 70(4) (2007), 523.

concept of free trade is not static but has varied in meaning across time and across political cultures.[273] Consequently, an automatic assumption that issues of health and environmental protection – and the ways that they interact with the liberalisation of trade flows – are not the core business of the WTO may thus not be valid, or at the very least, should be a matter open to debate. In contrast to the 'trade and' literature that tends to cast issues or approaches, such as the precautionary principle, as being outside of the WTO, such perspectives open up conceptual space to consider the extent to which 'non-trade' matters should or already do exist *within* the realm of the global trading system.

Conclusion

Taking the interdisciplinary insights regarding the nature of science and risk assessment developed in Chapter 3, this chapter has discussed how different risk regulatory approaches may emerge based upon 'variation in governmental preferences for differing types and quantities of scientific proof'.[274] The two regulatory paradigms that have come to dominate international debates and disputes over health and environmental risk – those of sound science and the precautionary principle – draw upon the respective regulatory trends and practices of science-based decision-making in the USA and the EU. Whereas the US notion of sound science emphasises the need for a high standard of scientific proof in order to put in place regulations addressing risk, the precautionary principle advocates that protective measures to address potentially severe health or environmental risks should not be delayed on the grounds of scientific uncertainty.

What differentiates the competing regulatory paradigms of sound science and the precautionary principle is largely not their degree of scientific credibility (as regulatory authorities in the different jurisdictions often rely upon much the same science in their decision-making processes). Rather it is underlying value concerns surrounding the risks of concern that most influence decisions as to what counts as sufficient proof of safety and under what circumstances. Both approaches are

[273] See, most notably, Howse, 'From Politics to Technocracy'.

[274] Sheila Jasanoff, 'Contingent Knowledge: Implications for Implementation and Compliance', in Edith Brown-Weiss and Harold Jacobson (eds.), *Engaging Countries: Strengthening Compliance with International Environmental Accords* (Cambridge, MA: MIT Press, 1998), p. 76.

thus science-based in the sense that scientific evidence is the departure point for assessments of risk (and increasingly so in recent articulations of the precautionary principle in EU law). However, the significance of available scientific evidence is evaluated in different ways in light of differing sensitivities to uncertainties and differing levels of emphasis placed on social and economic matters.

These differences can become a source of contention and dispute when national regulatory approaches come into conflict at the international level. The WTO has been a prominent site of such clashes given that much risk regulatory activity relates to risks attaching to products, the majority of which are traded internationally. International law and lawyers are still engaged in a debate as to how such normative conflicts should be approached and resolved. In the area of international trade law there are an increasing number of authors who recognise the need to revisit assumptions that global organisations such as the WTO are myopically focused on a narrow agenda of promoting free trade, opening the way for a debate about the purposes of the trading system and scope for political contestation regarding the other values or systems that it recognises. Nevertheless, perspectives that emphasise the self-contained nature of international legal regimes – and the WTO in particular – remain prevalent, both in discussions of international legal fragmentation and in related scholarship on trade linkage questions. Such perspectives have the tendency to exclude approaches or legal principles seen as being outside the mandate of the regime concerned.

This trend is observable in the SPS disputes, discussed in the next chapter, that have largely interpreted provisions of the SPS Agreement dealing with scientific evidence and risk assessment through a narrow lens that 'elevates the policing of trade-restrictive measures above the ability of national governments to address risk in the face of scientific uncertainty'.[275] The following chapter's consideration of risk regulation in the context of the SPS Agreement – encompassing both relevant decisions of the WTO dispute settlement organs and the activities of the political committee body under the Agreement – thus serves two interrelated functions. It stands as both a detailed case study of the use of science and risk assessment procedures in a prominent global

[275] Alan O. Sykes, 'Domestic Regulation, Sovereignty, and Scientific Evidence Requirements: A Pessimistic View', *Chicago Journal of International Law*, 3 (2002), 369.

governance institution and also as an evaluation of the potential for the narrower approach to science-based decision-making that has been characteristic of the SPS area to exercise influence over the way in which science is used in other international legal fora concerned with risk regulation.

5 Science and WTO regulation of SPS risk

Introduction

In practice, values, culture and context play a vital role, alongside scientific knowledge, in informing regulatory approaches for health and environmental risk. At the global level, however, this has not limited the appeal of science as a crucial resource for risk decision-making where international laws and institutions seek the acceptance of determinations as neutral and universally valid.[1] The Agreement on Sanitary and Phytosanitary Measures (SPS Agreement), negotiated during the Uruguay trade round that also led to the establishment of the WTO, reflects this faith in science. Its standards invoke scientific evidence and risk assessment as arbiters of the WTO-compatibility of trade-restrictive SPS risk regulatory measures, regardless of whether the measures concerned are discriminatory in nature.[2] Hence members' SPS measures that depart from the standards of recognised international expert bodies, such as the Codex Alimentarius Commission, must be founded on 'scientific principles', 'not maintained without sufficient scientific evidence'[3] and 'based on'[4] an adequate risk assessment if they are to avoid scrutiny through the processes of the WTO.

While introducing novel science-based requirements into global trade law, the SPS Agreement articulates these standards in a form 'so

1 Vern Walker, 'The Myth of Science as a "Neutral Arbiter" for Triggering Precautions', B.C. Int'l & Comp. L. Rev., 26 (2003), 197–8.
2 Joost Pauwelyn, 'The WTO Agreement on Sanitary and Phytosanitary (SPS) Measures as Applied in the First Three SPS Disputes EC – Hormones, Australia – Salmon and Japan Varietals', J. Int'l Economic Law, 2(4) (1999), 644.
3 Agreement on the Application of Sanitary and Phytosanitary Measures, opened for signature 15 April 1994, 1867 UNTS 493, in force 1 January 1995 (SPS Agreement), Article 2.2.
4 SPS Agreement, Article 5.1.

loose to be essentially unworkable in their own terms'.[5] Accordingly, many early analyses of the SPS Agreement predicted that it would have a benign, if not beneficial, impact on national and global risk regulation.[6] In practice, elaboration of the norms of the SPS Agreement has depended upon the workings of WTO institutions, most particularly the political arm represented by the SPS Committee comprising member state representatives, and the judicial branch constituted by the WTO dispute settlement system of panels and a standing Appellate Body.

Much of the previous analysis of the science requirements under the SPS Agreement – particularly in the legal literature – has focused on the interpretations developed by WTO judicial bodies. While in classical accounts of international law, WTO dispute settlement under the SPS Agreement might be seen as the working-out of interstate, consent-based commitments, SPS dispute resolution has more often been viewed as 'another layer of judicial review of domestic administrative action',[7] that functions as 'a mechanism of legislation and of governance'.[8] The findings of WTO judicial bodies in SPS disputes speak not just to the parties, but also exhibit significant 'spillover effects' in shaping risk regulatory activity within relevant policy spheres, national and global.[9]

[5] Joanne Scott, *The WTO Agreement on Sanitary and Phytosanitary Measures: A Commentary* (Oxford University Press, 2007), p. 44.

[6] E.g., Jeffery Atik, 'Science and International Regulatory Convergence', Northwestern J. Int'l L. & Business, 17 (1996–1997), 736; David Victor, 'The Sanitary and Phytosanitary Agreement of the World Trade Organization: An Assessment After Five Years', N.Y.U. J. Int'l Law & Politics, 32 (2000), 873; Robert Howse, 'Democracy, Science, and Free Trade: Risk Regulation on Trial at the World Trade Organization', Michigan L. Rev., 98 (2000), 2330; Steve Charnovitz, 'Improving the Agreement on Sanitary and Phytosanitary Standards', in Gary Sampson and W. Bradnee Chambers (eds.), *Trade, Environment, and the Millennium* (Tokyo: United Nations University Press, 2002), p. 223. See also, Donna Roberts and Laurian Unnevehr, 'Resolving Trade Disputes Arising from Trends in Food Safety Regulation: The Role of the Multilateral Governance Framework', *World Trade Review*, 4(3) (2005), 469, noting the coherence between SPS disciplines and new science-based approaches to food safety regulation in industrialised countries.

[7] Nico Krisch and Benedict Kingsbury, 'Global Governance and Global Administrative Law in the International Legal Order', European J. Int'l Law, 17(1) (2006), 3.

[8] Joel P. Trachtman, 'The Domain of WTO Dispute Resolution', Harv. Int'l L.J., 40 (1999), 336 and 339 for further comments.

[9] See Scott, *The WTO Agreement on Sanitary and Phytosanitary Measures*, p. 45. See also Donna Roberts, 'Preliminary Assessment of the Effects of the WTO Agreement on Sanitary and Phytosanitary Trade Regulations', *Journal of International Economic Law*, 1 (1998), 377, looking at action taken by domestic agencies to remove 'illegitimate' SPS measures. For an international example see Codex's risk assessment guidelines that bear the hallmarks of the SPS jurisprudence: Codex Alimentarius Commission, 'Working Principles for Risk Analysis for Application in the Framework of the Codex Alimentarius', in Codex Secretariat (ed.), *15th Procedural Manual* (2005), p. 101.

Moreover, at least for lawyers, the body of SPS jurisprudence – which now comprises panel and Appellate Body reports in six disputes (with a number of others pending)[10] – provides a readily accessible source for construing the meaning of the open-ended scientific evidence and risk assessment requirements established by the SPS Agreement.

To date, less attention has been directed in legal analyses to the role of the SPS Committee in elaborating norms under the SPS Agreement. As Joanne Scott argues in her seminal commentary on the SPS Agreement, this neglect is unjustified given the important 'quasi-legislative role' played by the Committee.[11] In addition, the Committee's unique mode of operation when dealing with members' SPS trade concerns and in elaborating relevant norms represents an alternative, perhaps more fruitful, mechanism for evolving the standards of the SPS Agreement than all-or-nothing, top-down dispute settlement proceedings.

Accordingly, this chapter includes a discussion of the workings of the SPS Committee and its contribution to developing practices of risk regulation under the SPS Agreement in addition to a detailed analysis of the SPS jurisprudence. The complex nature of the SPS disputes to date and their lengthy findings on issues pertinent to the use of science in risk regulation necessarily mean that the bulk of the chapter

[10] Five of these disputes have resulted in appeals to the Appellate Body, namely *EC – Measures Concerning Meat and Meat Products*, Report of the WTO Appellate Body, WT/DS26/AB/R & WT/DS48/AB/R, 16 January 1998 (*Hormones*); *Australia – Measures Affecting Importation of Salmon*, Report of the WTO Appellate Body, WT/DS18/AB/R, 20 October 1998 (*Salmon*); *Japan – Measures Affecting Agricultural Products*, Report of the WTO Appellate Body, WT/DS76/AB/R, 22 February 1999 (*Varietals*); *Japan – Measures Affecting the Importation of Apples*, Report of the WTO Appellate Body, WT/DS245/AB/R, 26 November 2003 (*Apples*); *United States – Continued Suspension of Obligations in the EC-Hormones Dispute*, Report of the WTO Appellate Body, WT/DS320/AB/R, 16 October 2008 (the report issued in DS321 brought by Canada is identical to the US report) (*Hormones II*). In the sixth dispute, *EC – Measures Affecting the Approval and Marketing of Biotech Products*, Reports of the Panel, WTO Docs WT/DS291/R, WT/DS292/R, WT/DS293/R, 29 September 2006 (*GMO*) the final panel reports issued on 29 September 2006 were not appealed to the Appellate Body. Pending SPS disputes include the following: DS270 *Australia – Measures Affecting the Importation of Fresh Fruit and Vegetables* (brought by the Philippines, panel established 29 August 2003); DS367 *Australia – Measures Affecting the Importation of Apples from New Zealand* (brought by New Zealand; interim panel decision released to parties on 31 March 2010 with final report expected by mid-2010) (*Tasman Apples*); DS389 *European Communities – Measures Affecting Poultry Meat and Poultry Meat Products from the United States* (brought by the USA, panel not yet established); DS391 *Korea – Measures Affecting the Importation of Bovine Meat and Meat Products from Canada* (brought by Canada, Request for Panel 10 July 2009); DS392 *United States – Certain Measures Affecting Imports of Poultry from China* (brought by China, Request for Panel 30 June 2009).

[11] Scott, *The WTO Agreement on Sanitary and Phytosanitary Measures*, p. 45.

is dedicated to the latter topic. This is not unwarranted given that (as the previous chapter discussed) the SPS jurisprudence is acquiring a broader relevance for developing notions of international risk governance consequent upon the WTO dispute settlement system's prominent place in the field of international legal adjudication. Added to this, the 2006 panel decision in the *GMO* case cast the net of coverage of the SPS Agreement very wide, potentially extending its disciplines to a range of indirect risks to human, animal or plant life or health.[12]

However, rather than exhaustively detailing the facts and rulings in each of the SPS cases to date,[13] the chapter focuses upon the general principles that can be drawn from the jurisprudence regarding key issues. These include the nature of scientific evidence considered in the SPS context, determinations regarding its 'sufficiency' and the requirements for an 'appropriate' risk assessment. The chapter departs from much of the existing SPS literature, which has concentrated on the more liberal elements of decisions (particularly those in the first *Hormones* case), to highlight instead the narrow ways in which science has generally been applied by the WTO dispute settlement bodies. Overall, these practices have provided the basis for a progressive tightening of the concepts of science and risk employed in the SPS jurisprudence, with the potential, over time, to generate a similarly restrictive culture of risk determination in the broader SPS regulatory sphere, as well as in other areas of international law.

Science and the SPS Agreement

Before turning to the elaboration of norms under the SPS Agreement that has been provided by the SPS Committee and the SPS jurisprudence, it is useful to set out a little of the history of the Agreement and its relevant requirements relating to the role of science and risk assessment. As was noted in the introduction, the rules laid down by the SPS Agreement are formulated in an open-ended fashion that facilitated

[12] Jacqueline Peel, 'A GMO by Any Other Name ... Might Be an SPS Risk!: Implications of Expanding the Scope of the WTO Sanitary and Phytosanitary Measures Agreement', European J. Int'l Law, 17(5) (2006), 1009.

[13] There is already an extensive body of scholarship which undertakes this task. See particularly Pauwelyn, 'The WTO Agreement on Sanitary and Phytosanitary (SPS) Measures'; Scott, *The WTO Agreement on Sanitary and Phytosanitary Measures*; Victor, 'The Sanitary and Phytosanitary Agreement of the World Trade Organization'; Rüdiger Wolfrum, Peter-Tobias Stoll and Anja Seibert-Fohr (eds.), *WTO – Technical Barriers and SPS Measures* (Leiden: Martinus Nijhoff, 2007), pp. 365–551.

state agreement during the negotiating process, albeit leaving to one side the meaning and means of implementing those requirements. The operationalisation of the SPS Agreement's standards has thus been the result of an evolving process of interpretation; one that has taken markedly different forms in the multilateral, consensus-based forum of the SPS Committee as opposed to the more formal, judicial realm of dispute settlement. Arguably in the latter forum, a lack of clarity in the text of the SPS Agreement has driven the dispute settlement bodies to endorse restrictive notions of science and objective risk assessment that, while they may serve the goal of fostering greater harmonisation of SPS measures,[14] also perpetuate the 'myth' that scientific knowledge and expert advice provide definitive criteria for international law to resolve risk disputes.[15]

The SPS Agreement: negotiating history and purpose

The SPS Agreement, dealing with measures applied to protect human, animal or plant life or health from pest, disease and food-related risks, was one of the suite of 'new' trade agreements negotiated during the Uruguay round.[16] The SPS Agreement came into force, together with the institutional structures of the WTO, on 1 January 1995. SPS issues were not foreign to the global trading regime under the GATT, the predecessor of the WTO. For instance, Article XX(b) of the GATT – which remains in force – allows members to adopt trade-restrictive, non-discriminatory measures which are deemed 'necessary to protect human, animal or plant life or health'.[17] However, over time the GATT rules came to be perceived as inadequate for governing the burgeoning area of SPS measures, many of which have a substantial regulatory impact beyond the adopting jurisdiction.[18]

[14] Atik, 'Science and International Regulatory Convergence', 755 notes that such harmonisation is not always in a direction determined by the logic of greater trading efficiencies. Science may instead promote regulatory convergence based on the 'happenstance of received scientific traditions'.

[15] Walker, 'The Myth of Science as a "Neutral Arbiter" for Triggering Precautions'.

[16] Sanitary and phytosanitary measures to which the SPS Agreement applies are formally defined in Annex A of the Agreement.

[17] The retention of GATT Article XX(b) raises complex questions of the relative priority of the GATT provisions vis-à-vis the SPS Agreement: see Gabrielle Marceau and Joel P. Trachtman, 'The Technical Barriers to Trade Agreement, the Sanitary and Phytosanitary Measures Agreement, and the General Agreement on Tariffs and Trade', *Journal of World Trade*, 36(5) (2002), 811.

[18] Discussions in the GATT on the need for clear rules to deal with SPS measures began as early as 1974: see Committee on Sanitary and Phytosanitary Measures,

Nonetheless, during the Uruguay round of trade talks, SPS issues were relatively uncontroversial, dwarfed by the broader negotiations over agriculture of which SPS controls were a subset.[19] This comparative obscurity of SPS matters facilitated a different dynamic in the subgroup responsible for negotiating the text of the new Agreement, in contrast to the highly politically charged atmosphere of the debates surrounding agricultural subsidies. Those negotiating the SPS text were mostly regulators and trade policy officials, many of who specialised in the area of quarantine and food safety control, and who saw themselves as fashioning a regime for an important, but nevertheless fairly narrow category of regulatory, non-tariff trade barriers.

Records and accounts of the negotiating history of the SPS Agreement suggest that participants brought to the negotiations at least two different understandings of the purpose of the new agreement. One early theme of the discussions in the SPS negotiating subgroup was the desirability of harmonising national SPS standards to facilitate trade in agricultural products. In this regard, Doaa Motaal recounts how negotiating parties initially focused on the use of international standards and reliance on the expertise of international standard-setting bodies, such as Codex, the International Office for Epizootics (OIE) and the International Plant Protection Convention (IPPC), to discipline the heterogeneity produced by multiple differing national standards.[20] She notes that '[a]t the outset of the negotiations, none of the contracting parties to GATT considered that requirements for scientific justification would be necessary'.[21] These concerns only emerged later in response to parties' realisation 'that the international harmonization of standards would not always be feasible (too time consuming to negotiate, too ambitious for standards in certain sectors, etc.), and that situations would anyway arise in which countries would need to exceed existing international standards'.[22]

Summary Report on the SPS Risk Analysis Workshop 19–20 June 2000, G/SPS/GEN/209, 3 November 2000, p. 2.

19 Thomas Cottier, 'Risk Management Experience in WTO Dispute Settlement', in Thomas Cottier (ed.), *The Challenge of WTO Law: Collected Essays* (London: Cameron May, 2007), p. 147. For an overview of the SPS negotiations in the Uruguay trade round see John Croome, *Reshaping the World Trading System: A History of the Uruguay Round* (The Hague: Kluwer Law International, 1999).

20 Doaa Motaal, 'The "Multilateral Scientific Consensus" and the World Trade Organization', J. World Trade, 38(5) (2004), 855.

21 *Ibid.*, 861. 22 *Ibid.*

Proposals for the use of scientific justification tests and risk assessment under the SPS Agreement were first introduced by negotiators from the USA, apparently to advance a different goal than that of regulatory harmonisation.[23] David Victor notes, for instance, the 'shadow' cast over the negotiations by a long-running dispute between the USA and the EC over the latter's bans on hormone-treated beef, which had its antecedents in EC measures first adopted in 1981.[24] This dispute (later to be decided under the SPS Agreement) had generated a great deal of suspicion on the part of the USA and other agricultural exporters over the potential for abuse of SPS measures. Many such exporters voiced the fear that 'as tariff barriers in agriculture came down, domestic agricultural lobbies would resort to sanitary and phytosanitary measures to keep food and agricultural products out of their markets'.[25] The introduction of requirements into the SPS text for scientific justification of measures and the performance of risk assessments hence seemed to be partly motivated by the desire to discipline members' domestic SPS standard-setting processes so as 'to promote international trade by limiting the use of SPS measures as disguised barriers to trade'.[26]

Not surprisingly, given the different objectives of negotiators, the final text of the SPS Agreement leaves open the question of its overall purpose.[27] On the one hand, the SPS Agreement seems to endorse regulatory harmonisation through reference to international SPS standards

[23] 'A Discussion Paper on Issues Related to the Negotiations Submitted by the United States', Negotiating Group on Agriculture, MTN.GNG/NG5/W/44, 22 February 1988. Proposals referring to the need to assess risk in establishing SPS measures were also submitted by the EC around the same time: Committee on Sanitary and Phytosanitary Measures, *Summary Report on the SPS Risk Analysis Workshop 19–20 June 2000*, G/SPS/GEN/209, 3 November 2000, p. 2.

[24] Victor, 'The Sanitary and Phytosanitary Agreement of the World Trade Organization', 871–2.

[25] Andrew Thompson, 'Australia-Salmon and Compliance Issues Surrounding the SPS Agreement: Sovereign Acceptance and Measure Adaptation', Law & Pol'y Int'l Bus., 33 (2002), 719; Roberts and Unnevehr, 'Resolving Trade Disputes Arising from Trends in Food Safety Regulation', 470.

[26] Victor, 'The Sanitary and Phytosanitary Agreement of the World Trade Organization', 875.

[27] Roberts and Unnevehr, 'Resolving Trade Disputes Arising from Trends in Food Safety Regulation', 483; Elizabeth Fisher, 'Beyond the Science/Democracy Dichotomy: The World Trade Organisation Sanitary and Phytosanitary Agreement and Administrative Constitutionalism', in Christian Joerges and Ernst-Ulrich Petersmann (eds.), *Constitutionalism, Multilevel Trade Governance and Social Regulation* (Portland: Hart Publishing, 2006), pp. 329–30.

devised by 'the relevant international organizations', namely Codex, the OIE and the IPPC. Article 3.1 of the SPS Agreement thus states:

> To harmonize sanitary and phytosanitary measures on as wide a basis as possible, Members shall base their sanitary or phytosanitary measures on international standards, guidelines or recommendations, where they exist, except as otherwise provided for in this Agreement, and in particular in paragraph 3.

However, a strict obligation of harmonisation is at odds with other provisions of the Agreement declaring the 'right' of members to take SPS measures 'necessary for the protection of human, animal or plant life or health'[28] in accordance with the level of SPS protection they deem 'appropriate'.[29] In addition, as Article 3.3 affirms, the level of SPS protection sought by a member may be one which is:

> higher ... than would be achieved by measures based on relevant international standards, guidelines or recommendations, if there is a scientific justification, or as a consequence of the level of sanitary or phytosanitary protection a Member determines to be appropriate in accordance with the relevant provisions of paragraphs 1 through 8 of Article 5.

Consistent with the idea that harmonisation around international norms is merely encouraged, rather than mandated, most of the remaining text of the SPS Agreement, including the provisions of Article 5, focuses on domestic standard-setting processes for SPS measures. Joanne Scott has described this as 'regulation by regulation ... defining the limits to legitimate diversity'.[30] In this respect, both procedural and substantive requirements are elaborated. Procedurally, members are required to act transparently with regard to their domestic standard-setting,[31] and to implement SPS regulations 'without undue delay'.[32]

As a substantive matter, there are obligations that relate to both the trade impacts of measures and their rationality as regulations necessary for protection against SPS threats. The former set of substantive obligations include those for members to ensure 'consistency' in their regulation of similar risks,[33] and to adopt SPS measures that are 'not

[28] SPS Agreement, Article 2.1.

[29] SPS Agreement, preamble recital 6, Articles 3.3, 4.1, 5.3–5.6, 9.1.

[30] Scott, *The WTO Agreement on Sanitary and Phytosanitary Measures*, p. 44.

[31] SPS Agreement, Article 7, Annex B.

[32] SPS Agreement, Article 8, Annex C.

[33] SPS Agreement, Articles 2.3, 5.5. These provisions raise many questions over the feasibility of comparative risk assessment, which lie beyond the scope of this book. For an analysis of such questions see Jeffery Atik, 'The Weakest

more trade-restrictive than required' to achieve a Member's appropriate level of SPS protection.[34] The latter set of obligations in Articles 2.2, 5.1 and 5.2 require national measures to be 'based on scientific principles' and 'not maintained without sufficient scientific evidence',[35] and to be 'based on' a risk assessment 'appropriate to the circumstances' that takes into account 'risk assessment techniques developed by the relevant international organizations',[36] as well as:

> available scientific evidence; relevant processes and production methods; relevant inspection, sampling and testing methods; prevalence of specific diseases or pests; existence of pest - or disease - free areas; relevant ecological and environmental conditions; and quarantine or other treatment.[37]

In Article 5.7 provision is made for a departure from the requirements of Articles 2.2 and 5 'where relevant scientific evidence is insufficient' in order for a member to adopt provisional SPS measures. Such measures must, however, be based on 'available pertinent information, including that from the relevant international organizations as well as from sanitary or phytosanitary measures applied by other Members'. Provisional measures are also subject to ongoing requirements for an adopting member to 'seek to obtain the additional information necessary for a more objective assessment of risk and review the sanitary or phytosanitary measure accordingly within a reasonable period of time'.

To government negotiators drafting the text of the SPS Agreement ambiguity as to its purpose undoubtedly seemed constructive, facilitating multiple interpretations of commitments by different parties depending on their particular priorities. To a large extent this flexibility is retained in the interpretations of the SPS Agreement that have been adopted by the SPS Committee, discussed further below. However, in the area of dispute resolution, the Dispute Settlement Understanding (DSU) negotiated during the Uruguay round,[38] gave to

Link: Demonstrating the Inconsistency of "Appropriate Levels of Protection" in *Australia-Salmon*', *Risk Analysis*, 24(2) (2004), 483.

34 SPS Agreement, Article 5.6. See also Articles 5.4 and 5.5 of the SPS Agreement.

35 SPS Agreement, Article 2.2.

36 SPS Agreement, Article 5.1.

37 SPS Agreement, Article 5.2. Further relevant factors are specified in Article 5.3 where the assessment and measures adopted concern risks to animal or plant life or health.

38 Marrakesh Agreement Establishing the World Trade Organisation, opened for signature 15 April 1994, 1867 UNTS 3, in force 1 January 1995, Annex 2 (Understanding on Rules and Procedures Governing the Settlement of Disputes) 1869 UNTS 401

WTO obligations – like those in the SPS Agreement – a much harder edge, exposing them to the possibility of binding interpretations issued by judicial decision-makers. As has often been noted, the system established by the DSU replaces the 'diplomatic' ethos of the ad hoc panels of the previous GATT dispute settlement system with one that is noticeably more legal and formal in character.[39] In this context, while the focus of decision-makers is upon making an 'objective assessment' of the facts and legal issues in dispute,[40] interpretations of relevant SPS rules are also issued, 'expressed in terms which allow for their extrapolation from context, and for their application in a *de facto* system of precedent'.[41]

This difference in approach became evident in the first WTO SPS dispute of *Hormones* (discussed in detail below), which raised directly the question of whether the SPS Agreement was designed to ensure harmonisation around international norms, or merely to weed out cases of disguised protectionism. The panel and Appellate Body reached divergent conclusions on this question, although both based their reasoning on an interpretation of the treaty text. The panel took the route of international harmonisation, emphasising the mandatory language of Article 3.1 and treating Article 3.3 as an 'exception', necessitating justification by the EC of its divergent standards.[42] However, the Appellate Body (in a move perhaps 'politically more astute' than the panel)[43] started its analysis with the requirements of Article 3.3 in which it saw 'an autonomous right' on the part of members to establish their

(DSU). The provisions of the DSU apply to disputes under the SPS Agreement as for other WTO Agreements: Article 11.1.

39 J. H. H. Weiler, 'The Rule of Lawyers and the Ethos of Diplomats: Reflections on WTO Dispute Settlement', in Roger B. Porter *et al.* (eds.), *Efficiency, Equity, and Legitimacy: the Multilateral Trading System at the Millennium* (Washington DC: Brookings Institution Press, 2001), p. 334.

40 DSU, Article 11.

41 Scott, *The WTO Agreement on Sanitary and Phytosanitary Measures*, p. 45. Although the judicial bodies of the WTO do not strictly follow the common law approach of *stare decisis*, they tend to display a high degree of fidelity to interpretations of WTO law developed in previous cases: Robert Howse, 'Adjudicative Legitimacy and Treaty Interpretation in International Trade Law: The Early Years of WTO Jurisprudence', in J. H. H. Weiler (ed.), *The EU, the WTO, and the NAFTA: Towards a Common Law of International Trade?* (Oxford University Press, 2000), p. 61.

42 *EC – Measures Concerning Meat and Meat Products*, Report of the Panel, WT/DS26/R & WT/DS48/ R, 12 July 1999 (*Hormones* Panel Report), [8.44]–[8.46].

43 Victor, The Sanitary and Phytosanitary Agreement of the World Trade Organization', 936.

own level of SPS protection, independent of international standards.[44] Consequently, the Appellate Body described the SPS Agreement as constraining:

> the use of [SPS] measures for arbitrary or unjustifiable discrimination between Members or as a disguised restriction on international trade, without preventing Members from adopting or enforcing measures which are both 'necessary to protect' human life or health and 'based on scientific principles', and without requiring them to change their appropriate level of protection.[45]

The Appellate Body's interpretation of Article 3 in the *Hormones* case has turned the course of subsequent SPS jurisprudence away from the assessment of national SPS measures against international benchmark standards (although this remains a major focus of activity in the SPS Committee).[46] Instead, the task of the WTO dispute settlement organs has largely become that of reviewing the internal workings of domestic SPS regulatory activity for its procedural and substantive adequacy so as to discern unnecessary – and by implication – protectionist measures. In the main this has been done by reference to apparently objective standards for distinguishing permissible from unjustified SPS measures. Hence, in the area of dispute settlement, the SPS Agreement's substantive requirements for scientific and risk-based justification of national measures have come to play 'a *key role* in turning the distinction between "protectionist" and "legitimate" regulations into an operationable legal construct'.[47]

Different understandings of science and SPS risk assessment

Like most legal texts – and particularly those which reflect compromises between their negotiating parties – the language used in the SPS Agreement is not always 'a model of clarity in drafting and communication'.[48] The text leaves many important questions

[44] *Hormones*, [104]. [45] *Ibid.*, [177].

[46] Articles 3.5 and 12.4 require the SPS Committee to 'develop a procedure to monitor the process of international harmonization and coordinate efforts in this regard with the relevant international organizations'. The Committee has since adopted and revised a procedure for this purpose: see Procedure to Monitor the Process of International Harmonization (G/SPS/11/REV.1 and G/SPS/40). The procedure is discussed in Scott, *The WTO Agreement on Sanitary and Phytosanitary Measures*, pp. 65–69.

[47] Oren Perez, *Ecological Sensitivity and Global Legal Pluralism: Rethinking the Trade and Environment Conflict* (Portland: Hart Publishing, 2004), p. 117 (emphasis as in the original).

[48] *Hormones*, [175]. The Appellate Body was commenting on the text of Article 3.3 of the Agreement.

unresolved, not least so with respect to its requirements for the scientific justification of SPS measures. The crucial concepts of 'scientific principles' and '(in)sufficient' scientific evidence in Articles 2.2 and 5.7 are left undefined. Similarly, Article 5's provisions on risk assessment provide little indication of what amounts to an 'appropriate' assessment for SPS purposes, beyond articulating the factors and methodologies to be taken into account, and requiring that an 'evaluation' of risks takes place.[49] Further, although the SPS Agreement makes reference to members' determining their 'appropriate level of sanitary or phytosanitary protection' and adopting SPS measures for the purpose of achieving that level,[50] there is no mention of risk management in the text,[51] raising questions as to whether this function (or elements of it) are incorporated into the 'assessment' of SPS risks.

Such ambiguities are not surprising given the inherent difficulties encountered in fixing upon any one notion of science or risk assessment in a regulatory context. As was highlighted in Chapters 3 and 4, mainstream positivist notions of science and technical risk perspectives are both highly contestable, something which has facilitated the development of competing regulatory paradigms that take different views regarding the use of relevant scientific evidence, the importance to be assigned to areas of uncertainty and the role of social and economic factors in risk decision-making. Questions over the sufficiency of the scientific support for regulations and the adequacy of risk assessment processes thus tend to be judged by decision-makers depending on the risk regulatory framework – sound science-based or uncertainty-focused – that has been adopted. This is evident from the arguments put by disputing parties in SPS cases regarding the proper interpretation to be given to terms such as 'sufficient scientific evidence' and risk assessment. For instance, complainants have argued that appropriate scientific support for SPS measures should be judged on the basis of adherence to the 'scientific method',[52] and an adequate

[49] SPS Agreement, Annex A, para. 4 (containing the Agreement's definitions of risk assessment).

[50] See particularly SPS Agreement, Article 5.6 and Annex A(5). The Appellate Body has also described it as the 'prerogative' of a WTO Member to determine the level of protection the member deems appropriate: *Salmon*, [199].

[51] Apparently this was because countries negotiating the SPS Agreement judged that it was inappropriate for the WTO to be prescriptive regarding risk management with international organisations like Codex providing a better forum in this respect: Roberts and Unnevehr, 'Resolving Trade Disputes Arising from Trends in Food Safety Regulation', 488.

[52] *Hormones* Panel Report, IV.24 (US position).

risk assessment in terms of evaluations of probability drawing on firm scientific findings.[53] For other members, however, science-based regulation merely signifies a distinction from the 'non-scientific' domains of religion and superstition.[54]

In its decision in the latest iteration of the *Hormones* dispute (*Hormones II*), the Appellate Body appeared to acknowledge the role of such 'framing' considerations in carrying out risk assessment, remarking:

> where the [member's] chosen level of protection is higher than would be achieved by a measure based on an international standard, this may have some bearing on the scope or method of the risk assessment. In such a situation, the fact that the WTO Member has chosen to set a higher level of protection may require it to perform certain research as part of its risk assessment that is different from the parameters considered and the research carried out in the risk assessment underlying the international standard.[55]

Consequently, the Appellate Body disagreed with the panel's finding in the case that 'the determination of whether scientific evidence is sufficient to assess the existence and magnitude of a risk must be disconnected from the intended level of protection'.[56] Nonetheless, it added the caveat that the chosen level of protection should not 'predetermine' this assessment, which 'must remain, in essence, a rigorous and objective process'.[57]

The open-ended meaning of the scientific and risk assessment requirements in the text of the SPS Agreement also helps to explain why a number of prominent commentators have seen the Agreement as supporting a plurality of risk regulatory styles. For instance, Jeffrey Atik (writing prior to the first *Hormones* decision) pointed out that the SPS Agreement does not call for measures to be based on the best or truest science and hence should 'admit, at least implicitly, the possibility of multiple, mutually-exclusive, sciences'.[58] The result of this flexibility, he argued, is that the science-based provisions of the SPS Agreement 'represent a substantial

[53] *Australia – Measures Affecting Importation of Salmon*, Report of the Panel, WT/DS18/R, 12 June 1998 (*Salmon* Panel Report), [4.140] (Canadian position).

[54] *Hormones* Panel Report, IV. 25 (EC position).

[55] *Hormones II*, [685]. See also [534].

[56] *US – Continued Suspension of Obligations in the EC – Hormones dispute*, Report of the Panel, WT/DS320/R, 31 March 2008 (US Hormones II Panel Report), [7.612]; *Canada – Continued Suspension of Obligations in the EC – Hormones dispute*, Report of the Panel, WT/DS321/R, 31 March 2008 (Canada Hormones II Panel Report), [7.590.685].

[57] *Hormones II*, [686]. See also [534].

[58] Atik, 'Science and International Regulatory Convergence', 748.

restoration of rulemaking authority to national institutions', which 'categorically empower the nations to regulate in the SPS sphere' even where regional or global treatment of a risk might be 'optimal'.[59] Along similar lines, are the views of risk specialists Douglas Crawford-Brown and his co-authors, although they believe the potential for multiple science-based arguments to be put forward in support of SPS measures is a factor limiting the effectiveness of the Agreement.[60] One does not need to go so far as to take the 'social constructivist' view of science, they argue, in order to see the possibility of 'rational flexibility in scientific arguments, allowing both the development and application of risk estimates to be captured and used strategically to foster trade barriers'.[61]

Just as the SPS Agreement, on its face, is permissive of multiple understandings of science, so it seems it could be read to accommodate 'different approaches to coming to a conclusion on risk'.[62] On this basis, risk assessment under the Agreement might be seen to connote a qualitative, deliberative exercise in which scientific evidence is considered alongside a range of social and cultural inputs relating to risk.[63] Accordingly, the adequacy of a national SPS risk assessment process and the role that science plays in that process could well be judged in light of the degree of adherence to deliberative standards.[64] In this conception, science would be a necessary, rather than decisive, input into SPS risk assessment, designed not to trump democratic judgments about risk but instead to ensure that they result from an appropriately structured, deliberative process.[65] This notion of risk assessment has much in common with some of the models for precautionary regulation that have been developed in the EU.[66]

59 *Ibid.*, 739, 743.

60 Douglas Crawford-Brown, Joost Pauwelyn and Kelly Smith, 'Environmental Risk, Precaution, and Scientific Rationality in the Context of WTO/NAFTA Trade Rules', *Risk Analysis*, 24(2) (2004), 461.

61 *Ibid.*, 462. 62 *Ibid.*

63 David Winickoff *et al.*, 'Adjudicating the GM Food Wars: Science, Risk, and Democracy in World Trade Law', Yale J. Int'l L., 30 (2005), 81.

64 Fisher, 'Beyond the Science/Democracy Dichotomy', pp. 335–6. See also Robert Howse, *The WTO System: Law, Politics and Legitimacy* (London: Cameron May, 2007), pp. 66–7 contending that 'when read properly, the rules can be understood mostly as requiring a procedure of public justification for [SPS] regulations, including the gathering of scientific evidence, which may enhance democracy, by allowing fuller public debate of the issues, and better public information'.

65 Howse, 'Democracy, Science, and Free Trade', 2335, 2341.

66 E.g., Ortwin Renn *et al.*, *The Application of the Precautionary Principle in the European Union* (Stuttgart: Precaupri Project, European Commission, 2003).

Equally, however, the SPS Agreement is open to a 'divergent' interpretation on the question of appropriate risk assessment, including the acceptance of risk regulatory paradigms that posit different roles for science in the decision-making process.[67] Elizabeth Fisher describes one such alternative paradigm as 'rational-instrumental' in its orientation.[68] The characteristics of this paradigm, according to Fisher, are an emphasis on discretion-constraining analytical methodologies (such as quantitative risk assessment) and the harnessing of science 'so as to ensure the efficient pursuit of goals generated by the democratic process'.[69]

Arguably, therefore, there exists substantial scope for the SPS Agreement to recognise a diversity of regulatory approaches for the use of science in risk assessment and to function as a site for debating the merits of alternative approaches.[70] The following sections of the chapter examine the extent to which this potential has been realised in the political forum of the SPS Committee and the judicial forum of SPS dispute settlement, respectively. We might anticipate, however, that preserving scope for plurality in risk regulation would be a more difficult task in the dispute settlement arena where decision-makers are expected to deliver a final and authoritative resolution of parties' disputes.

Risk regulation and the role of the SPS Committee

The SPS Committee is one part of the extensive, albeit largely invisible, political infrastructure of the WTO.[71] It comprises representatives of WTO member states, most usually diplomats or specialists drawn from national ministries in SPS-covered fields, who meet on a regular basis with assistance provided by a Secretariat. The Committee was established by the SPS Agreement in order to 'carry out the functions necessary to implement the provisions of [the] Agreement and the furtherance of its objectives, in particular with respect to harmonization'.[72] The Committee serves as a regular forum for consultations and negotiations among member states on specific SPS issues,[73] and also plays an important role in liaising with relevant international standard-setting

[67] Fisher, 'Beyond the Science/Democracy Dichotomy', pp. 334–5.
[68] *Ibid.*, p. 335. [69] *Ibid.* [70] *Ibid.*, p. 336.
[71] Scott, *The WTO Agreement on Sanitary and Phytosanitary Measures*, p. 45.
[72] SPS Agreement, Article 12.1.
[73] SPS Agreement, Article 12.2.

bodies 'with the objective of securing the best available scientific and technical advice for the administration of [the] Agreement and in order to ensure that unnecessary duplication of effort is avoided'.[74]

Beyond specifying that the Committee should reach decisions 'by consensus', the SPS Agreement offers no further guidance as to the appropriate mode of operation of the body.[75] As Joanne Scott remarks, the SPS Committee is thus 'an experiment in institutional self-invention'.[76] The procedures which it has evolved differ in nature from those that characterise the realm of WTO dispute settlement. For instance, they are more participatory and cooperative; forward-looking in terms of dealing with regulatory proposals rather than finalised measures; and more nuanced in outcome, often leading to 'mutual adjustment of regulatory expectation and regulatory performance, and to collaboration in problem solving'.[77] A consultative approach – both between members, and between the Committee and international organisations – is also a feature of the work of the Committee, which assists in coordinating the diverse elements of the SPS governance framework.

For commentators such as Scott, the SPS Committee clearly offers an alternative, less confrontational model for dealing with SPS risk regulatory matters than dispute settlement and one, moreover, which might achieve more sustainable resolutions of regulatory differences that exist between member states. She comments that the back-and-forth of contestation and reasoned justification in the Committee seems productive of behavioural change and learning on the part of members, as well as inculcating a heightened sense of empathy with the situations of others that can lead to policy change not readily explicable on the basis of self-interest.[78]

Discussion of members' specific trade concerns

The SPS Committee can be conceptualised as a trans-governmental network deploying 'soft power', along the lines discussed in Chapter 2. As such, one of its primary functions is as a forum for information exchange and peer review, which serves to disseminate shared views on good practice in risk regulation as well as providing an accountability check on what counts as an acceptable domestic SPS measure.[79] One

[74] SPS Agreement, Article 12.3.
[75] SPS Agreement, Article 12.1.
[76] Scott, *The WTO Agreement on Sanitary and Phytosanitary Measures*, p. 48.
[77] *Ibid.*, pp. 74–5. [78] *Ibid*, p. 75. [79] *Ibid.*, pp. 47, 54.

of the most important aspects of its information exchange function is the procedures the Committee has developed around the notification and consideration of members' 'specific trade concerns'.

Under the SPS Agreement members are subject to notification and transparency requirements regarding their SPS measures.[80] The SPS Committee has issued recommendations regarding notification procedures that build on these requirements.[81] A large number of notifications are received under the SPS Agreement regarding new or altered SPS measures, with 4,376 notifications circulated as of May 2005.[82] These provide an opportunity for a member's trading partners to raise objections in the SPS Committee to a proposed measure, designated 'specific trade concerns'.[83]

In the ten years between 1995 and 2004, 204 specific trade concerns were raised in the SPS Committee, with just over a quarter being reported as resolved in the same period.[84] The vast majority of these concerns related to animal health, especially emergency national measures adopted in response to concerns over the transmission of mad cow disease. While developed countries are the initiators and targets of the majority of complaints, in contrast to SPS dispute settlement, developing countries also play an active role in raising specific trade concerns.[85]

The substance of specific trade concerns raised in respect of a member's SPS measures may be procedural or substantive. In the latter category, for example, are complaints about the adequacy of the scientific basis underlying a regulation or the need for updating of a risk assessment. Scott describes the process that occurs in the consideration of specific trade concerns as follows:

> The raising of a specific trade concern acts as a catalyst for dialogue, often involving give and take across the course of several meetings. Complaining

[80] SPS Agreement, Article 7 and Annex B.

[81] Committee on Sanitary and Phytosanitary Measures, 'Recommended Procedures for Implementing the Transparency Obligations of the SPS Agreement (Article 7)', G/SPS/7/Rev.2, 2 April 2002.

[82] Committee on Sanitary and Phytosanitary Measures, 'Review of the Operation and Implementation of the Agreement on the Application of Sanitary and Phytosanitary Measures', Report adopted by the Committee on 30 June 2005, G/SPS/36, 11 July 2005, p. 5.

[83] Roberts and Unnevehr, 'Resolving Trade Disputes Arising from Trends in Food Safety Regulation', 480.

[84] Committee on Sanitary and Phytosanitary Measures, 'Recommended Procedures for Implementing the Transparency Obligations of the SPS Agreement (Article 7)', p. 17.

[85] *Ibid.*, pp. 19–20.

Members are at pains to exemplify the consequences of regulatory proposals for them. Regulating Members are called upon to provide further information and clarification of their proposals, and to elucidate the evidence upon which a measure is to be based. Contestation is matched by efforts at justification. Summary minutes of the meetings record the nature and content of Members' epistemic claims and arguments.[86]

Importantly, discussions are constrained by reference to the standards of the SPS Agreement, as well as those of international standard-setting bodies. Members often use the procedure 'as a way of turning up the political heat, without necessitating costly and acrimonious recourse to the "courts"', although formal dispute settlement always remains in the background as an option.[87]

Although raising of a specific trade concern focuses attention on the SPS measures introduced by a particular member, Scott remarks that accountability operates 'two-way'. Not only does the member whose measure is targeted become more sensitised to the external impacts of its regulatory proposals, but also the complaining member is subjected to oversight which may identify deficiencies in its capacity to guarantee the safety of its exports.[88] Thus the result of the process might either be adjustment in the importing member's regulations, or alternatively enhancement in the capacity of the exporting member to enable it to meet the applicable standards or better demonstrate compliance.[89]

While as a risk governance mechanism the specific trade concerns procedure operated by the SPS Committee undoubtedly has many commendable features (dynamism, reflexivity, opportunities for learning and so on), it would appear to have been most effective in resolving relatively straightforward trade concerns. These kind of disputes encompass situations where a member's measures are overtly discriminatory (as for example an exemption from an Australian ban on sauces containing benzoic acid granted to New Zealand but not to other producing countries)[90] or where differences can be resolved by way of an updated risk assessment (as in the case of an EC reassessment of the toxicity of a potential food contaminant in soy sauce that led to a finding that the risks involved were not as high as initially thought).[91] On the other hand, disputes that involve divergences over 'risk management' – for

[86] Scott, *The WTO Agreement on Sanitary and Phytosanitary Measures*, p. 52.

[87] *Ibid.*, p. 58. [88] *Ibid.*, pp. 56–7. [89] *Ibid.*, p. 58.

[90] Roberts and Unnevehr, 'Resolving Trade Disputes Arising from Trends in Food Safety Regulation', 485–6.

[91] Scott, *The WTO Agreement on Sanitary and Phytosanitary Measures*, pp. 47–8.

instance whether a precautionary approach is employed and the role of 'other legitimate factors' in decision-making – have proven more intractable.[92]

Normative elaboration of the SPS text

In addition to its functions regarding information exchange and peer review of SPS measures, the SPS Committee also plays a role in elaborating the norms of the SPS Agreement. In this respect, the Committee operates as a quasi-legislative body capable of clarifying SPS standards, albeit in a 'soft law' form.[93] Since its establishment, the Committee has adopted a number of acts, ranging from guidelines (most notably its guidelines on practical implementation of the consistency requirement in Article 5.5 of the SPS Agreement)[94] to recommendations (such as those regarding the notification procedures discussed above) and more formal decisions (such as that on the implementation of the 'equivalence' requirement in Article 4 of the SPS Agreement).[95]

Once again, the process of normative elaboration pertaining in the SPS Committee differs noticeably from the development of interpretations of the SPS Agreement worked out in the context of adversarial dispute settlement. Scott remarks that the process exhibits dynamism and reflexivity, is generally deferential to member states, and focuses on providing assistance rather than augmenting or restricting the rights and obligations of members articulated by the SPS Agreement.[96] To date the Committee has not exercised its norm elaboration function in respect of the scientific evidence or risk assessment requirements of the SPS Agreement.[97] One might speculate that were it to do

[92] Roberts and Unnevehr, 'Resolving Trade Disputes Arising from Trends in Food Safety Regulation', 488.

[93] Scott, *The WTO Agreement on Sanitary and Phytosanitary Measures*, p. 72.

[94] Committee on Sanitary and Phytosanitary Measures, 'Guidelines to Further the Practical Implementation of Article 5.5', G/SPS/15, 18 July 2000.

[95] Committee on Sanitary and Phytosanitary Measures, 'Decision on the Implementation of Article 4 of the Agreement on the Application of Sanitary and Phytosanitary Measures', G/SPS/19, 26 October 2001.

[96] Scott, *The WTO Agreement on Sanitary and Phytosanitary Measures*, p. 70.

[97] While there is no express call in the SPS Agreement for the SPS Committee to play this function in respect of the requirements of Articles 2.2 and 5.1 (unlike Article 5.5), and there has been no such request for elaboration issued by the WTO's General Council, there would seem to be nothing to prevent the Committee issuing recommendations that build upon the risk assessment requirements of the SPS Agreement in a similar fashion to the work that it has done in respect of the transparency requirements in Article 7.

so, this could result in evolution of the SPS rules in ways that would enhance their flexibility and permit of procedures for ongoing review and re-evaluation.[98]

However, in order to develop the science-based requirements of the SPS Agreement, the Committee would need to determine how it would respond to the detailed elaboration of those requirements that has already emerged from the SPS jurisprudence, discussed in the next section of the chapter. Past practice of the Committee suggests that the rulings in SPS disputes would exercise a substantial influence over any recommendations issued by the Committee in this regard.[99] While the SPS Agreement leaves unresolved the question of the relationship between the Committee and the dispute settlement system, 'there is evidence that the committee, in its determinations, looks to the decisions of the dispute settlement bodies, encompassing their findings within the text of [its] measures.'[100] By contrast, the dispute settlement bodies have not relied in a substantive way on the work of the SPS Committee in developing interpretations of provisions under the SPS Agreement. Indeed, the strengthening of the WTO dispute settlement system that occurred with the conclusion of the DSU in 1995 may be said to have 'created a bias in WTO law in favour of courts'.[101] Consequently, rulings of panels and the Appellate Body remain the major source for understanding the meaning of the science-based requirements of the SPS Agreement.

Science in the SPS jurisprudence

The contours of SPS dispute settlement, and the interpretations of the SPS Agreement developed through this process, were significantly shaped by the first dispute of *Hormones*. This dispute has had a long and rancorous history,[102] involving the two major players in the WTO

[98] See Rosie Cooney and Andrew T. F. Lang, 'Taking Uncertainty Seriously: Adaptive Governance and International Trade', European J. Int'l Law, 18(3) (2007), 523 for a strong plea for the institution of adaptive governance of this kind in the SPS forum.

[99] For instance, the Committee's Article 5.5 guidelines largely follow the Appellate Body's rulings in *Hormones* and *Salmon*.

[100] Scott, *The WTO Agreement on Sanitary and Phytosanitary Measures*, p. 72.

[101] *Ibid.*, p. 74.

[102] However, the dispute may now be coming to an end with a provisional deal reached by the USA and the EU. Under the deal, the EC ban on US hormone-treated beef remains in place but duty-free access for non hormone-treated US beef has been increased. In return the USA will reduce punitive trade sanctions on EU exports such as Roquefort cheese and Italian mineral water. Further WTO litigation between the parties on the matter has also been suspended. See Embassy of

system, the USA and the EC, as well as Canada. The outcome of the first round of the dispute (*Hormones*) was a determination in favour of the complainants in as much as the challenged EC measures were found not to be based on a risk assessment as required by Article 5.1 of the SPS Agreement. The failure of the EC to remove its impugned measures led the USA and Canada to seek approval for trade sanctions against EC products, which still remain in effect.

In 2003, after seeking seventeen scientific opinions over the period 1999–2002, the EC introduced revised measures that, albeit somewhat less stringent, still had the effect of excluding the complainants' beef products from the EU market.[103] In November 2004 the EC initiated the second round of the dispute, asking the WTO dispute settlement bodies to order the removal of trade sanctions on the basis that the EC's new measure complied with the SPS Agreement.[104] The appeal of the panel's findings to the Appellate Body in *Hormones II* resulted in some important clarifications of the original *Hormones* rulings. Nonetheless, the core question at the heart of the dispute – whether the EC measures are based upon an SPS-compliant risk assessment – remains unresolved as significant deficiencies in the panel's assessment of the scientific evidence left the Appellate Body unable to 'complete the analysis' on the substantive legal issues.[105]

the US (London), Statement on U.S.-EU Beef Hormone Agreement, 14 May 2009, available at www.usembassy.org.uk/euro013.html.

[103] Directive 2003/74/EC of the European Parliament and of the Council of 22 September 2003 amending Council Directive 96/22/EC concerning the prohibition on the use in stockfarming of certain substances having a hormonal or thyrostatic action and of beta-agonists, Official Journal, L 262, 14 October 2003. Under the new EC Directive only one hormone – oestradiol-17ß – was banned outright. For the other five hormones (testosterone, progesterone, trenbolone acetate, zeranol and MGA) provisional bans were introduced which the EC sought to justify under Article 5.7. These claims, and the relevant Appellate Body rulings in *Hormones II*, are discussed in the section on precautionary regulation.

[104] United States – Continued Suspension of Obligations in the EC – Hormones Dispute, Request for Consultations by the European Communities, WT/DS320/1, G/L/713, 10 November 2004; United States – Continued Suspension of Obligations in the EC – Hormones Dispute, Request for the Establishment of a Panel by the European Communities, WT/DS320/6, 14 January 2005. The essence of the EC's case was the scientific opinions sought by the EC since 1999 comprised an adequate risk assessment for SPS purposes: *Hormones II*, [736].

[105] *Hormones II*, [736]. Consequently, the Appellate Body recommended that the parties be requested to initiate compliance proceedings under Article 21.5 of the DSU without delay: [737]. Under the provisional deal reached by the USA and EU (described above at note 102) the parties have agreed, however, to suspend further WTO litigation for the meantime.

The measures at issue in the first phase of the *Hormones* dispute were EC restrictions under Directive 96/22/EC on the import of beef produced using hormone treatment with any of the six hormones listed by the Directive.[106] The basis for introduction of the EC ban was possible – but scientifically unproven – risks to health posed by the consumption of hormone residues that might be present in such beef. Previous attempts by the parties to resolve the dispute drawing on the provisions of the GATT had been unsuccessful.[107] Hence, when the USA and Canada initiated a dispute over the EC's measures under the SPS Agreement in 1996, it was viewed by many as an important test case for the novel, science-based disciplines of the Agreement.[108]

A re-reading of the Hormones dispute

Following the Appellate Body's first decision in the *Hormones* case in 1998, numerous analyses of the judgment were written, equally praising and criticising the interpretations of SPS provisions that had been put forward.[109] In general, commentators focused their attention on several central rulings of the Appellate Body, which were seen to endorse

[106] The facts of the first *Hormones* dispute have been exhaustively detailed in other commentaries and so will only be outlined here. For a comprehensive summary of the facts and background to the dispute see Dale McNiel, 'The First Case Under the WTO's Sanitary and Phytosanitary Agreement: The European Union's Hormone Ban', Virginia J. Int'l Law, 39 (1998), 89.

[107] Negotiations were also sought under the Agreement on Technical Barriers to Trade, 12 April 1979, 1186 UNTS 276, in force 1 January 1980, the predecessor of the Agreement on Technical Barriers to Trade, 15 April 1994, 1868 UNTS 120, in force 1 January 1995 (TBT Agreement).

[108] E.g., Vern Walker, 'Keeping the WTO from Becoming the "World Trans-science Organisation": Scientific Uncertainty, Science Policy and Fact-Finding in the Growth Hormones Dispute', Cornell Int'l. L.J., 31 (1998), 251; David Wirth, 'International Trade Agreements: Vehicles for Regulatory Reform?', *University of Chicago Legal Forum*, (1997), 331.

[109] E.g., Michele Carter, 'Selling Science under the SPS Agreement: Accommodating Consumer Preference in the Growth Hormones Controversy', Minnesota J. Global Trade, 6 (1997), 625; David Hurst, 'Decisions of the Appellate Body of the World Trade Organization: Hormones: European Communities – Measures Affecting Meat and Meat Products', European J. Int'l L., 9 (1998), 182; Jan McDonald, 'Big Beef Up or Consumer Health Threat?: The WTO Food Safety Agreement, Bovine Growth Hormone and the Precautionary Principle', *Environmental & Planning Law Journal*, 15(2) (1998), 115; George Rountree, 'Raging Hormones: A Discussion of the World Trade Organization's Decision in the European Union-United States Beef Dispute', Georgia J. Int'l & Comp. L., 3 (1999), 607; Iain Sandford, 'Hormonal imbalance? Balancing free trade and SPS measures after the decision in Hormones', Victoria Uni. Wellington Law Review, 29(2) (1999), 389; Ryan Thomas, 'Where's the Beef? Mad Cows and the Blight of the SPS Agreement', Vanderbilt J. Transnational

a broad notion of SPS risk assessment.[110] In the SPS commentary that has burgeoned since the *Hormones* dispute (and indeed, also in subsequent SPS case law), these rulings have taken on a life of their own as quasi-maxims of the Appellate Body's interpretative approach that are rarely subjected to critical analysis. These principles, drawn from the *Hormones* decision and largely reiterated by the Appellate Body in *Hormones II*, can be summarised as follows:

1. **Burden of Proof**: Under the SPS Agreement, the complainants in a dispute are responsible for presenting evidence and legal arguments sufficient to demonstrate that challenged SPS measures are inconsistent with obligations under each relevant article of the Agreement (including Articles 2.2 and 5.1). Only after such a *prima facie* determination has been made by a panel does the onus shift to the defending party to bring forward evidence and arguments to disprove the complaining party's claims.[111]
2. **Prudence and Precaution**: A panel charged with determining whether sufficient scientific evidence exists to warrant a particular SPS measure should 'bear in mind that responsible, representative governments commonly act from perspectives of prudence and precaution where risks of irreversible, e.g. life-terminating, damage to human health are concerned'.[112]
3. **No minimum magnitude of risk**: There is no basis in the SPS Agreement for the imposition of a quantitative requirement for a 'minimum magnitude of risk' to be established via risk assessment.[113] Hence, by implication, members may evaluate SPS risks either quantitatively or qualitatively.[114]

Law, 32 (1999), 487; David A. Wirth, 'European Communities Restrictions on Imports of Beef Treated with Hormones – Nontariff Trade Barriers – Control of Food Additives – Scientific Basis for Restrictions – WTO Dispute Settlement Mechanisms – Scope of Review', American J. Int'l L., 92 (1998), 755; Reinhard Quick and Andreas Blüthner, 'Has the Appellate Body Erred? An Appraisal and Criticism of the Ruling in the WTO *Hormones* Case', *Journal of International Economic Law*, 2 (1999), 603.

110 This view has been taken by some of the most astute commentators on WTO law and the SPS Agreement, such as Robert Howse. See Howse, *The WTO System*, p. 67, contending: 'The Appellate Body went to great lengths to emphasize that it would not second-guess a WTO Member's regulatory choices, provided there was some scientific evidence on the record concerning the risks in question, and even stated that a Member could act on the basis of "non-mainstream" science, thereby ensuring that "science" does not become an orthodoxy precluding democratic contestability in the area of risk regulation.'

111 *Hormones*, [109]. 112 *Hormones*, [124]; *Hormones II*, [680].

113 *Hormones*, [186]; *Hormones II*, [569].

114 This was confirmed by the Appellate Body in *Salmon*, [124]. See also *Hormones II*, [530].

4. **Real world risks**: Risk assessment under Article 5.1 extends to matters beyond those 'susceptible of quantitative analysis by the empirical or experimental laboratory methods commonly associated with the physical sciences'.[115] While science 'plays a central role in risk assessment', the Appellate Body 'has cautioned against taking too narrow an approach to a risk assessment'.[116] Consequently, the risk to be evaluated in an SPS risk assessment 'is not only risk ascertainable in a science laboratory operating under strictly controlled conditions, but also risk in human societies as they actually exist, in other words, the actual potential for adverse effects on human health in the real world where people live and work and die'.[117]
5. **No minimum procedural requirement**: There is no requirement for a member to demonstrate that it performed and 'actually took into account' a risk assessment at the time of enacting its measures. A particular SPS measure might find objective justification in a risk assessment carried out by another member or an international organisation.[118]
6. **Reliance on divergent or minority scientific opinion**: A risk assessment is not required to come to a 'monolithic conclusion' that coincides with the scientific conclusion or view implicit in a member's SPS measures and need not 'embody only the view of a majority of the relevant scientific community'. Responsible and representative governments may act in good faith on the basis of what, at the time, may be a 'divergent opinion coming from qualified and respected sources' and still meet, in some cases, the requirement that SPS measures be based on a risk assessment. This is especially so 'where the risk involved is life-threatening in character and is perceived to constitute a clear and imminent threat to public health and safety'.[119]
7. **Relevance of socio-cultural dimensions of risk**: Although the Appellate Body has not provided a clear demarcation of the factors that may be considered in an SPS risk assessment, the list of factors provided in Article 5.2 is not considered to be a closed one.[120] Factors such as public 'anxieties' generated by general scientific studies establishing potential health risks, the dangers of abuse or misuse

[115] *Hormones*, [187]. [116] *Hormones II*, [527]. See also [542].

[117] *Hormones*, [187]; *Hormones II*, [527]. In *Hormones II* the Appellate Body criticised the panel for failing to engage with evidence relating to abuse or misuse in the administration of hormones discussed in the scientific opinions sought by the EC and by the experts advising the panel. It held the panel's summary dismissal of this evidence resulted in the panel incorrectly applying Article 5.1 and the SPS definition of risk assessment, as interpreted by the Appellate Body: [553].

[118] *Hormones*, [189]–[190]; *Hormones II*, [530].

[119] *Hormones*, [194]; *Hormones II*, [529]. [120] *Hormones II*, [535].

> and difficulties of control in the administration of substances such as growth-promoting hormones, and 'the intense concern of consumers' may be considered in the context of a risk assessment and provide, in some circumstances, evidence that measures have a legitimate rather than a protectionist purpose.[121]

Taken in isolation, these seven principles would seem to provide a flexible interpretative framework for understanding the SPS Agreement's notion of risk assessment; one that is capable of accommodating both precautionary regulation that favours minority scientific views about risk, as well as deliberative risk assessment processes which place a premium on social risk perceptions. However, such readings of the *Hormones* decision pay insufficient attention to other critical rulings in the case (discussed below) that provide the genesis for a much stricter application of notions of science and risk assessment in the SPS context.

In *Hormones II* the Appellate Body provided important clarifications regarding aspects of the SPS risk assessment process that may ameliorate the stringency of its approach in the first case, at least to some extent. It also articulated a new standard of review for panels examining expert evidence and members' risk assessments under the SPS Agreement, with potentially far-reaching implications for the evaluation of the scientific basis underlying WTO members' SPS measures. Nonetheless, although there is some evidence in the *Hormones II* decision that the pendulum may be swinging back in favour of a more deferential stance towards members' risk regulatory practices, Joanne Scott's early assessment of the *Hormones* ruling is still largely warranted: 'Its bark may be muted, but its bite is strong'.[122]

Rational relationship requirement

The efficacy of many of the Appellate Body's more liberal findings regarding the scope of risk assessment in *Hormones* turned on its interpretation of an important phrase in Article 5.1; that requires

121 *Hormones*, [245]; *Hormones II*, [535]. In *Hormones* these factors were highlighted in the Appellate Body's analysis of Article 5.5 and the question of whether differences in the treatment of similar risk situations provided evidence that the EC measures were in fact a disguised restriction on trade. In *Hormones II* the context for discussion was the factors assessable in risk assessment, although only 'the abuse or misuse and difficulties of control in the administration of hormones' was mentioned.

122 Joanne Scott, 'On Kith and Kine (and Crustaceans): Trade and Environment in the EU and WTO', in J. H. H. Weiler (ed.), *The EU, the WTO, and the NAFTA: Towards a Common Law of International Trade?* (Oxford University Press, 2000), p. 155.

members' regulatory measures to be 'based on' an assessment of SPS risks. The importance of this phrase lies in its consequences for the nature and stringency of WTO evaluation of a member's SPS measures.[123] At one end of the interpretative spectrum is a fairly loose relationship between SPS measures and a risk assessment as, for example, where a member has adopted some but not all of the conclusions of a risk assessment,[124] or merely considered its results in the course of the standard-setting process. This might occur in some precautionary or deliberative regulatory settings where national authorities wish to place more emphasis on uncertainties or public concerns in devising risk standards than on the findings of expert risk assessment.[125] At the other end of the spectrum is a requirement for a substantial degree of coherence between the (scientific) conclusions of a risk assessment and the measures adopted as a result.[126] Taking this approach, a member's measure would need to demonstrate consistency with particular risk assessment findings in order to pass WTO scrutiny.

In its analysis of the 'based on' requirement in Article 5.1, the Appellate Body in *Hormones* rejected, in part, the conclusions of the panel, which had found both a 'substantive aspect' and the existence of a 'minimal procedural requirement' for a member 'to submit evidence that at least it actually took into account a risk assessment when it enacted or maintained its sanitary measure'.[127] The Appellate Body took issue with the latter finding, citing the lack of a 'textual basis' in Article 5.1 for any 'minimum procedural requirement'.[128] According to the Appellate Body, the preferable interpretation was one construing the phrase 'based on' to refer 'to a certain *objective relationship* between two elements, that is to say, to an *objective situation* that persists and is observable between an SPS measure and a risk assessment'.[129]

How such an objective situation was to be discerned was not made clear in *Hormones* by the Appellate Body, which contented itself with

[123] Hurst, 'Decisions of the Appellate Body of the World Trade Organization'.

[124] *Hormones*, [163] considering the meaning of the term 'based on' in Article 3.1.

[125] Howse, 'Democracy, Science, and Free Trade', 2336–7.

[126] E.g., in *European Communities – Trade Description of Sardines*, Report of the Appellate Body, WT/DS231/AB/R, 26 September 2002, [245], Article 2.4 of the TBT Agreement (directing members to use international standards 'as a basis for' their own measures) was construed as requiring 'a very strong and very close relationship between [the] two things'.

[127] *Hormones* Panel Report, [8.113].

[128] *Hormones*, [189].

[129] *Ibid.*, [189] (emphasis in original).

the general statement that '[d]etermination of the presence or absence of that relationship can only be done on a case-by-case basis, after account is taken of all considerations rationally bearing upon the issue of potential adverse effects'.[130] Nonetheless, it hinted that the scientific evidence requirements of Article 2.2 were relevant to this assessment, given that Article 5.1 is to be read 'in conjunction with and as informed by Article 2.2 of the SPS Agreement'.[131] This suggested that a member's measures would fail Article 5.1's 'based on' test where they did not bear an objective relationship to the *scientific* findings of a risk assessment (for example, if the member concerned pointed instead to uncertainties or public risk concerns to justify regulation).

The Appellate Body's insistence that the 'based on' requirement in Article 5.1 is a substantive one, requiring 'a rational relationship between the measure and the risk assessment',[132] has several important consequences for the nature of the evaluation carried out by WTO judicial bodies in a dispute settlement setting. The first is to take emphasis away from the processes (deliberative, precautionary or otherwise) followed by a WTO member in adopting its SPS measures in favour of determining whether a rational/objective relationship between the measure and a risk assessment can be established at the time a measure is challenged before the WTO dispute settlement organs.[133] This tends to direct WTO decision-makers' scrutiny away from the contextual factors influencing a particular risk assessment or the good governance features of a member's standard-setting process towards an analysis of objective evidence of risk.

The second consequence of the Appellate Body's approach – related to the first – is that a substantive test encourages panels 'to entangle themselves in evaluations of science' in an attempt to discern whether a rational relationship exists.[134] Indeed, this was the very kind of analysis in which the *Hormones* panel engaged when attempting to evaluate the substantive aspect of the 'based on' requirement in respect of the EC's measures.[135] Not only does this approach raise concerns about the competence of legal decision-makers to engage in in-depth analysis of scientific conclusions,[136] but, as the Appellate Body itself noted, it may

[130] *Ibid.*, [194]. [131] *Ibid.*, [193]. [132] *Ibid.*
[133] Hurst, 'Decisions of the Appellate Body of the World Trade Organization'.
[134] Andrew Guzman, 'Food Fears: Health and Safety at the WTO', Virginia J. Int'l L., 45 (2004), 1, 38.
[135] *Hormones* Panel Report, [8.117].
[136] Pauwelyn, 'The WTO Agreement on Sanitary and Phytosanitary (SPS) Measures', 661.

also encourage scientific risk findings to 'be assigned relevance to the exclusion of everything else'.[137]

Relationship between risk assessment and scientific evidence

Clarification of the necessary relationship between the risk assessment relied upon by a WTO member and any SPS measure adopted as a result is only one part of the interpretive puzzle presented by the scientific requirements of the SPS Agreement. Another important question concerns the requisite relationship between the risk assessment on which a member's SPS measures must be based and scientific evidence regarding those risks. The text of Article 5, which was the focus of the Appellate Body's findings in *Hormones* and *Hormones II*, suggests this relationship is a fairly flexible one, given that 'available scientific evidence' is only one of several factors members are to 'take into account' in the assessment of SPS risks.[138] On the other hand, Article 2.2, designated one of the 'basic obligations' of members under the SPS Agreement, might be read in a much stricter manner to demand a close coherence between national SPS measures and available scientific evidence regarding the risks of concern. As David Wirth has pointed out, this depends upon whether Article 2.2 is interpreted as authorising WTO panels to undertake a penetrating examination of the adequacy of the scientific data underlying a measure, or as simply indicating 'a need for a minimal level of scientific evidence', which might be satisfied by panels asking 'whether the empirical data are minimally adequate to support the national government's scientific conclusions'.[139]

In *Hormones* the question of the relationship between Articles 2.2 and 5.1 was addressed only indirectly. However, the Appellate Body's comments have had significant consequences for the approach taken to the evaluation of members' measures in subsequent SPS disputes. The Appellate Body remarked on the importance of 'constantly' reading the two provisions together since 'Article 2.2 informs Article 5.1: the elements that define the basic obligation set out in Article 2.2 impart meaning to Article 5.1'.[140] Albeit rather cryptic, this

[137] *Hormones*, [193].

[138] SPS Agreement, Article 5.2.

[139] David Wirth, 'The Role of Science in the Uruguay Round and NAFTA Trade Disciplines', Cornell Int'l. L.J., 27 (1994), 856–7.

[140] *Hormones*, [180]. In *Hormones II* the Appellate Body stated that the requirements of Article 2.2 are 'made operative in other provisions of the *SPS Agreement*, including Article 5.1': [674].

statement is difficult to reconcile with the Appellate Body's more flexible rulings in *Hormones* regarding the permissible scope of SPS risk assessment.[141]

Although in the context of the applicable standard of review (discussed further below), rather than on the question of the interrelationship between Articles 2.2 and 5.1, the Appellate Body in *Hormones II* also stressed the need for the reviewing panel to identify the 'scientific basis' upon which an SPS measure is adopted.[142] Taken together with the direction to read Articles 2.2 and 5.1 in concert, this would seem to emphasise the importance of a sufficient scientific evidentiary basis for *all* risks of concern. It is difficult, however, to imagine how this requirement could be satisfied for some categories of risk, such as the 'real world' risks related to the maladministration of hormones, which featured prominently in the EC's revised risk assessment at issue in *Hormones II*.

An important implication arising from reading Articles 2.2 and 5.1 together is that violation of the more general provisions of the former article will support a conclusion that the requirement for risk assessment in Article 5 also has not been met.[143] Following the first *Hormones* decision, some astute observers therefore advised complaining parties in SPS cases 'to follow the "direct" route of applying Article 2 SPS instead of the "complex and indirect" route of Article 5 SPS'.[144] The Appellate Body in *Hormones* also encouraged panels in this direction, expressing its 'surprise' that the *Hormones* panel began its analysis with Article 5 of the SPS Agreement rather than the more 'logically attractive' approach of first focusing on Article 2.[145] Logical attractiveness aside, the question of whether to focus SPS scrutiny on the provisions of Article 2 or 5 is one that implicates more than the legal niceties of treaty interpretation. Approaching the task of review as 'an exercise of evidentiary assessment',[146] determined by the provisions of Article 2.2, may end up overshadowing the nuances of a member's risk assessment, including attempts to rely on uncertainties or non-scientific information in the decision-making process.

141 Quick and Blüthner, 'Has the Appellate Body Erred?', 637.

142 *Hormones II*, [591]. 143 See also *Salmon*, [137].

144 Quick and Blüthner, 'Has the Appellate Body Erred?', 629.

145 *Hormones*, [250].

146 Fisher, 'Beyond the Science/Democracy Dichotomy', p. 346.

Ascertainable risk rather than theoretical uncertainty

The issue of the extent to which uncertainties may be looked to to found risk concerns in a member's risk assessment was further muddied by the Appellate Body's rulings in *Hormones* concerning the distinction between 'ascertainable risk' and 'theoretical uncertainty'.

As we have seen in previous chapters, the dividing line between what is ascertainable as a matter of science and what falls into the realm of uncertainty is inherently fuzzy. This is because 'science can never provide a certainty, i.e. exclude once and for all that a specific substance can ever have adverse health effects'.[147] In standard scientific research certainty is approximated by applying the conventions of scientific proof that maintain a hypothesis of no harm until a substantial body of evidence has accumulated which contradicts that hypothesis.[148] This approach is thought to facilitate the integrity of scientific research by increasing the likelihood that current hypotheses approximate as closely as possible to the truth, while still allowing for them to be displaced if better theories later emerge.[149] More pragmatically, it reflects the reluctance of research scientists ever to declare a line of inquiry closed. As one of the advising experts in *Hormones* remarked:

> I know of no scientist who has ever said on any issue (because we are all looking for funding) I know enough, please don't provide any more money on this issue.[150]

In a regulatory context, however, an insistence on high standards of scientific proof before accepting the possibility of harm (or risk, broadly defined) may be counterproductive. The result may be that harms are not scientifically verified until substantial health or environmental damage has already occurred.[151] This has motivated adoption of a precautionary approach by some authorities whereby information regarding potential adverse effects is treated as sufficient for regulatory action even where it could not be said to meet the standard of proof applied in

147 *Hormones* Panel Report, [8.152] noting the expert advice in this regard.

148 For an accessible introduction to the scientific method see Peter Riggs, *Whys and Ways of Science: Introducing Philosophical and Sociological Theories of Science* (Melbourne University Press, 1992).

149 David Fisk, 'Environmental Science and Environmental Law', J. Envt'l L., 10(1) (1998), 3.

150 *Hormones* Panel Report, Annex, [856] (Dr Ritter).

151 Lene Buhl-Mortensen, 'Type-II Statistical Errors in Environmental Science and the Precautionary Principle', *Marine Pollution Bulletin*, 32(7) (1996), 529.

conventional science. Some of the most difficult cases are those where regulators are presented with a new theory of harm causation that has not been sufficiently tested to yield reliable, publishable scientific results, or for which all elements of the putative causal chain have not yet been substantiated with plausible data. In such circumstances the question arises as to whether regulatory authorities must wait until definitive evidence emerges before introducing protective measures, or whether they may act on the basis of a theoretical risk.

In both *Hormones* and *Hormones II* this was the situation that faced national and international standard-setting bodies. By 2003, when the EC was introducing its second hormones directive, the strongest conclusion drawn by its scientific advisors regarding the most well-characterised hormone, oestradiol-17β, was that recent scientific literature demonstrated the genotoxicity of this substance.[152] A genotoxic substance is one that damages DNA; damage which – if not repaired – may give rise to mutations in the genetic code.[153] Genetic mutations that occur in genes responsible for the production of growth promoters may cause abnormal growth patterns that eventually lead to the development of cancer. However, evidence that a substance is genotoxic is not equivalent to scientific proof of potential adverse health effects, such as cancer, in the absence of other data verifying the link between DNA damage and the development of cancerous cells.[154]

Based on the available scientific evidence regarding oestradiol-17, Codex has established an international food standard for the hormone when used as a veterinary drug that designates maximum residue limits that are seen as posing no harm to health.[155] The EC, on the other hand, takes the view that the scientific theory underlying the international standard is itself flawed as it discounts the possibility of a genotoxic mechanism for cancer causation. On the basis of alternative

152 Scientific Committee on Veterinary Measures relating to Public Health, 'Opinion of the Scientific Committee on Veterinary Measures Relating to Public Health on Review of Previous SCVPH opinions of 30 April 1999 and 3 May 2000 on the potential risks to human health from hormone residues in bovine meat and meat products', (2002), p. 14.

153 *US – Hormones II* Panel Report, Annex G, 849. The SCVMPH 2002 opinion, *ibid.*, also discusses evidence that oestradiol-17ß gives rise to DNA mutations in cultured mammalian cells. However, this *in vitro* (i.e. experimental) evidence is not equivalent to a finding that oestradiol-17ß causes DNA mutations *in vivo* (i.e. in whole, live animal subjects).

154 *US – Hormones II* Panel Report, Annex G, [210].

155 Codex Alimentarius Commission, Maximum Residue Limits for Veterinary Drugs in Foods (2009).

theories emerging in science, albeit ones that have not yet yielded conclusive evidence of health effects, the EC has decided to follow a more cautious approach, banning meat and meat products from cattle treated with oestradiol-17β.

While conventional research science purports to define hard boundaries between theory and science, it is important to emphasise that this exercise is not itself based on science. As Oren Perez observes, 'science does not offer a definite algorithm for "*closing*" the *incompletable* universe in which law operates'.[156] Rather it is the operation of policy or broader values concerns that bring closure to the inherently open-ended knowledge spectrum provided by science. In scientific research the overriding policy concern with ensuring the integrity of data has given rise to a bias towards hypotheses verified by a substantial body of evidence. In a regulatory context, however, there may be more pressing policy considerations, such as public calls for a conservative approach to the protection of human health. Motivated by such concerns, decision-makers may choose to treat the absence of definitive scientific evidence of risk as indicative, not of the fact that no (unacceptable) risk exists, but rather of the remaining potential for adverse effects to manifest in the future.[157]

In *Hormones* the Appellate Body took a position on the dividing line between the kinds of risk that are a proper foundation for SPS measures and theoretical harms, although it appeared to regard this choice as one dictated by science rather than as being influenced by policy considerations. Its comments were made in the context of reviewing the panel's opposition of 'a requirement of an "identifiable risk" to the uncertainty that theoretically always remains since science can *never* provide *absolute* certainty that a given substance will not *ever* have adverse health effects'.[158] The Appellate Body stated its agreement with the panel 'that this theoretical uncertainty is not the kind of risk which, under Article 5.1, is to be assessed'.[159] Accordingly, the Appellate Body ruled that in order to be of regulatory concern, an SPS risk must

[156] Perez, *Ecological Sensitivity and Global Legal Pluralism*, p. 128.

[157] Anne Orford, 'Beyond Harmonization: Trade, Human Rights and the Economy of Sacrifice', *Leiden Journal of International Law*, 18(2) (2005), 179, 195, noting that the SPS Agreement imposes 'no requirement that the rationality of this decision *not* to regulate be established, or that the reasoning involved in reaching this decision be made public, supported by adequate documentation, or based on scientific principles'.

[158] *Hormones*, [186] (emphasis as in original).

[159] *Ibid.*

be 'an ascertainable risk' because 'if a risk is not ascertainable, how does a Member ever know or demonstrate that it exists?'[160] The implication from this statement is that uncertainties about what the risks are cannot, by themselves, ground regulatory action in the absence of a demonstration of risk.

In *Hormones II* the Appellate Body reiterated its previous findings regarding the need for the evaluation of 'ascertainable' risk rather than 'theoretical uncertainty'.[161] It also provided some clarification as to what it saw as the difference between the two concepts, remarking that:

> it is ... difficult to understand the concept of risk as being devoid of any indication of potentiality. A risk assessment is intended to identify adverse effects and evaluate the possibility that such adverse effects might arise. This distinguishes an ascertainable risk from theoretical uncertainty.[162]

Further, the Appellate Body agreed with a Canadian submission that 'to examine the "potential" for adverse effects is to ask whether those adverse effects could ever occur'.[163] These comments suggest that an ascertainable risk in the Appellate Body's mind is one more aligned with a technical perspective on risk assessment than with precautionary regulatory approaches. On this basis, only a possible harm for which some evidence could be assembled to evaluate its potential occurrence would be treated as an ascertainable risk. Evidence pointing to possible harms but not allowing for an assessment of the potential for the harm to materialise would not be sufficient. Moreover, the Appellate Body made clear that if 'there is no ascertainable risk ... no SPS measure can be taken'.[164]

For some, the implicit value judgment underlying the Appellate Body's insistence on a regulatory prerequisite of ascertainable risk is undesirable as it creates 'a world in which hypothetical risk must be endured, regardless of the nature of the risk-generating activity and the social worth attaching to it'.[165] Equally, though, there are many others who would endorse this approach as sound, both as a means to safeguard against arbitrary decision-making,[166] and also because it

[160] *Ibid.* [161] *Hormones II*, [530]. [162] *Ibid.*, [569].
[163] *Ibid.*, [572].
[164] *Ibid.*, [531].
[165] Scott, 'On Kith and Kine', p. 157.
[166] Gary Marchant and Kenneth Mossman, *Arbitrary and Capricious: The Precautionary Principle in the European Union Courts* (Washington DC: AEI Press, 2004).

will almost always be possible to find a scientist who disagrees with the existing consensus and is prepared to put forward an alternative theory. While the Appellate Body's apparent assumption that the dividing line it has drawn is one based in science (a representation also often made by scientists), 'boundary work' of this kind is invariably policy-driven.[167] In fact, empirical studies in domestic regulatory settings suggest that boundary-drawing between scientific and policy realms tends to be most effective (that is, considered persuasive and legitimate) where value dimensions are acknowledged and some room left for decision-makers and their advisors 'to negotiate the location and meaning of the boundaries'.[168]

The alternative approach of leaving definition of the boundary between an 'ascertainable risk' and 'theoretical uncertainty' in the hands of scientists limits the opportunities for negotiation of this dividing line on a case-by-case basis. One consequence is to reduce the scope for alternative choices that allow more room for the consideration of uncertain risk, something which is intrinsic to precautionary risk regulation.[169] Risk assessment approaches that place significant weight on 'unknown and uncertain elements' may thus become untenable,[170] at least insofar as they seek to justify long-term risk regulatory measures on this basis. In such situations members' regulatory options may be limited to provisional measures adopted in accordance with Article 5.7, the scope of which is discussed further below.

Necessity for specific scientific studies

Related to the question of the balance struck between considerations of uncertainty and science in risk assessment is that of the weight to be given to different types of scientific information in the decision-making process. Here, once again, the regulatory approach or paradigm adopted by decision-makers will play an influential role in the assessment of what evidence is considered to be a sufficient indication of risk. Pursuant to the precautionary principle risk regulatory paradigm, discussed in the previous chapter, regulators tend to consider,

[167] Sheila Jasanoff, *The Fifth Branch: Science Advisors as Policymakers* (Cambridge, MA: Harvard University Press, 1990), p. 236.

[168] *Ibid.*

[169] Laurence Boisson de Chazournes and Makane Moïse Mbengue, 'GMOs and Trade: Issues at Stake in the EC Biotech Dispute', *Review of European Community and International Environmental Law*, 13(3) (2004), 296.

[170] *Salmon*, [130].

and give credence to, a broad range of evidence regarding risk in order to minimise the possibility for adverse surprises and to ensure a high level of health or environmental protection.[171] For instance, they might look to minority scientific views, new theories of harm causation or evidence that is indicative of a general risk from exposure to a category of substances.[172] If, on the other hand, regulators operate under a regulatory approach that requires them to discharge stringent scientific burdens before introducing new risk measures, they are likely to weight preferentially material which is more relevant or specific to the risks at hand.[173]

Other important factors influencing the weight given to different types of scientific information in risk assessment are the risks at stake and the importance placed on preventing their materialisation. For example, regulators are often cautious in excluding evidence of health risks – even if only general in nature – because most societies attach great significance to the preservation of human health.[174] This indicates the connection, in practice, between risk management decisions about acceptable risk – in SPS terms, the appropriate level of protection – and the scope of the evidence considered in risk assessment. As highlighted earlier, at least in some parts of its judgment in *Hormones II*, the Appellate Body acknowledged this link.

However, the dominant approach that the Appellate Body has taken in the *Hormones* litigation to the question of appropriate scientific evidence of SPS risk is expressed by what has become known as the 'specificity requirement'. In the first *Hormones* case the Appellate Body's designation of a specificity requirement came in its findings regarding whether the EC's measures could be said to be based on a risk assessment. In support of its measure, the EC had put forward various scientific studies and individual scientific opinions pertaining to the general

171 This was essentially the argument put by Japan in *Apples* regarding the need to take account of indirect evidence of risk: *Japan – Measures Affecting the Importation of Apples*, Report of the Panel, WT/DS245/R, 15 July 2003 (*Apples* Panel Report), [4.56].

172 One of the experts advising the panel in *Hormones II*, Dr Cogliano, noted that this was the kind of evidence relied upon in assessing the health risks from cigarette smoking. Hence it was the 'totality of the evidence about smoking' that was crucial rather than any quantitative evaluation of the effects of smoking on health: *US – Hormones II* Panel Report, Annex G, pp. 883–4.

173 E.g., in *Apples*, the USA took the approach that only 'something serving as proof that was valid according to the objective principles of the scientific method' should qualify as scientific evidence for SPS purposes: *Apples* Panel Report, [4.57].

174 M. Gregg Bloche and Elizabeth Jungman, 'Health Policy and the WTO', *Journal of Law, Medicine and Ethics*, 31 (2003), 529.

cancer risk associated with increasing levels of hormones in the body. In regard to this evidence the Appellate Body held that although the EC's studies did 'indeed show the existence of a general risk of cancer ... they do not focus on and do not address the particular kind of risk here at stake – the carcinogenic or genotoxic potential of the residues of those hormones found in meat derived from cattle to which the hormones had been administered for growth promotion purposes'.[175] It concluded that the 'general studies, are in other words, relevant but do not appear to be sufficiently specific to the case at hand'.[176]

The Appellate Body's rulings on this issue in *Hormones* echoed the findings of the panel that, in turn, seemed to have been strongly shaped by views of the experts appointed to advise the panel. The experts had been critical of emerging studies presented on behalf of the EC because the studies involved used hormone dosages that were much higher than the lower doses likely to be present as residues in meat (that is, they were less specific).[177] The experts were also reluctant to displace the risk findings of specific studies, which had the weight of a majority of the relevant scientific community behind them, in favour of new, more general, risk conclusions.[178] Such advice from experts is not surprising given that both their scientific training, as well as social conventions operating in the scientific community regarding appropriate methodologies and peer review requirements, encourage them to approach scientific evidence in this way. But it does not necessarily follow that what is accepted as the best approach for science or the scientific community is also the only legitimate approach in the broader context of risk regulation.

In *Hormones II* the Appellate Body reaffirmed the validity of the specificity requirement in SPS law,[179] albeit while recognising some greater flexibility in its application. Much as in the first dispute, the panel in *Hormones II* held that the scientific opinions relied upon by the EC did not amount to a risk assessment because of the failure to undertake a specific evaluation.[180] The panel observed that:

> a risk assessment in this instance required not a general evaluation of the carcinogenic potential of entire categories of hormones, but rather should

[175] *Hormones*, [200]. [176] *Ibid.*

[177] *Hormones* Panel Report, VI.85, VI.92.

[178] *Ibid.*, [8.133]. [179] *Hormones II*, [530].

[180] *US – Hormones II* Panel Report, 7.537, 7.578; *Canada – Hormones II* Panel Report, 7.509, 7.548.

include an examination of residues of those hormones found in meat derived from cattle to which the hormones had been administered for growth promotion purposes.[181]

On appeal, the Appellate Body upheld the test articulated by the panel as being compatible with the definition of a risk assessment in Annex A(4) of the SPS Agreement and with its previous findings in *Hormones*.[182] Thus it ruled:

> The definition of a risk assessment in paragraph 4 of Annex A, as interpreted by the Appellate Body, required the European Communities to conduct a risk assessment that addresses the specific risk at issue. The particular risk being evaluated by the European Communities in this case was the potential for neurobiological, developmental, reproductive, and immunological effects, as well as immunotoxic, genotoxic and carcinogenic effects from the residues of oestradiol-17β found in meat derived from cattle to which this hormone was administered for growth-promoting purposes. Although the European Communities is correct in arguing that it was not required to demonstrate that these adverse health effects would actually arise, it was nevertheless required to demonstrate these adverse effects could arise from the presence of residues of oestradiol-17β in meat from treated cattle.[183]

Where the Appellate Body granted a little more latitude was in respect of the EC's obligation 'to evaluate whether a causal connection exists between the consumption of meat from cattle treated with oestradiol-17β and the possibility of adverse health effects'.[184] The Appellate Body acknowledged that, in the case of substances potentially toxic to human health, it would be unethical to insist on a 'specific' evaluation of risks through testing the effects of actual human consumption of the substances.[185] In addition, the Appellate Body found that there was no need for the EC to establish 'a direct causal relationship' as 'it was sufficient for the European Communities to demonstrate that the additional human exposure to residues of oestradiol-17β in meat from treated cattle is one of the factors contributing to the possible adverse health effects'.[186] This latter ruling of the Appellate Body should help to ease the stringency of the specificity requirement in situations of

181 *US – Hormones II* Panel Report, 7.511; *Canada – Hormones II* Panel Report, 7.483.

182 *Hormones II*, [558]. 183 *Ibid.*, [559]. 184 *Ibid.*, [562].

185 *Ibid.*, [563]. By implication, therefore, the Appellate Body recognised the need for continuation of the toxicological practice of assessing human health risks on the basis of tests in animals, which approximate the physiological reactions of human beings.

186 *Ibid.*

cumulative risk as, for example, where hormone residues in consumed beef add to levels of hormones and other substances already present in the body to give rise to health effects. Accordingly, '[w]here multiple factors may contribute to a particular risk, a risk assessor is not required to differentiate the individual contribution made by each factor'.[187]

The Appellate Body also pointed to the requirement in Article 5.1 that SPS measures be based on a risk assessment 'as appropriate to the circumstances' as indicating the need for the underlying scientific inquiry to 'take due account of particular methodological difficulties posed by the nature and characteristics of the particular substance and risk being evaluated'.[188] Ultimately, however, the specificity requirement still stands, meaning that risk assessors are not excused from evaluating whether there is a connection between a particular substance being evaluated and the possibility that adverse health effects may arise.[189]

Like the requirement for an ascertainable risk, the need for specificity in the scientific evidence linking 'the harm concerned and the precise agent that may possibly cause the harm'[190] tends to give preference in the risk assessment process to data that demonstrates a risk over studies raising uncertainties. This places substantial obstacles in the way of precautionary regulation that seeks to introduce measures in advance of the emergence (or availability) of conclusive scientific evidence of harm. This was the case for the EC's measure in *Hormones* pertaining to residues of the synthetic hormone melengestrol acetate (MGA) – a substance for which little specific data was available at the time of the dispute given its relative novelty. The Appellate Body held that this measure was unsupported by an adequate risk assessment.[191] Moreover, while the complainants initially bore the burden of proving a violation of Article 5.1, it seemed that this burden was easily discharged by asserting the absence of a specific risk assessment, even though it was clear in the circumstances that it was the complainants who had better access to the relevant scientific information.[192]

Similar hurdles are likely to be faced by members who seek to rely on minority scientific opinion or real world concerns about risk,

[187] *Ibid.* [188] *Ibid.* [189] *Ibid.* [190] *Ibid.*, [530]. [191] *Hormones*, [201].

[192] The Appellate Body noted that the USA and Canada 'declined to submit any assessment of MGA upon the ground that the material they were aware of was proprietary and confidential in nature': *ibid.*

notwithstanding the Appellate Body's endorsement of these sources as appropriate considerations in SPS risk assessment. There are strong indications in the Appellate Body's *Hormones* rulings that the acceptance of these alternative types of evidence in risk assessment is also conditional upon their meeting the specificity requirement. For example, although the EC put forward a handful of studies in *Hormones* identifying the misapplication of growth-promoting hormones as a 'real world' problem,[193] the Appellate Body found that the rather general nature of these studies meant that they at best represented the beginnings of an assessment of such risks.[194] Likewise, the Appellate Body's treatment of a 'single divergent opinion' expressed by one of the advising experts in the case,[195] suggested that minority scientific opinion, even from qualified and respected sources, will rarely displace accumulated contrary evidence, unless it can be backed up with specific scientific studies relating to the case at hand.[196] Since minority scientific opinion relied upon in precautionary regulation is frequently based on inconclusive but suggestive findings or expert views extrapolating from known facts, rather than specific studies proving a causal link, such opinion is unlikely to provide an adequate basis for risk assessment and permanent SPS measures in most cases.

Standard of review and the evaluation of expert evidence

In examining the relationship between a member's SPS measures, risk assessment and the underlying scientific evidence, an overriding consideration is the applicable standard of review.[197] As we saw in

193 David Driesen, 'What is Free Trade?: The Real Issue Lurking Behind the Trade and Environment Debate', *Virginia Journal of International Law*, 41 (2001), 297.

194 *Hormones*, [207].

195 This was the opinion of Dr George Lucier who, in response to the panel's questions about the adequacy of Codex limits for natural hormone residues in meat, stated: 'For every million women alive in the United States, Canada, Europe today, about a 110,000 of those women will get breast cancer ... by my estimates one of those 110,000 would come from eating meat containing oestrogens as a growth promoter, if used as prescribed' (*Hormones* Panel Report, Annex, [819]). If 'realistic', the Appellate Body noted this assessment equated to an estimated 371 women out of an EU population of 371 million getting cancer: *Hormones*, [198], fn. 182.

196 *Hormones*, [198]. Steve Charnovitz has commented that it was not entirely clear whether the Appellate Body regarded Dr Lucier's risk estimate as expressing only an insignificant (one in a million) risk, or whether it considered the absence of specific scientific support as rendering the estimate scientifically 'unsound': Steve Charnovitz, 'The Supervision of Health and Biosafety Regulation by World Trade Rules', Tulane Envtl. L.J., 13 (2000), 282.

197 See, generally, Matthias Oesch, 'Standards of Review in WTO Dispute Resolution', J. Int'l Economic Law, 6 (2003), 635.

Chapter 4, the question of the appropriate standard of review to be applied by courts evaluating risk regulatory measures and expert scientific opinion has generated a range of approaches on both sides of the Atlantic. Judicial endorsement of precautionary regulation – during the 1970s in the USA and later in the 1990s in the EU – has often been associated with the exercise of less stringent review by the courts and greater deference to the risk conclusions reached by regulators on the basis of uncertain scientific evidence. However, this more deferential standard of review is in constant tension with approaches that call for a 'harder look' at the science underlying risk regulatory measures, whether this is to ensure that the evidence relied upon meets the substantive criteria of rigorous science, or that it has been deployed according to transparent and objective processes.[198]

A prevalent view prior to the first *Hormones* decision was that WTO decision-makers' review of national SPS measures under the SPS Agreement would apply a lenient standard of review, allowing national regulators substantial autonomy in the introduction and maintenance of SPS risk regulation. For instance, US trade officials involved in the SPS negotiations stated confidently that 'the requirement in the S&P Agreement that measures be based on scientific principles and not be maintained "without sufficient scientific evidence" would not authorise a dispute settlement panel to substitute its scientific judgment for that of the government maintaining the sanitary and phytosanitary measure'.[199]

Between *de novo* review and total reference

Not surprisingly, in the first *Hormones* case it was the EC, rather than the USA, which argued for a deferential standard of review to be applied in WTO evaluation of the findings of risk assessment. The EC contended that deference should be extended, not only to a member's scientific assessment and management of risk, but also 'when reviewing a Member's decision to adopt a particular science policy or

[198] Debate over the 'hard look' doctrine in the USA, as well as more recent literature in the EU on judicial review of risk regulation, recognises that there may be little practical difference between a substantive hard look at the science of risk regulation and rigorous procedural review. See Catherine Button, *The Power to Protect: Trade, Health and Uncertainty in the WTO* (Oxford: Hart Publishing, 2004), pp. 130–58.

[199] Uruguay Round Agreements Act, Statement of Administrative Action, reprinted in H.R. DOC. NO. 103-316, at 656, 746 (1994).

a Member's determination that a particular inference from the available data is scientifically plausible'.[200] As Vern Walker explains, science policies are decision-making rules 'not justified on purely scientific grounds', which are routinely applied by risk assessors in cases where uncertainties require a choice 'among alternative models or inputs'.[201] Examples include conservative assumptions about risk, extrapolations from the available data based on professional judgment and the application of safety factors in deriving health or environmental standards from toxicological research.[202] The EC's contention that members' science policies and inferences should be accorded deferential review amounted, in effect, to a call for the WTO dispute settlement organs not to interfere with the way members choose to balance competing considerations of science and uncertainty in risk decision-making.

This call was not heeded by the Appellate Body in the first *Hormones* case, which instead saw its task as one of determining the 'finely drawn balance' between the relative 'jurisdictional competences' of the WTO and its members on the basis of an analysis of the text of the SPS Agreement.[203] Appropriate textual guidance, in its view, was to be found in Article 11 of the DSU, which the Appellate Body saw as articulating 'with great succinctness but with sufficient clarity the appropriate standard of review for panels in respect of both the ascertainment of facts and the legal characterization of such facts under the relevant agreements'.[204] Article 11 of the DSU in fact does not deal with standards of review but instead places an obligation on a dispute settlement panel to make 'an objective assessment of the matter before it, including an objective assessment of the facts of the case and the applicability of and conformity with the relevant covered agreements'. The Appellate Body construed this provision to mean that when a panel examines disputed facts (including conflicting evaluations of the available scientific evidence), the applicable standard of review is 'neither *de novo* review, as such, nor "total deference", but rather the "objective assessment of the facts"'.[205] Whereas the standard of *de novo* review was said to connote 'complete freedom to come to a different view than the competent authority of the Member whose act or determination

200 *Hormones*, [14].

201 Walker, 'Keeping the WTO from Becoming the "World Trans-science Organisation"', 260–1. Walker served as a consultant to the EC in the case.

202 See also Rory Sullivan and Amanda R. Hunt, 'Risk Assessment: the Myth of Scientific Objectivity', *Environmental And Planning Law Journal*, 16(6) (1999), 522.

203 *Hormones*, [115]. 204 *Ibid.*, [116]. 205 *Ibid.*, [117].

is being reviewed',[206] the alternative of total deference would seem to amount to adopting, without question, the same evaluation of risk and the underlying scientific evidence as that put forward by a defending member.[207]

As former Appellate Body members Claus-Dieter Ehlermann and Nicolas Lockhart have commented, the findings in *Hormones* at least 'fleshed out what is not permitted under the WTO standard of review'.[208] Nonetheless, it remained unclear what the Appellate Body contemplated by a review that neither 'redoes an investigation into the facts that has already been done by a national authority',[209] nor defers to that authority's risk assessment approach. Only instances involving a very serious departure from the objective assessment standard were discussed by the Appellate Body. For instance, in *Hormones* the Appellate Body considered that a panel would fail to meet the standard of 'objective assessment' if there were 'deliberate disregard of, or refusal to consider' evidence submitted to the panel, or the 'wilful distortion or misrepresentation of the evidence put before a panel', constituting 'an egregious error that calls into question the good faith of a panel'.[210] Since it would be a rare occurrence where the good faith of an independent panel might be called into question, this statement suggested that the Appellate Body would not exercise any real supervisory powers over the way in which panels approached, considered and weighed the scientific evidence and expert opinion they received in the course of a hearing.[211]

In SPS cases that generally involve the presentation and evaluation of vast amounts of technical material, the level of oversight likely to be exercised by the Appellate Body over panels' assessment of scientific facts is of no little matter. Moreover, the panel's fact-finding exercise has acquired more significance in light of the Appellate Body's rulings in *Hormones* regarding the need for evaluation, as a substantive matter, of the existence of a rational relationship between a risk assessment and members' measures. Beyond the scientific views presented by experts on the disputing parties' delegations,[212] panels receive assistance with

[206] *Ibid.*, [111]. [207] See also *Apples*, [165].

[208] Claus-Dieter Ehlermann and Nicolas Lockhart, 'Standard of Review in WTO Law', J. Int'l Economic Law, 7 (2004), 501.

[209] *Ibid.*, 501. [210] *Hormones*, [133].

[211] Button, *The Power to Protect*, p. 174.

[212] However, these experts are generally not seen as having sufficient objectivity; hence it is often better for parties to get their experts appointed as panel

this evaluation via their powers to appoint independent experts.[213] Neither the DSU nor the SPS Agreement offers panels meaningful guidance as to how they might formulate appropriate questions for experts or determine the weight to be given to competing scientific views. One consequence, as Robert Howse and Petros Mavroidis have noted, is that panels may put questions to the advising experts that extend beyond the experts' areas of scientific expertise.[214]

Another possible consequence – the likelihood of which increases as more emphasis is placed upon assessing the scientific underpinnings of SPS measures – is that panels feel obliged to defer to the 'epistemic superiority' of experts and, therefore, accord significant weight to those experts' views.[215] This is problematic where expert views about risk embed within them particular framings of the risk problems at hand that may not be widely shared in the broader community. An 'objective assessment of the facts' by a panel necessarily reliant on the advice of independent scientists who possess greater technical expertise may thus potentially become a vehicle for particular value judgments to enter the evaluative process under the guise of science.

In *Hormones II* the Appellate Body took the opportunity to clarify the standard applicable in the review of risk assessments under the SPS Agreement, as well as related standards pertaining to the treatment of expert evidence about risk. Discussion of the latter came in the context of an evaluation of whether the EC had been afforded 'due process' as a result of the panel's decision to consult two experts closely associated with the production of risk assessments for hormone residues underlying relevant Codex standards. The EC's submissions also revisited the question of the applicable standard of review in SPS cases, arguing that the panel's mandate was limited to determining whether there was any 'reasonable scientific basis' for the SPS measures concerned.[216] On

advisors: Joost Pauwelyn, 'The Use of Experts in WTO Dispute Settlement', *International and Comparative Law Quarterly*, 51 (2002), 333–4.

213 DSU, Article 13.2; SPS Agreement, Article 11.2.

214 Robert Howse and Petros Mavroidis, 'Europe's Evolving Regulatory Strategy for GMOs – The Issue of Consistency with WTO Law: Of Kine and Brine', Fordham Int'l L.J., 24 (2000), 317, 348. For example, the panel in the *Hormones* case asked scientists questions relating to the effectiveness of the EC's regulatory policies, as well as the cost implications of different techniques for testing for hormone residues. *Hormones* Panel Report, VI.189ff., VI. 203ff.; one of the experts, Dr Andre, responded that these questions did not relate to matters of scientific expertise.

215 Pauwelyn, 'The Use of Experts in WTO Dispute Settlement', 349, 355.

216 *Hormones II*, [587].

this basis the EC took issue with the panel's approach to the expert evidence that essentially sought to determine whether particular findings were generally accepted by the relevant scientific community.[217]

Due process in the selection and consultation of advising experts

Of particular concern to the EC was the panel's selection and heavy reliance on the views of Drs Boobis and Boisseau, both of whom had participated in a substantial manner in the risk assessment conducted by the scientific committee known as the Joint FAO/WHO Expert Committee on Food Additives (JEFCA), which provides independent scientific expert advice to the Codex Alimentarius Commission to assist with its standard-setting role in the field of food safety.[218] An essential part of the EC's case was to call into question the adequacy of JEFCA's risk assessment process for growth-promoting hormones and the international standards established on that basis. The Appellate Body recognised that it was problematic for the panel to consult experts with close institutional links to JEFCA and direct involvement in the risk assessments performed by that committee for the hormones at issue in the dispute.[219] 'The natural inclination of someone placed in that situation', the Appellate Body opined, 'would be to compare the [JEFCA and EC] risk assessments, rather than to assess whether the science relied upon by the European Communities can support the conclusions it reached, and to favour or defend JECFA's approach.'[220]

According to the Appellate Body, the panel was under an obligation to afford the parties to the dispute 'due process' to ensure that the proceedings were conducted with fairness and impartiality and that one party was not unfairly disadvantaged with respect to the other parties in the dispute.[221] Given that '[s]cientific experts and the manner in which their opinions are solicited and evaluated can have a significant bearing on a panel's consideration of the evidence and its review of a domestic measure, especially in cases … involving highly complex scientific issues', the Appellate Body recognised that appointment and consultation of experts who are not independent or impartial can compromise a panel's ability to act as an independent adjudicator in an SPS case.[222] In light of their

[217] *Ibid.*, [607], [610]. [218] See further, Chapter 6.

[219] *Hormones II*, [479]. [220] *Ibid.*, [469]. [221] *Ibid.*, [433].

[222] *Ibid.*, [436]. The Appellate Body stressed that the obligation to afford due process was not circumscribed to the expert selection stage and does not end with the appointment of experts but continues to apply throughout the panel's questioning and consultations with experts: [473].

close association with JEFCA and its risk assessments related to hormone use, the Appellate Body ruled that it was improper for the panel to have asked Drs Boobis and Boisseau to evaluate the EC's risk assessment, and incompatible with applicable due process obligations.[223] Consequently, the Appellate Body found that it was difficult to sustain the panel's findings on scientific and risk assessment issues in the case, which relied heavily upon the responses of the two experts in question.[224]

To be fair to the *Hormones II* panel, it had put forward a number of plausible reasons for consulting Drs Boobis and Boisseau despite their links to JEFCA. For instance, the panel argued that it was entitled to rely on the experts' objectivity as scientists, their recognition as internationally renowned experts by way of their appointment to JEFCA, the special assistance they could provide to the panel in understanding the work of JEFCA, and the presumption in the SPS Agreement itself that Codex standards meet the Agreement's scientific evidence and risk assessment requirements. Clearly the panel had also faced substantial 'practical difficulties' in selecting advising experts with the requisite level of expertise who would be acceptable to the parties.[225]

Apparently taking what might be described as a more constructivist view of science, the Appellate Body recognised that Drs Boobis and Boisseau's institutional links to the JEFCA epistemic community might well influence their views as to what amounted to an appropriate risk assessment of the hormones at issue in the dispute. It was also quite forthright about the 'decisive role' experts consulted by a panel can play in SPS cases, which invariably involve highly complex scientific questions.[226] These considerations led the Appellate Body to weight the importance of requirements of independence and impartiality in consultations with experts over the undeniable practical issues that panels face in obtaining suitable expert advice in SPS disputes.

Emergence of a procedural standard of review?

In *Hormones II* the EC's concerns with respect to the expert evidence related not just to the panel's decision to consult scientists of questionable independence and impartiality, but also to the way in which the panel relied upon the expert evidence in reaching its findings. The panel's practice in this regard did not differ substantially from what has become the norm in SPS disputes. Reflecting their lack of confidence

[223] *Ibid.*, [469]. [224] *Ibid.*, [484].
[225] A fact acknowledged by the Appellate Body: *ibid.*, [480].
[226] *Ibid.*, [480].

as non-scientists to engage deeply with the scientific evidence,[227] the panel tended simply to survey and summarise the opinions of the experts on a particular issue and then reach a conclusion based upon the view expressed by 'a majority in the spectrum of the scientific experts consulted by the Panel'.[228]

In contrast to previous cases, the Appellate Body in *Hormones II* decided to tackle the appropriateness of the panel's assessment of scientific evidence, using the question of the applicable standard of review as the vehicle for this exercise. Criticising the panel for having 'reviewed the scientific experts' opinions and somewhat peremptorily decid[ing] what it considered to be the best science',[229] the Appellate Body went on to articulate what it saw as the appropriate standard and approach to the review of consistency of a member's SPS measure with Article 5.1 of the SPS Agreement. It ruled that the review power of a panel pursuant to Article 11 of the DSU 'is not to determine whether the risk assessment undertaken by a WTO Member is correct, but rather to determine whether that risk assessment is supported by coherent reasoning and respectable scientific evidence and is, in this sense, objectively justifiable'.[230]

The Appellate Body then went on to spell out, in some detail, the correct methodology for a panel to follow where it is reviewing a member's risk assessment, particularly one that encompasses divergent or minority scientific perspectives on the risks in question. This methodology can be distilled into three main steps, as follows:[231]

1. **Step 1: Identification of the scientific basis of the measure**. The first task of a panel reviewing the consistency of an SPS measure with Article 5.1 of the SPS Agreement is to identify the 'scientific basis' upon which the SPS measure was adopted. This need not reflect the majority view within the scientific community but may instead embrace divergent opinions or minority perspectives.
2. **Step 2: Verifying that the scientific basis of a measure comes from a respected and qualified source**. Having identified the scientific basis of the measure, the panel must then ensure that this material has 'the necessary scientific and methodological rigour to be considered reputable science'. This does not mean that the views concerned have to have been accepted by the broader scientific community so long as they are 'considered to be legitimate science according

[227] *US Hormones II*, Panel Report, [7.553]; *Canada Hormones II*, Panel Report, [7.521].
[228] *Hormones II*, [597]. See also [598], [602].
[229] *Ibid.*, [612]. [230] *Ibid.*, [590]. [231] *Ibid.*, [591].

to the standards of the relevant scientific community'. A panel must further assess whether the reasoning articulated on the basis of the scientific evidence is objective and coherent or, put another way, whether the particular conclusions drawn by the member assessing risk find sufficient support in the scientific evidence relied upon.

3. **Step 3: Determining whether the results of the risk assessment sufficiently warrant the SPS measure**. The final step in the panel's review process is to determine whether the requisite objective relationship exists between the identified scientific basis and the SPS measure adopted by the member. Once again, the scientific basis cited as warranting a particular SPS measure need not reflect the majority expert view, provided it comes from a qualified and respected source.

The experts advising the panel may, and indeed are expected, to play a major role in the panel's review of an SPS measure. However, the role of the experts is commensurate with the limited mandate of the panel. Consultations with the experts thus 'should not seek to test whether the experts would have done a risk assessment in the same way and would have reached the same conclusions as the risk assessor'.[232] Instead the Appellate Body specified that the panel may seek the assistance of experts with the following tasks:[233]

- Identification of the scientific basis of the SPS measure at issue;
- Verification that the scientific basis comes from a qualified and respected source, irrespective of whether it represents minority or majority scientific views;
- Reviewing whether the reasoning articulated on the basis of the scientific evidence is objective and coherent;
- Determining whether the particular conclusions drawn by a member find sufficient support in the scientific evidence; and
- Advising on the relationship between a risk assessment and the SPS measure in order to assist the panel in determining whether the risk assessment sufficiently warrants the particular SPS measure.

The clarification of the applicable SPS standard of review offered by the Appellate Body in *Hormones II* suggests that this standard is procedural in nature. In this sense, panels should not defer entirely to a member's process and conclusions of risk assessment, but nor does the Appellate Body endorse panels (or their advising experts) conducting their own risk assessment.[234] Instead, the Appellate Body's preferred methodology apparently charts the middle ground between the extremes of

[232] *Ibid.*, [592]. [233] *Ibid.* [234] *Ibid.*, [614].

deference and substantive review, calling only for the panel to evaluate the coherence of the reasoning offered by the member and the reputability of the sources of evidence relied upon.[235]

The procedural overtones of the Appellate Body's rulings on the applicable standard of review in *Hormones II* seem to open the door a little wider to recognition of a greater diversity of risk assessment approaches in the SPS context. It is as yet too early to determine whether legal clarification of the requisite standard of review will produce substantially different outcomes in panels' evaluation of scientific evidence in subsequent SPS cases.[236] Indeed, some caution would be warranted in this regard given that the Appellate Body's new review methodology still contains a substantial emphasis on scientific factors (for example, the 'scientific basis' for measures, which must display 'scientific and methodological rigour' sufficient to be considered 'legitimate science' by reference to 'the standards of the relevant scientific community').[237] Although panels are instructed not to disregard scientific evidence that comes from a minority of scientists, they will still be relying heavily on their advising experts to help them identify what evidence should or should not be considered 'legitimate science', an evaluation which

[235] The Appellate Body's approach bears some similarities to the standard for the admission of scientific testimony and expert evidence set by the US Supreme Court in the case of *Daubert* v. *Merrell Dow Pharmaceuticals*, 113 S. Ct. 2786 (1993). In its decision, the Supreme Court overturned the previous *Frye* standard that called for scientific evidence to be admitted only where it was sufficiently established to have general acceptance in the field to which it belonged. The new standard adopted by the Supreme Court relied instead on a flexible set of factors designed to ensure that admitted scientific evidence is both relevant and reliable. Since the Supreme Court's decision, however, the *Daubert* test would not seem to have proved any easier for courts to apply than the previous *Frye* standard: Michael Saks and David Faigman, 'Expert Evidence After Daubert', Annual *Review of Law and Social Science*, 1 (2005), 105; David Faigman *et al.*, 'Check Your Crystal Ball at the Courthouse Door, Please: Exploring the Past, Understanding the Present, and Worrying About the Future of Scientific Evidence', *Cardozo Law Review*, 15 (1994), 1799. Moreover, the Supreme Court's emphasis on the 'scientific validity' of evidence has tended to push reviewing courts to impose stricter 'sound science' requirements on the admission of expert opinion.

[236] *Tasman Apples*, Panel's Second Substantive Meeting with the Parties: Closing Statement of Australia, 2 July 2009, [2]–[3], available at www.dfat.gov.au/trade/negotiations/disputes/. 'Leaked' findings from the panel's interim report in *Tasman Apples* would suggest Australia's arguments for the adoption of a more lenient standard of review were not accepted. See Adam Bennett, 'NZ Apples to Take a Bite Out of Australian Fruit Market', *NZ Herald*, 13 April 2010, available at www.nzherald.co.nz/nz/news/article.cfm?c_id=1&objectid=10638033.

[237] See also WorldTradeLaw.net LLC, 'Appellate Body Reports: Canada/United States – Continued Suspension of Obligations in the EC – Hormones Dispute', (2008), p. 30.

may well gravitate towards peer-reviewed studies or data that have widespread acceptance in the relevant scientific community.

Regardless of whether the Appellate Body's new standard of review produces substantive differences in the evaluation of members' risk assessments in accordance with the requirements of the SPS Agreement, difficult theoretical questions remain surrounding what constitutes 'legitimate science'. To take a pertinent example considered in the *Hormones II* case, are the assumptions (or science policies) used by risk assessors to overcome irremediable gaps in the available scientific data properly considered part of the 'scientific basis' founding an SPS measure? This was an issue for the panel when evaluating the different treatment given by the EC and JEFCA to scientific evidence concerning the genotoxicity of growth-promoting hormones. The available data on this question comes from high-dose studies in short-lived animals such as rats, but for the purposes of setting human health standards this data must be extrapolated to the very low doses found in meat products.

Significant scientific controversy surrounds the question of whether extrapolation should be based on an assumption that substances such as growth-promoting substances exhibit a risk threshold, with exposures below that threshold being deemed safe,[238] or whether the existence of some evidence suggesting the genotoxicity of these substances should lead to the conclusion that no threshold can be determined. Depending upon which assumption is preferred (something that will largely turn on risk management considerations relating to what is an acceptable risk), risk assessors may extrapolate from high dose data in different ways, leading to different assessments about the 'safe' level of exposure.[239]

As one of the advising experts in *Hormones II* pointed out, the question remains 'an area of legitimate scientific disagreement that has gone on for many years' with the differences in the scientific community largely turning on 'a matter of professional scientific judgement' and 'the assumptions that scientists bring to risk assessment'.[240] In this light, the Appellate Body's chastisement of the *Hormones II* panel for

[238] For regulatory authorities that take this approach, an extra level of safety is provided by dividing the 'safe' dose obtained in experimental studies by an arbitrary 'safety factor', e.g. 1,000, to yield the permissible exposure level for the substance.

[239] *US Hormones II*, Panel Report, Annex G, p. 880.

[240] *US Hormones II*, Panel Report, Annex G, p. 953 (Dr Cogliano).

failing to identify the scientific basis for the EC's 'conclusion that a threshold could not be established for oestradiol-17'[241] is exposed as far more than a straightforward matter of application of the correct standard of review in evaluating a member's risk assessment.

Building on the Hormones legacy

Unlike the *Hormones* litigation, which concerned the politically charged and culturally divisive area of food safety,[242] the SPS disputes that followed in the wake of the first *Hormones* case – *Salmon*, *Varietals* and *Apples* – dealt with less high-profile issues of quarantine protection.[243] In *Hormones* the Appellate Body suggested that quarantine risk concerns fall into a different category than those regarding food safety, interpreting the definitions of 'risk assessment' in Annex A of the SPS Agreement to connote different stringencies of risk 'evaluation'. Hence it found that the requirement for food and feed safety risks to be evaluated in light of 'the *potential* for adverse effects on human or animal health' needed demonstration only of a 'possibility' of harm, which was a less demanding standard than the showing of 'likelihood' specified for quarantine risks associated with the introduction of pests and diseases.[244] Quite apart from this (questionable) distinction drawn by the Appellate Body, the cases following the *Hormones* decision (with the notable exceptions of the *GMO* and *Hormones II* disputes) have been considered by many to be 'winner cases', which involved 'relatively clear violations of the SPS Agreement'.[245]

While the quarantine cases litigated under the SPS Agreement have dealt with long-running disputes over stringent requirements imposed by countries against a background of accumulated scientific studies suggesting negligible risk, the science-based SPS disciplines do not

[241] *Hormones II*, [607]. Indeed, the Appellate Body went on to quote a statement by one of the experts indicating that potential for adverse effects is a matter of inference given that animal models are limited, offer poor correspondence with humans and in light of a lack of epidemiological studies of matched populations consuming meat from untreated and hormone-treated cattle: [612].

[242] Marsha Echols, 'Food Safety Regulation in the European Union and the United States: Different Cultures, Different Laws', Colum. J. Eur. L., 4 (1998), 525.

[243] Quarantine risks are also the subject of the *Tasman Apples* dispute: *Australia – Measures Affecting the Importation of Apples*, WTO Doc WT/DS367/1 (2007) (Request for Consultations – New Zealand).

[244] *Hormones*, [184].

[245] Victor, 'The Sanitary and Phytosanitary Agreement of the World Trade Organization', 897–8.

appear to have offered any ready solution to these disputes. In each case, the members whose measures were subject to challenge had carried out some kind of risk evaluation and purported to base their measures on these evaluations. In addition, each stressed important political and economic dimensions of the risks at issue – tied to the members' disease-free status – that provided plausible reasons for favouring uncertain, possible risks over specific scientific evidence of harm in the process of risk assessment. While such exercises might have been motivated by protectionism, it became evident that broad notions of science and risk assessment would not offer workable detection tools in these circumstances. In order to use the science-based disciplines of the SPS Agreement for this purpose, the WTO judicial bodies engaged, perhaps unconsciously, in a reconstruction of these concepts, working with building blocks supplied by the Appellate Body's decision in *Hormones*.

Narrowing the notion of risk assessment

Soon after the issue of the *Hormones* decision, Canada's challenge to Australian quarantine restrictions on the import of uncooked salmon provided the Appellate Body with an opportunity to develop further the notion of risk assessment relevant for SPS purposes. The Australian measures at issue in the *Salmon* case were underpinned by an import risk analysis that was, in many ways, a model risk assessment.[246] Rather than the content of the Australian risk assessment, however, it was the process of developing risk management conclusions that had most invited suspicion because of the substantial differences that existed between initial versions of the analysis and a Final Report produced in 1996.[247]

Whereas initial drafts of the Australian risk assessment recorded 'no evidence' of the spread of fish diseases via the importation of fresh Canadian salmon for human consumption,[248] the Final Report stressed significant 'gaps' in the available scientific evidence, leading

[246] This was described as including 'identification of potential disease agents, analysis of disease risks, scientific review of data for salmonid diseases, socio-economic assessment of the potential impact of salmonid disease introduction and identification of options for risk reduction and risk management': *Australia – Measures Affecting Importation of Salmon*, Report of the Panel, WT/DS18/R, 12 June 1998 (*Salmon* Panel Report), [2.27].

[247] Howse, 'Democracy, Science, and Free Trade', 2347–8.

[248] *Salmon* Panel Report, [2.28].

to the conclusion of 'a possibility that up to 20 disease agents exotic to Australia might be present in Pacific salmon products'.[249] In *Salmon*, however, the question before the Appellate Body was not the merits of the Australian risk assessment process, but rather the substantive issue of whether the Final Report amounted to a risk assessment for the purposes of the SPS Agreement.

As has been discussed earlier, the SPS Agreement provides little specific guidance as to the determinants of an appropriate SPS risk assessment, beyond indicating that it must involve an 'evaluation' of risks. In the context of risks associated with 'entry, establishment or spread of a pest or disease within the territory of an importing Member', the required evaluation is of the 'likelihood' of such events 'according to the sanitary or phytosanitary measures which might be applied, and of the associated potential biological and economic consequences'.[250] In its *Salmon* decision, the Appellate Body treated the term 'likelihood' as pivotal to the understanding of what was required for pest/disease risk assessments under the SPS Agreement. Rather than a mere '*possibility* of entry, establishment or spread of diseases and associated biological and economic consequences', it concluded that:

> A proper risk assessment of this type must evaluate the '*likelihood*', i.e., the '*probability*', of entry, establishment or spread of diseases and associated biological and economic consequences as well as the '*likelihood*', i.e., '*probability*', of entry, establishment or spread of diseases according to the SPS measures which might be applied.[251]

In the Australian Final Report there was a mix of statements regarding risk probabilities and possibilities, reflecting the fact that in some cases the Australian authorities had chosen to place significant weight on uncertainties or gaps in the available scientific data.[252] The *Salmon* panel, for its part, had been content to accept 'some' evaluation of risk probabilities as an adequate basis for a risk assessment.[253] However, the Appellate Body, relying on the text of the relevant definition of risk assessment in Annex A of the SPS Agreement, took a more stringent approach. The definition, the Appellate Body noted 'refers to "the evaluation of the likelihood" and not to some evaluation of the likelihood'.[254] Consequently the Appellate Body concluded that the

[249] *Ibid.*, [2.30].
[250] SPS Agreement, Annex A, para. 4, sentence 1.
[251] *Salmon*, [123] (emphasis added).
[252] *Salmon* Panel Report, [8.82]. [253] *Ibid.*, [8.80]. [254] *Salmon*, [124].

Australian Final Report did not contain 'the "evaluation of the likelihood of entry, establishment or spread" of the diseases of concern "and of the associated potential biological and economic consequences" as required by paragraph 4 of Annex A of the SPS Agreement' because 'some evaluation of the likelihood is not enough'.[255]

The Appellate Body's findings in *Salmon* built upon the distinction, initially drawn in *Hormones*, between evaluating potential adverse effects and likely harms to create a demanding standard of probabilistic risk assessment for pest and disease-related risks. While there may be some textual basis for this distinction, the normative rationale behind a more stringent risk assessment standard for quarantine risks is not as apparent and, in any event, the variation is capable of producing some odd results. In general, food safety risks might be thought to be of greater public saliency than quarantine risks (at least in industrialised countries). However, in the case of a specific quarantine risk (say that related to the potential introduction of mad cow disease through imported beef) it is difficult to justify imposing a more stringent standard of risk assessment than would apply in the context of a food safety risk (such as the possible health effects of consuming beef containing hormone residues).

The approach can also result in different stringencies of scientific support being required for a measure depending on whether it is construed as being directed to health or environmental risks. In the *GMO* dispute, discussed further below, this saw the panel applying different tests in judging the various studies (some food safety-related, some ecologically based) put forward in support of the EC measures at issue, notwithstanding expert advice that health risks were less well substantiated by the available scientific evidence.[256]

While the Appellate Body's rulings in *Salmon* have principally had the effect of narrowing the notion of risk assessment relevant for pest and disease-related risks, rejection of the adequacy of 'some evaluation' of harms also has implications for food safety risk assessment. In both cases it seems that risk evaluation must be *complete* in order to pass SPS review. As the Appellate Body highlighted in *Salmon*, this means that where there are gaps in the scientific evidence available for risk assessment they cannot simply be filled by relying on 'unknown and uncertain elements', particularly in light of the fact that 'the elements that define the basic obligation set out in Article 2.2 impart meaning to Article 5.1'.[257] Hence, it seems that reading Articles 2.2 and

[255] *Ibid.*, [128]. [256] *GMO*, [8.5]. [257] *Salmon*, [130].

5.1 together – as was endorsed by the Appellate Body in *Hormones* – may indeed have the consequence of imparting a narrower, more scientific orientation to the risk assessment process by requiring all risk conclusions to be well supported by the available scientific evidence.

Rational relationship between science and risk measures

While the *Hormones* and *Salmon* litigation focused on member's risk assessments and their adequacy for SPS purposes, the case of *Varietals* signalled an important shift in the focus of the SPS jurisprudence. For the first time the Appellate Body conducted an explicit analysis of the scientific evidence requirements of Article 2.2, including the critical question of the level of scientific support necessary to conclude that measures are 'not maintained without sufficient scientific evidence'.

Arguments under Article 2.2 had been vigorously pursued by the complainant in the case – the USA – in respect of Japanese measures that required testing of the efficacy of chemical quarantine treatment for each new variety of an agricultural product before it could be imported into Japan. The US objections to the Japanese measures were based on its conviction that varietal testing requirements were unnecessary to protect against plant disease risks, a conclusion that appeared to be well supported, both by the available scientific evidence and international quarantine practice. Nevertheless, the scientific evidence did reveal some variation – albeit small – between the chemical efficacy of quarantine treatments when applied to different varieties of an agricultural product. The expert advice provided to the panel suggested that these variations were unlikely to be biologically significant, but did not rule out the possibility of risk altogether in light of the lack of knowledge concerning the extent to which chemical sorption levels differ with different varieties.[258] As one advising expert pointed out, this revealed questions over the legitimacy of Japan's phytosanitary measures in the case to be 'more one[s] of risk management as the scientifically definable differences would normally be small and difficult to determine due to variability'.[259]

In approaching the interpretation of Article 2.2 in *Varietals*, the Appellate Body took a highly textual approach, focusing squarely on the ordinary meaning and context of the terms used in the provision.

[258] *Japan – Measures Affecting Agricultural Products*, Report of the Panel, WT/DS76/R, 27 October 1998 (*Varietals* Panel Report), [8.37]–[8.40].

[259] *Ibid.*, [6.06].

Its central findings related to the word 'sufficient' in Article 2.2, which it characterised as a 'relational concept' requiring 'the existence of a sufficient or adequate relationship between two elements, *in casu*, between the SPS measure and the scientific evidence'.[260] Applying a number of its interpretations developed in the first *Hormones* case,[261] the Appellate Body held that not only did the relationship between the SPS measure and the scientific evidence need to be sufficient or adequate, but also 'rational and objective' in order to ensure the measure was 'not maintained without sufficient scientific evidence'.[262]

The Appellate Body found that the requirement for a 'rational' relationship – much like that applicable in the context of the 'based on' test in Article 5.1 – should be 'determined on a case-by-case basis' in light of 'the particular circumstances of the case', which include 'the characteristics of the measure at issue and the quality and quantity of the scientific evidence'.[263] While this suggests some flexibility in determining the degree of coherence required between scientific evidence and an SPS measure, the Appellate Body in *Varietals* frowned upon Japan's argument in favour of a precautionary interpretation of Article 2.2, and stressed the dangers of '[a]n overly broad and flexible interpretation of that obligation' if the provisional measures exception in Article 5.7 was not to be rendered 'meaningless'.[264]

The shift evident in the Appellate Body's decision in *Varietals* toward paying greater attention to the relationship SPS measures bear to scientific evidence, rather than simply to a risk assessment that takes such evidence into account, is a subtle but nonetheless important one.[265] It may be, for example, that the review methodology discussed by the Appellate Body in *Hormones II* is confined to evaluations under Article 5, thereby not extending to a panel's examination of the sufficiency of scientific evidence pursuant to Article 2.2.[266] The significance of a direct enquiry into the scientific basis of SPS measures was highlighted by the

260 *Varietals*, [73]. 261 *Ibid.*, [75]–[77]. 262 *Ibid.*, [84].

263 *Ibid.* 264 *Ibid.*, [80]–[81].

265 Oliver Landwehr, 'Decisions of the Appellate Body of the World Trade Organization: Japan – Measures Affecting Agricultural Products', European J Int'l L., 10 (1999), 803.

266 Worldtradelaw.net, 'Appellate Body Reports: Canada/United States – Continued Suspension of Obligations in the EC – Hormones Disputes', (2008), 30–1. In the *Tasman Apples* dispute Australia has pleaded that the deference to the initial risk assessment required by Article 5.1 should also be shown under Article 2.2 (*First Written Submissions*, [197]–[208]), which New Zealand has resisted (*Second Written Submissions*, [2.82]–[2.83]).

Varietals panel's approach to evaluating the requirement for a rational relationship, which it assumed was an essentially scientific question, governed by expert advice.[267] In contrast to its strictures issued to the *Hormones II* panel, the Appellate Body in *Varietals* did not fault the panel for requiring a close connection between the Japanese measure and the available scientific evidence, even though the panel seemed to be searching for an 'actual causal link between the differences in the test results and the presence of varietal differences'.[268]

While the evaluative approach taken by the *Varietals* panel may not seem overly problematic in a case 'which seemingly involve[d] nothing other than a straight-forward application of the SPS Agreement',[269] it could be seen to set a precedent which favours the scientific analysis of risk questions over other approaches, such as those that place importance on uncertainties or use different value-frames in evaluating information about risk. Moreover, WTO reviewers' narrow focus on the scientific basis of challenged measures in the case did not ultimately prove to be a helpful approach in discerning the legitimacy of the SPS measures at stake. Instead, this facilitated deconstruction of the scientific evidence in a way that suggested the existence of significant expert disagreement at odds with the fairly settled international quarantine practice.[270]

The actual result of scientific review achieved in the *Varietals* case seems to bear out Sheila Jasanoff's warning that although legal adversarial procedures are 'a wonderful instrument for deconstructing "facts"', they are far less effective 'in reconstructing the community held beliefs that reasonably pass for scientific truth'.[271] Insistence on the use of science as a measure of the rationality of risk regulation may thus have the undesirable effect of promoting 'endless deconstruction' of even well-settled scientific findings.[272] Avoiding this eventuality may

[267] *Varietals* Panel Report, [8.35].

[268] *Varietals*, [83].

[269] Jeffrey Dunoff, 'Lotus Eaters: The Varietals Dispute, the SPS Agreement, and WTO Resolution', in George Bermann and Petros Mavroidis (eds.), *Health Regulation in the WTO* (Cambridge University Press, 2006), p. 153.

[270] Indeed, on the basis of the scientific advice received from its advising experts, the panel ended up validating a version of varietal testing as an available alternative measure for the purposes of Article 5.6: *Varietals* Panel Report, [8.74]–[8.75].

[271] Sheila Jasanoff, 'What Judges Should Know About the Sociology of Science', *Judicature*, 77(2) (1993), 80.

[272] Sheila Jasanoff, 'Technologies of Humility: Citizen Participation in Governing Science', *Minerva*, 41(3) (2003), 240.

only be possible where judicial bodies artificially reconstruct science in a more certain image than it supplies in practice.

Intimations of *de minimis* risk

The case of *Apples*, in which Japanese phytosanitary requirements were once again at issue, illustrates how readily an investigation of the evidentiary basis for risk regulation can serve as a vehicle for constructing narrow notions of science and acceptable risk. The arguments of the complainant in the case (the USA) in respect of Japanese measures for fire blight control alleged that there was 'no scientific evidence' that harvested, mature US apples exported to Japan could serve as a pathway for introduction of fire blight.[273] The USA supported this contention by deconstructing the putative pathway for transmission of the disease from apples harvested in American orchards to apples growing in Japan, pointing to the lack of scientific evidence available for a probabilistic evaluation of risk in relation to each and every step in that pathway. (A similar approach was taken by New Zealand in respect of Australian restrictions on apple imports, which are currently before the WTO dispute settlement system in the case of *Tasman Apples*. Like the USA, New Zealand argued there is 'no scientific support' for Australia's contention that mature, symptomless apples are a pathway for transmitting various plant diseases and pests, including fire blight, meaning that the Australian measures are inconsistent with Article 2.2 of the SPS Agreement).[274]

In its rulings the panel in *Apples* essentially adopted the evaluative approach advocated by the complainant, carrying out its own objective step-by-step assessment of the relationship the available scientific evidence bore to Japan's allegations of risk. The panel's conclusion that the overall risk of disease transmission was 'negligible', and hence that Japan's stringent phytosanitary requirements were 'clearly disproportionate',[275] signalled its acceptance of scientific assessments as being determinative of the level of phytosanitary risk. These findings intimated (for the first time in an SPS case) a *de minimis* risk requirement under the SPS Agreement, effectively excluding from the scope of permissible SPS risk regulation risks scientifically assessed as negligible.[276]

273 *Apples* Panel Report, [4.21].

274 New Zealand, First Written Submissions, 41–87. It would appear these arguments have been successful before the panel: see note 236.

275 *Ibid.*, [8.198].

276 Cf. Joost Pauwelyn, 'Does the WTO Stand for "Deference to" or "Interference with" National Health Authorities When Applying the Agreement on Sanitary and

Despite the potentially far-reaching effects of the panel's rulings, however, their factual orientation served largely to screen them from Appellate Body review. In the *Tasman Apples* case it will be interesting to see the nature of the panel's evaluation of the Australian measures and whether this is influenced by the Appellate Body's rulings in *Hormones II* on the applicable standard of review with respect to Article 5.1.

The course of the panel's reasoning in the *Apples* case in reaching its conclusion of negligible risk demonstrates the ways in which the SPS jurisprudence (at least in the quarantine field) has evolved since *Hormones* to embrace a narrow focus on scientific evidence as diagnostic of SPS risk.[277] The majority of the panel's report was taken up with a detailed review of the available scientific studies, including questions of their methodological soundness,[278] and whether or not they qualified as 'scientific evidence' for the purposes of Article 2.2.[279]

Nonetheless, much as was the case in *Varietals*, an analysis of hard scientific data was not capable of answering the question put to the panel of whether it could infer from a lack of such evidence the absence of an SPS risk. As the experts advising the panel observed, this inference was difficult to draw as a scientific matter because the events that might give rise to a risk, if they ever did occur, would do so only rarely and hence were not readily amenable to scientific study.[280] In this context, the advising scientists – following standard scientific methodologies – gave their opinion that the potential for disease transmission via trade in apple fruit amounted to a negligible risk; not able to be ruled out with certainty but very small, somewhere in the vicinity between zero and a one in one million probability of harm.[281] At the same time they expressed their discomfort with the idea that a scientific assessment of negligible risk should lead to the elimination of Japan's phytosanitary controls 'in one step'.[282] Despite this, the panel assumed that it was possible to equate scientific and regulatory notions of SPS risk

Phytosanitary Measures (SPS Agreement)?', in Thomas Cottier and Petros Mavroidis (eds.), *The Role of the Judge in International Trade Regulation: Experience and Lessons for the WTO* (Ann Arbor: University of Michigan Press, 2003), p. 175.

277 This narrow approach was also evident in *Japan – Measures Affecting the Importation of Apples; Recourse to Article 21.5 of the DSU by the United States*, Report of the Panel, WT/DS245/RW, 23 June 2005 (*Apples Recourse* Panel Report), which examined the revised measures implemented by Japanese authorities following the Appellate Body report in the case.

278 E.g., *Apples* Panel Report, [8.127].

279 *Ibid.*, [8.92], [8.93], [8.95].

280 *Ibid.*, Annex 3, [39] (Dr Smith).

281 *Ibid.*, [8.149].

282 *Ibid.*, [8.173].

and, indeed, seemed to believe it would be impossible to come to any finding about risk in the absence of sufficient scientific evidence being submitted.[283] It concluded that the overall risk of fire blight transmission through the import of US apples was negligible and thus could not reasonably support a highly risk-averse measure, such as the stringent Japanese phytosanitary controls.

What emerged from the panel's analysis in *Apples* was an approach to the assessment of SPS risk that called for a scientific demonstration of risk, preferably based upon an evaluation of evidence directly 'probative' of that 'fact'.[284] The panel's approach evidenced a shift away from evaluating the relationship between (specific) scientific evidence (alone or as part of a risk assessment) and a particular measure, to an enquiry directed to whether a risk arises that can be established *on the basis of the available scientific evidence*. For risks which cannot be scientifically substantiated, it seems these will be considered *de minimis* for SPS purposes and hence as failing to provide a rational or objective basis for members' measures.[285] In turn, this implies that there are limits on the appropriate risk levels members can choose in respect of (quarantine-related) SPS risks, and that these limits can be determined by reference to science. Yet, as the advice provided by the panel's experts emphasised, the inference of 'no reasonable risk' from the available scientific evidence was one requiring policy as well as scientific, judgments, effectively 'a weighing of the costs of precaution against the costs of risk-taking'.[286]

Although the panel's treatment of the scientific evidence in the *Apples* case was subject to an appeal by Japan, the Appellate Body had little to say on the evaluative 'methodology' employed by the panel, beyond noting that it did 'not exhaust the range of methodologies available to determine whether a measure is maintained "without sufficient scientific evidence" within the meaning of Article 2.2'.[287] Left unresolved was the question of how far future panels should be permitted to go in demanding a strong scientific basis for claims of SPS risk in order to found strict risk management measures. Arguably, the Appellate Body's body of jurisprudence up to and including *Apples* is one that contains definite signposts ushering panels in the direction

[283] *Ibid.*, [8.175].

[284] *Ibid.*, [8.98]. See also *Apples Recourse* Panel Report, [8.45].

[285] Button, *The Power to Protect*, p. 48.

[286] Walker, 'The Myth of Science as a "Neutral Arbiter" for Triggering Precautions', 201.

[287] *Apples*, [164].

of a narrowly science-based evaluation of SPS risk. These findings build progressively, from the *Hormones* emphasis upon the logical attractiveness of an Article 2.2 analysis, to *Salmon*'s restrictive reading of risk assessment in light of the scientific requirements of Article 2.2, to the *Varietals*' endorsement of a rational relationship test under Article 2.2, and the Appellate Body's reluctance in *Apples* to weigh in on the crucial question of how panels should evaluate and apply scientific evidence under the Agreement. Even if the Appellate Body in a later case – say *Tasman Apples* – develops the standard of review for Article 2.2 along the lines of its rulings in *Hormones II*, a strong emphasis on evaluating the 'scientific basis' of members' SPS measures is likely to remain.

Precautionary risk regulation: the interpretation of Article 5.7

The increasing stringency of the required scientific basis in order for national SPS measures to satisfy Articles 2.2 and 5.1 has given added saliency to the question of the scope for precautionary regulation under the SPS Agreement. One possibility for such regulation is the adoption of provisional SPS measures by members, in accordance with Article 5.7, '[i]n cases where relevant scientific evidence is insufficient'. In *Hormones* the Appellate Body stated that the precautionary principle 'finds reflection' in Article 5.7 and may also have relevance for the SPS Agreement beyond the provisions of that article, for instance where panels are assessing the sufficiency of the scientific evidence underlying members' permanent SPS measures regarding irreversible risks.[288] As discussed further in the final section of the chapter, this hints that, at least in the face of some risks, WTO reviewers may be prepared to relax requirements for the scientific justification of measures in order to recognise the risk priorities favoured by democratic societies. Some commentators have argued that a similar result can be achieved by interpreting the term 'insufficient' in Article 5.7 in light of social science perspectives on risk assessment so as to encompass situations where relevant scientific evidence does not address the 'risks that a particular community actually cares about'.[289]

Up until the *Apples* decision, most appellate-level analysis of the scope for precautionary SPS regulation focused on the requirements for maintaining provisional measures once taken, rather than the

[288] *Hormones*, [124]; *Hormones II*, [680].
[289] Winickoff *et al.*, 'Adjudicating the GM Food Wars', 113.

pre-requisites for adoption of a provisional measure in the first place.[290] In *Varietals*, the Appellate Body noted that Article 5.7 sets out four requirements, that:

1. a provisional measure is imposed in respect of a situation where 'relevant scientific information is insufficient';
2. the measure is adopted 'on the basis of available pertinent information';
3. the member that adopted the measure 'seek[s] to obtain the additional information necessary for a more objective assessment of risk'; and
4. 'review[s] the ... measure accordingly within a reasonable period of time'.

It stated that these requirements are 'clearly cumulative in nature', but only issued further clarifications in respect of the latter two conditions.[291] For example, the Appellate Body noted that the obligation to collect 'additional information' on risk requires that the information sought must be 'germane' to conducting 'a more objective assessment of risk', and that in circumstances where 'collecting the necessary additional information would be relatively easy', a delay of even a few years in reviewing the provisional measures could be considered unreasonable.[292] The Appellate Body's only comment in *Varietals* that went to the substantive scope of the Article 5.7 was that the provision 'operates as a *qualified* exemption from the obligation under Article 2.2'.[293] This suggests that an overly broad and flexible interpretation of Article 5.7 will be discouraged by the Appellate Body, although in *Hormones II* the Appellate Body indicated there might be some latitude for a more lenient assessment of compliance with the article's requirements in 'emergency situations'.[294]

As a number of commentators have noted, from a substantive viewpoint the critical concept in determining the operative scope of Article 5.7 is that of the insufficiency of relevant scientific evidence (and its differences from what is encompassed by the notion of 'available

[290] This distinction in Article 5.7 between requirements that must be met before adoption of a provisional measure and conditions for maintaining the measure once it has been taken was affirmed by the Appellate Body in *Hormones II*, [676].

[291] *Varietals*, [89]. [292] *Ibid.*, [92], [93].

[293] *Ibid.*, [80]. Cf the panel's ruling in the *GMO* case, [7.2969], which instead construed Article 5.7 as a 'right' of WTO members.

[294] *Hormones II*, [680].

pertinent information').[295] Arguably, the concept of insufficient scientific evidence is one that should encompass situations of pervasive scientific uncertainty, especially since the Appellate Body has found that an evaluation of the sufficiency of scientific evidence embraces considerations of both 'the quality and quantity of the scientific evidence'.[296] However, interpretations of the insufficiency concept developed in *Apples*, and even *Hormones II*, suggest a narrower operation, limited to the kinds of uncertainties emphasised in expert risk assessments rather than incorporating broader notions of ignorance and indeterminacy.

Insufficiency judged in relation to the task of risk assessment

It was the *Apples* decision that provided the first important clarification of the concept of insufficient scientific evidence in the context of Article 5.7. The *Apples* panel had interpreted the notion narrowly, seeing Article 5.7 as primarily 'designed to be invoked in situations where little, or no, reliable evidence was available on the subject matter at issue'.[297] The Appellate Body, on the other hand, adopted a seemingly broader approach, judging the concept of insufficiency in light of the wider task of risk assessment under Article 5 of the SPS Agreement. It ruled that '"relevant scientific evidence" will be "insufficient" within the meaning of Article 5.7 if the body of available scientific evidence does not allow, in quantitative or qualitative terms, the performance of an adequate assessment of risks as required under Article 5.1 and as defined in Annex A to the SPS Agreement'.[298] This finding clarifies the relationship between Articles 2.2 and 5.1, on the one hand, and Article 5.7 on the other. Consequently, where relevant scientific evidence is sufficient to perform a risk assessment to the standards specified by the SPS Agreement, a member may take a SPS measure only where the measure is based on a risk assessment (in accordance with Article 5.1) and meets the obligations set out in Article 2.2. If the relevant scientific evidence is insufficient to perform a risk assessment, a member may adopt a provisional SPS measure, subject to meeting the cumulative obligations set out in Article 5.7.[299]

[295] Button, *The Power to Protect*, 74; Charnovitz, 'The Supervision of Health and Biosafety Regulation by World Trade Rules', 291–2.

[296] *Varietals*, [84]. See also the *amicus curiae* brief submitted to the *GMO* panel by a number of US non-governmental organisations arguing that uncertainty is a factor that affects the quality and quantity of scientific evidence concerning a risk, [36] available from www.ciel.org/Publications/ECBiotech_AmicusBrief_2June04.pdf.

[297] *Apples* Panel Report, [8.219]. [298] *Apples*, [179]. [299] See also *Hormones II*, [674].

In *Apples* the Appellate Body emphasised that its test of insufficiency, judged against the task of risk assessment, would not necessarily exclude from the ambit of Article 5.7 'cases where the available evidence is more than minimal in quantity but has not led to reliable or conclusive results'.[300] Nonetheless, it left open the crucial questions of what amounts to an 'adequate' assessment of risks, as well as the circumstances in which relevant evidence will be considered 'reliable' for the purposes of risk evaluation. Critically, the flexibility afforded by this interpretation of insufficiency depends on the preparedness of WTO decision-makers to recognise the context-dependent nature of questions surrounding the adequacy of risk assessment (which will be influenced by risk management considerations, such as the desired level of SPS protection)[301] and the perceived reliability of the relevant scientific evidence (which will be affected by the allowance made for areas of uncertainty).

In this latter respect, other findings of the Appellate Body in *Apples* militated against a broad and flexible application of the insufficiency concept in Article 5.7. Japan's contentions regarding the scope for consideration of scientific uncertainty under Article 5.7 were summarily dismissed by the Appellate Body relying on the text of Article 5.7. It ruled that that the two concepts of insufficiency of relevant scientific evidence and scientific uncertainty 'are not interchangeable'.[302] However, if scientific uncertainty is not a trigger for precautionary action under Article 5.7, then questions of the reliability of the available evidence and the adequacy of risk assessment are likely to be assessed primarily from a scientific perspective, rather than in light of alternative risk framings. Generally it is divergences over the relevance of different sources of uncertainty that provide the means by which public risk framings enter risk disputes, given that such divergences serve to expose the values and regulatory assumptions that often underlie scientific risk assessments.[303] Absent members being able to rely on a broad consideration of scientific uncertainty in their assessment of SPS risks, we might expect that Article 5.7 in practice will permit little beyond the adoption of provisional measures in circumstances where

[300] *Apples*, [185].
[301] As mentioned earlier, this link was acknowledged by the Appellate Body in *Hormones II*, [694].
[302] *Apples*, [184].
[303] Winickoff *et al.*, 'Adjudicating the GM Food Wars', 113.

(a majority of) experts judge there to be insufficiencies that adversely affect the reliability of evidence concerning a particular risk.

Insufficiency as a spectrum

The recent rulings of the Appellate Body in *Hormones II*, overturning restrictive interpretations of the insufficiency concept developed by the panel, might seem on their face to permit a broader scope of operation for Article 5.7.[304] A close reading of the Appellate Body's findings, however, suggests that Article 5.7 still falls short of endorsing risk regulation based on the precautionary principle in the context of the SPS Agreement.

Unlike its earlier directive at issue in the first *Hormones* case, the EC's revised hormones directive sought to rely directly on Article 5.7 in order to sustain a provisional ban on meat treated with any of the five hormones: testosterone, progesterone, trenbolone acetate, zeranol and MGA. The panel found that this provisional ban failed to meet the requirements of Article 5.7 on the basis that the EC had not shown that the relevant scientific evidence regarding the health effects of residues of these hormones in meat was insufficient.

An important factor influencing the panel's assessment was the fact that JEFCA had performed risk assessments for all five hormones and adopted international standards relating to their use as veterinary drugs. Relying on the presumption of consistency with the SPS Agreement in Article 3.2 for international standards adopted by relevant international organisations such as Codex, the panel essentially treated the JEFCA risk assessments as a legal benchmark for assessing the sufficiency or otherwise of the available scientific evidence regarding the five hormones at issue.[305] The inference drawn by the panel was that the scientific evidence could not be 'insufficient' where JEFCA had been able to conduct a risk assessment.[306] In reaching this conclusion, the panel set forth its understanding of what would be needed to displace the scientific evidence upon which the international standards

[304] Andrew T. F. Lang, 'Provisional Measures under Article 5.7 of the WTO's Agreement on Sanitary and Phytosanitary Measures: Some Criticisms of the Jurisprudence So Far', *Journal of World Trade*, 42(6) (2008), 1097 and 1105.

[305] The Appellate Body overruled the panel's interpretation, noting that it was incorrect to use JEFCA's risk assessments as a legal benchmark for assessing insufficiency: *Hormones II*, [708]. In this sense, the evidence underlying JEFCA's risk assessments had probative value for determining whether relevant scientific evidence was insufficient but was not dispositive of the matter: *ibid.*, [697].

[306] *Hormones II*, [708].

relied, rendering that evidence insufficient for Article 5.7 purposes. It ruled:

> there must be a *critical mass* of new evidence and/or information that calls into question the fundamental precepts of previous knowledge and evidence so as to make relevant, previously sufficient, evidence now insufficient. In the present case where risk assessments have been performed and a large body of quality evidence has been accumulated, this would be possible only if it put into question existing relevant evidence *to the point that* this evidence is no longer sufficient to support the conclusions of existing risks assessments.[307]

The Appellate Body's findings on Article 5.7 in *Hormones II* largely focused on the 'critical mass' test developed by the panel, eventually reversing the panel's approach as setting too inflexible and too high a threshold for determination of the issue of insufficiency.[308] By the same token, the Appellate Body did not accept the binary interpretation put forward by the EC whereby SPS measures are either 'based on' a risk assessment or the relevant scientific evidence is insufficient, justifying the adoption of provisional measures under Article 5.7.[309] Instead the Appellate Body used the concept of 'a spectrum' to illuminate the question of the sufficiency or insufficiency of scientific knowledge for SPS risk assessment in the context of a constantly evolving body of scientific knowledge (see Figure 5.1).[310]

At one extreme of this spectrum, according to the Appellate Body, 'lies the incremental advance of science'; '[w]here these scientific advances are at the margins, they would not support the conclusion that previously sufficient evidence has become insufficient'.[311] Given that members are permitted to rely on divergent or minority views from qualified and respected sources in risk assessment, the Appellate Body indicated that mere scientific controversy, or the possibility of conducting further research or of analysing additional information, by themselves, do not render relevant scientific evidence 'insufficient' for the purposes of Article 5.7.[312]

At the other extreme of the spectrum 'lie the more radical [and infrequent] scientific changes that lead to a paradigm shift'.[313] The application of Article 5.7 is not limited to such situations as, for instance,

[307] *US – Hormones II* Panel Report, [7.648]; *Canada – Hormones II* Panel Report, [7.626] (original emphasis; footnote omitted).

[308] See *Hormones II*, [705]–[707], [712], [725] and [731].

[309] *Ibid.*, [681]. [310] *Ibid.*, [703].

[311] *Ibid.* [312] *Ibid.*, [677], [702]. [313] *Ibid.*, [703].

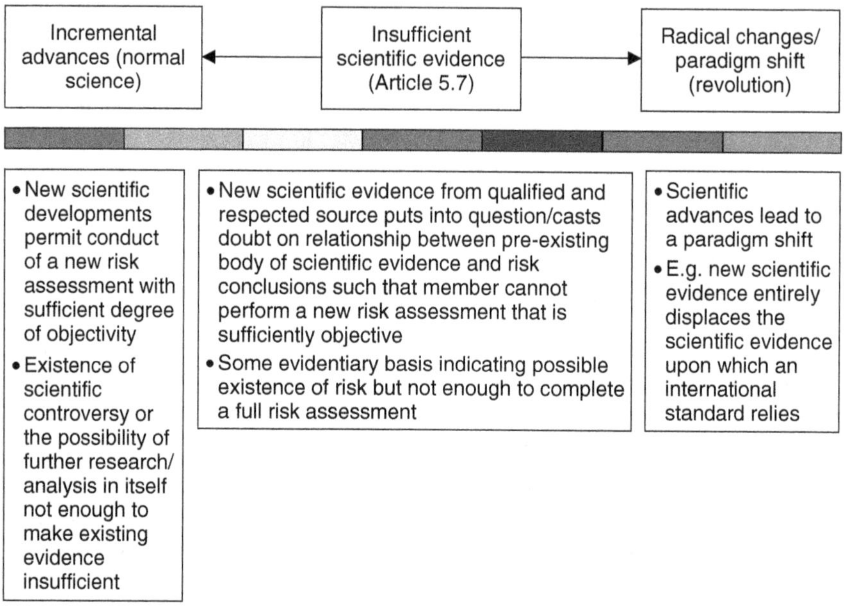

Figure 5.1: Insufficiency spectrum

where new scientific evidence emerges that entirely displaces the scientific theories upon which previous research relies.[314] Rather 'WTO Members should be permitted to take a provisional measure where new evidence from a qualified and respected source puts into question the relationship between the pre-existing body of scientific evidence and the conclusions regarding risks', in other words, 'where new scientific evidence casts doubt as to whether the previously existing body of scientific evidence still permits of a sufficiently objective assessment of risk'.[315]

In its analytical approach, and even in the terminology used, the Appellate Body's 'spectrum' concept adopts a constructivist, Kuhnian understanding of the evolution of knowledge in science. Incremental advances or advances at the margins of an existing research paradigm are what Kuhn described as 'normal science' activity. The other extreme of the spectrum – a paradigm shift – equates to Kuhn's notion of a revolution that leads to the acceptance of a new paradigm in the relevant research field. The Appellate Body positions Article 5.7 insufficiency

[314] *Ibid.*, [725]. [315] *Ibid.*, [703].

in the zone between these two extremes, suggesting provisional measures will be permissible 'where some evidence of a risk exists but not enough to complete a full risk assessment'.[316]

In practice, discerning the difference between 'some evidence' of risk and 'enough' to complete a full and sufficiently objective risk assessment is likely to prove a complex and fraught task.[317] Kuhn's analysis suggests that the transition from normal science to a new paradigm is not a linear or predictable process.[318] Indeed, it may only be possible to identify a paradigm shift once it has occurred and has been acknowledged as such by the scientific community; prior to that time, knowledge generated outside the boundaries of the dominant paradigm is likely to be rejected by the vast majority of scientists. Consequently, for a non-scientifically trained panel, the issue of whether relevant scientific evidence is sufficient or insufficient for the purposes of Article 5.7 will raise difficult matters of judgment, with the probability that the panel will be highly reliant on its advising experts to understand the current nature of scientific research in a particular field. If these experts are chosen to represent a wide range of opinions and disciplines, then the panel will be more likely to become aware of scientific views that challenge the existing consensus or research paradigm. On the other hand, if the experts are chosen on a narrower basis, or if the views of only some of the experts are given an adequate hearing by the panel, this will tend to reinforce the consensus perspective pertaining in the area of 'normal science' within the field.

[316] *Ibid.*, [678].

[317] In this regard, the Appellate Body criticised the panel's failure to explore further the question of the relevance of a study put forward by the EC documenting more sensitive detection methods capable of identifying lower endogenous levels of oestradiol in pre-pubertal children: *Hormones II*, [721]–[730]. The utility of this evidence had been questioned by some of the experts advising the panel given a lack of validation of the new detection methods. Ultimately the Appellate Body did not determine whether this new evidence created conditions of insufficiency for the purposes of Article 5.7 as it ruled that 'numerous flaws' in the panel's analysis, together with the highly contested nature of the facts in the case made it impossible to 'complete the analysis' with respect to consistency of the EC's measures with Article 5.7: *ibid.*, [735].

[318] See also Oren Perez's perceptive comment that scientific propositions invoked in the context of the SPS Agreement are – by the very nature of the scientific endeavour – 'insufficient', meaning that the problem facing the law is not to distinguish between conditions of 'full' and 'insufficient' knowledge but between different levels of insufficiency: Oren Perez, 'Anomalies at the Precautionary Kingdom: Reflections on the GMO Panel's Decision', *World Trade Review*, 6(2) (2007), 275.

Despite the apparent flexibility of its 'spectrum' approach, there are other aspects of the Appellate Body's rulings on Article 5.7 in *Hormones II* that suggest a narrower understanding of the concept of 'insufficient' scientific evidence. For instance, the Appellate Body described the condition of insufficiency of scientific evidence as 'not a perennial state, but rather a transitory one, which lasts only until such time as the imposing Member procures the additional scientific evidence which allows the performance of a more objective assessment of risk'.[319] This view aligns strongly with the positivist/technical perspective on scientific uncertainty, which sees it as something remediable over time through further scientific research. However, as discussed in Chapter 3, this kind of scientific uncertainty is often dwarfed by more pervasive issues of ignorance and indeterminacy that are not readily amendable to resolution through further research or the application of uncertainty management techniques. The Appellate Body's interpretation of Article 5.7 would thus seem to restrict substantially the scope for members to adopt provisional measures to address inadequacies in the available scientific evidence created by issues of ignorance or indeterminacy.

There is also a narrowing perceptible when it comes to the understanding of the kind of 'available pertinent information' that might be relied on by members adopting provisional measures in order to put into question the relationship between the relevant scientific evidence and the conclusions of the majority of experts in relation to a particular risk. There is nothing in the text of the SPS Agreement to indicate that the new information put forward by a member to challenge the adequacy of existing risk assessments has to come from a scientific source. Conceivably, therefore, 'available pertinent information' might take the form of non-scientific material, such as documented experience with the abuse, misuse or maladministration of hormone-based veterinary drugs, or views obtained in public consultation processes that show greater community sensitivity to uncertainties where adverse health effects for pre-pubertal children are at stake. However, in overruling the EC's argument that the relevant scientific evidence should be deemed insufficient where SPS measures cannot be said to be based on a risk assessment, the Appellate Body noted:

> there may be situations where there is no pertinent *scientific* information available indicating a risk such that an SPS measure would be unwarranted even on a provisional basis.[320]

[319] *Hormones II*, [679]. [320] *Ibid.*, [681] (emphasis added).

The insertion of the word 'scientific' into the phrase 'pertinent information' may have been inadvertent on the part of the Appellate Body. However, it opens the door to arguments that Article 5.7 is concerned only with provisional measures adopted on the basis of some *scientific* evidence of risk. If this is the case, it would tend to restrict the scope for precautionary risk regulation in the SPS field, particularly that which seeks in circumstances of scientific uncertainty to rely on non-scientific sources to identify risks of concern.

Panel report in the GMO case

Analysis of the contribution of the SPS jurisprudence to developing international legal notions of science and risk assessment would not be complete without discussion of the 2006 decision of a WTO panel in the *GMO* case (also known as the *Biotech* case).[321] Like the *Hormones* litigation, the *GMO* case concerned a high stakes dispute over risks of significant public saliency that have given rise to divergent risk regulatory approaches at the national and international levels. Consequently, expectations surrounding the outcomes of the *GMO* case were high, with some seeing the dispute settlement exercise as crucial to 'the WTO's very legitimacy as an institution of global governance'.[322] Since the panel's decision, a substantial literature has developed, analysing the findings of the panel report, as well as their broader implications for the field of biotechnology and GMO agriculture.[323]

321 The different naming of the dispute is reflective of deeper differences over the social perception of genetic modification and its products.

322 Winickoff *et al.*, 'Adjudicating the GM Food Wars', 82.

323 See, e.g., Denise Prevost, 'Opening Pandora's Box: The Panel's Findings in the EC-Biotech Products Dispute', Legal Issues of Economic Integration, 34(1) (2007), 67; Simon Lester, 'International Decision: European Communities – Measures Affecting the Approval and Marketing of Biotech Products', A.J.I.L., 101 (2007), 453; Robert Howse and Henrik Horn, 'European Communities – Measures Affecting the Approval and Marketing of Biotech Products', *World Trade Review*, 8(1) (2009), 49; Freya Baetens, 'Safe Until Proven Harmful? Risk Regulation in Situations of Scientific Uncertainty: The GMO Case', *Cambridge Law Journal*, 66(2) (2007), 276; Ilona Cheyne, 'Life after the Biotech Products Dispute', Environmental Law Review, 10 (2008), 52; Caroline E. Foster, 'Prior Approval Systems and the Substance-Procedure Dichotomy Under the WTO Agreement on Sanitary and Phytosanitary Measures', *Journal of World Trade*, 42(6) (2008), 1199; Gregory Shaffer, 'A Structural Theory of WTO Dispute Settlement: Why Institutional Choice Lies at the Centre of the GMO Case', *New York University Journal of International Law and Politics*, 41 (2008), 1; Andrew Thomison, 'A New and Controversial Mandate for the SPS Agreement: The WTO Panel's Interim Report in the EC – Biotech Dispute', *Columbia Journal of*

Despite this, the decision of the parties (the USA, Canada and Argentina as complainants; the EC as the defendant) not to appeal the panel's legal interpretations of the SPS Agreement leaves the *GMO* panel report in a precedential grey zone. Whereas all the other SPS cases litigated to date have been subjected to appellate review, with the resulting Appellate Body findings being extensively cited in subsequent decisions, the *GMO* panel report was only referred to with approval on a few occasions by the panel in *Hormones II* and not at all by the Appellate Body in its appeal decision.[324] In part this might reflect the very particular course taken by the legal arguments in the *GMO* case, which only incidentally touched on the major scientific evidence and risk assessment requirements of the SPS Agreement. However, it is also consistent with the tendency in WTO dispute settlement for the rulings of the superior court of the system – the Appellate Body – to be treated as the most authoritative interpretations of the WTO legal texts.[325]

Accordingly, while the following sections detail the principal issues and legal findings of the panel in the *GMO* case as an important ruling in its own right (and one, moreover, that addressed questions of *environmental* risk under the SPS Agreement for the first time), it remains unclear to what extent the Appellate Body might follow the reasoning of the *GMO* panel in any subsequent case. Already, some of the panel's findings – particularly with respect to the insufficiency of scientific evidence for the purposes of Article 5.7 – seem to stand at odds with new Appellate Body rulings in the *Hormones II* case.

Background to the GMO case

The *GMO* case concerned (another) long-running transatlantic dispute over risk regulation, bringing into conflict the sound science-based American approach to the risk assessment and authorisation of products

Environmental Law, 32 (2007), 287; Noah Zerbe, 'Risking Regulation, Regulating Risk: Lessons from the Transatlantic Biotech Dispute', *Review of Policy Research*, 24(5) (2007), 407; Christiane Conrad, 'PPMs, the EC-Biotech Dispute and Applicability of the SPS Agreement: Are the Panel's Findings Built on Shaky Ground?', *Hebrew University International Law Research Paper*, No. 8–06 (2006), available at http://papers.ssrn.com/sol3/papers.cfm?abstract_id=920742.

324 *US – Hormones II* Panel Report, [7.429], [7.431], [7.433], [7.609], [7.626], [7.633]; *Canada – Hormones II* Panel Report, [7.420], [7.422], [7.424], [7.587], [7.604], [7.611].

325 Michael J. Trebilcock and Robert Howse, *The Regulation of International Trade*, 3rd edn (London: Routledge, 2005), p. 135.

of biotechnology and the precautionary EC regime for GMOs. The regulatory differences between North America and Europe concerning GMOs have been extensively analysed elsewhere, and largely reflect the entrenched positions of supporters versus opponents of the technology.[326] The main points of difference between the two camps relate to uncertainties over the nature and extent of risks associated with GMOs, and the potential socio-economic implications of GMO agriculture.

Regulatory authorities, such as those in the EU and some of its member states that have taken a cautious approach in the evaluation and authorisation of GMOs, have generally pointed to factors such as the lack of long-term studies of the health and ecological risks of GMOs that might allow an assessment of their potential inter-generational effects, the practical difficulties of segregating GMO from non-GMO agriculture in production and distribution chains, and divergent ways of framing the risks of concern depending on the extent to which social, ethical and economic considerations are taken into account. Against this, biotechnology proponents and governments (such as that in the USA) that have adopted a permissive stance to the authorisation of GMO products point to the lack of scientific evidence confirming health or environmental risks from GMOs in countries where their release and use has been authorised for over a decade, the minimal physical differences of GMOs compared with the parent organisms from which they are derived, and the substantial benefits for both the sustainability of agriculture and the security of food production promised by widespread uptake of GMO crops.[327]

326 For a selection of the literature see: Les Levidow and Susan Carr, 'Normalizing Novelty: Regulating Biotechnological Risk at the U.S. EPA', *Risk: Health, Safety and Environment*, 11 (2000), 9; Les Levidow *et al.*, 'European Biotechnology Regulation: Framing the Risk Assessment of a Herbicide-Tolerant Crop', *Science, Technology and Human Values*, 22(4) (1997), 472; Jacqueline Peel, Rebecca Nelson and Lee Godden, 'GMO Trade Wars: The Submissions in the *EC-GMO* Dispute in the WTO', *Melbourne Journal of International Law*, 6(1) (2005), 141; Joanne Scott, 'European Regulation of GMOs: Thinking about Judicial Review in the WTO', in Jane Holder (ed.), *Current Legal Problems 2004* (Oxford University Press, 2005), p. 117; George E. C. York, 'Global Foods, Local Tastes and Biotechnology: the New Legal Architecture of International Agriculture Trade', *Columbia Journal of European Law*, 7 (2001), 423; Sheila Jasanoff, 'Between Risk and Precaution – Reassessing the Future of GM Crops', *Journal of Risk Research*, 3(3) (2000), 277; Rosemary Lyster, 'Sustainability, Regulatory Dilemmas and GMOs: the US and the EU compared', *Asia Pacific Journal of Environmental Law*, 8(3/4) (2004), 111; Howse and Mavroidis, 'Europe's Evolving Regulatory Strategy for GMOs', 317.

327 The various arguments for and against GMOs have been helpfully canvassed in a number of government reports. See, e.g., Australian Senate Community Affairs Committee, *A Cautionary Tale: Fish Don't Lay Tomatoes (A Report on the*

Unsurprisingly, when the *GMO* dispute was initiated by the three complainants in May 2003, a much narrower construction of the issues was favoured, focusing heavily (although not entirely in the case of the latter two members)[328] on questions of the SPS compatibility of the EC's regulatory scheme for GMOs.[329] The target of the complainants' challenge was the way in which the EC carried out decision-making under its GMO regulatory scheme, rather than the scheme itself. Indeed the complainants were at pains to point out that they were not contesting the right of the EC to require an approval process for GMO products and would have been content had the EC simply applied the process on its statute books. The focus of the complainants' concerns was thus significant delays, starting in October 1998 and continuing up to the time the dispute was brought before the panel, in the processing of applications for the approval of new GMO products.[330]

The complainants argued that delays in the European regulatory process evidenced a general, de facto moratorium instituted by the EC authorities to block further approvals of GMOs, including twenty-seven particular products identified by the complainants. The complainants also challenged measures being maintained by several EC member states in accordance with 'safeguard clauses' under relevant regulations of the EC's GMO scheme.[331] In regard to these safeguard measures, the complainants noted that EC-level scientific committees had assessed new scientific information put forward by member states in support of their measures and determined that no changes were warranted to favourable EC risk assessments previously issued for the products concerned. A mainstay of the complainants' arguments, both in respect of the alleged moratorium and the challenges to member states' safeguard measures, was that each entirely lacked a sufficient

Gene Technology Bill 2000) (Canberra: Parliament of Australia, 2000); National Research Council, *Environmental Effects of Transgenic Plants: the Scope and Adequacy of Regulation* (Washington DC: NRC, 2002); Royal Society of Canada, *Elements of Precaution: Recommendations for the Regulation of Food Biotechnology in Canada* (Ottawa: Canadian Government, 2001).

328 Canada and Argentina also presented claims under the GATT and the TBT Agreement; however, in light of its findings under the SPS Agreement the panel did not proceed to consider the validity of these claims.

329 See Chapter 4 for an overview of this scheme.

330 Applications approved after the *GMO* panel convened were found to have ended the EC's moratorium: *GMO*, [7.1303].

331 Safeguard measures were adopted pursuant to Article 16 of Directive 90/220 (which has since been replaced by Article 23 of Directive 2001/18) and Regulation 258/97, Article 12.

scientific basis, rendering them inconsistent with the science-based disciplines of the SPS Agreement, as well as its procedural requirements for members to maintain transparent assessment processes and proceed to decisions without undue delay.

On its face, the panel ruling in the *GMO* case would seem to have fallen significantly short of the complainants' hopes of WTO-endorsement of their claims that the EC maintained a ban on GMO products without scientific basis. The panel upheld the complainants' allegations of a general de facto moratorium affecting GMO products and causing delays in the processing of specific product applications,[332] but declined to make any finding as to the consistency of the moratorium, or its product-specific manifestations, with the provisions of Articles 2.2 and 5.1 of the SPS Agreement. The panel achieved this result by distinguishing between the EC's overall pre-marketing approval scheme – which it found was an SPS measure – and the implementation of that scheme – which it held to be simply a 'procedural' decision 'relating to the application, or operation, of the existing EC approval procedures'.[333]

Based on this reasoning, member states' safeguard measures were assessable under Articles 2.2 and 5.1 (as would have been the EC GMO scheme itself had it been challenged by the complainants).[334] On the other hand, the moratorium, as something less than an SPS measure, was not evaluated against these provisions. Rather the panel focused its attention on the previously unexplored requirements of Annex C(1)(a) of the SPS Agreement that speak of members ensuring 'with respect to any procedure to check and ensure the fulfilment of sanitary or phytosanitary measures that … such procedures are undertaken and completed without undue delay'. The panel found that the EC's moratorium maintained between June 1999 and August 2003 had resulted in undue delay generally in the approval process and also in twenty-four of the twenty-seven product cases cited by the complainants.[335]

While the panel's concern with aspects of the EC's regulatory *process* represented a departure from the substantive scientific focus characteristic of other SPS case law, there were nonetheless a number of its rulings that indicated substantial congruence with the overall course of that jurisprudence. Indeed, the final part of the panel's report, which concerned its evaluation of member states' safeguard measures,

[332] *GMO*, [7.1272]. [333] *Ibid.*, [7.1378].
[334] *Ibid.*, [8.4]. [335] *Ibid.*, [8.6].

was notable for containing some of the narrowest applications to date of the notion of SPS risk assessment.[336]

In addition, the panel's finding that a broad range of concerns relating to GMOs are covered by the provisions of the SPS Agreement could have far-reaching implications if it remains undisturbed in future cases, as it broadens the scope of the Agreement beyond food safety and quarantine risks to a wide range of environmental and biodiversity-related harms.[337] A striking feature of the panel's analysis in this part of its report was its emphasis on decontextualised, dictionary-based evaluations of terms and definitions in the SPS Agreement, disregarding both the extensive scientific evidence assembled in the case and the heated socio-political context of the decision.[338]

Scope of measures covered by the SPS Agreement

In most SPS disputes the question of whether a measure falls within the scope of the SPS Agreement has been relatively uncontroversial. Disputes such as *Salmon*, *Varietals* and *Apples* have all involved quarantine measures primarily directed to preventing the introduction of pests and diseases of agricultural significance. While the measures challenged in the *Hormones* litigation concerned potential food safety risks posed by hormone residues in meat, this situation was most probably in the minds of negotiators when they added a specification to the relevant definition of SPS measures in the Agreement, including 'veterinary drug residues and extraneous matter' within the scope of food 'contaminants'.[339]

By contrast, it is highly unlikely that those drafting the text of the SPS Agreement in the early 1990s had any inkling that it might provide a forum for the analysis of GMO risk concerns a decade later. Arguably the breadth of potential harms discussed in the scientific and social scientific literature regarding GMOs is inadequately captured

[336] Following leaking of the panel's interim, normally confidential, rulings in the dispute, the panel sought to ameliorate or at least clarify aspects of its reasoning that had attracted significant criticism from civil society groups in a letter to parties. This letter is reproduced in Annex K of the panel report.

[337] See Peel, 'A GMO by Any Other Name'.

[338] As Howse and Horn, 'Measures Affecting the Approval and Marketing of Biotech Products', 53 note, this encompasses not just regulatory differences between Europe and America but also 'a broader debate about the desirability, costs, and benefits more precisely, of GMOs as a strategy for development and food security in developing countries'.

[339] SPS Agreement, Annex A, fn. 4.

by the four categories of SPS measures defined in the SPS Agreement. These are spelt out in Annex A, paragraph 1 of the Agreement and are directed to (a) 'risks arising from the entry, establishment or spread of pests, diseases, disease-carrying organisms or disease-causing organisms'; (b) 'risks arising from additives, contaminants, toxins or disease-causing organisms in foods, beverages or feedstuffs'; (c) 'risks arising from diseases carried by animals, plants or products thereof, or from the entry, establishment or spread of pests'; or (d) 'other damage within the territory of the Member from the entry, establishment or spread of pests'.[340]

In view of the narrow notions of (in)sufficient scientific evidence and appropriate risk assessment developed in the SPS jurisprudence to date, the question of the possible scope of the SPS Agreement is an important one. If the Agreement is interpreted so as to extend to a wide range of health and environmental risk measures with trade effects, then the sphere of influence of the SPS disciplines is also broadened, with implications for other fora of international risk governance. In its assessment of the scope of the SPS Agreement, however, the panel in the *GMO* case seemed entirely untroubled by the ramifications of its rulings for SPS coverage of environmental risks, and the way that this might affect relationships between international trade and environmental rules more generally. Instead the panel adopted a textual style of analysis, taking the definition of SPS measures in Annex A of the SPS Agreement as a starting point and amplifying that by reference to the *Oxford English Dictionary*. Looking largely to the ordinary meaning of the terms used in Annex A, the panel developed far-reaching interpretations of the nature of SPS risks that brought within the ambit of the SPS Agreement a wide range of environmental, health, agricultural and economic flow-on effects of GMO use and food production.

A central plank of the panel's analysis was that the frequently used phrase 'animal or plant life or health' in the SPS Agreement was 'meant to be comprehensive in coverage'.[341] Therefore, it found that risks to animal and plant life and health encompassed concerns relating to the effects of GMO crops on micro-flora and micro-fauna (such as soil organisms), as well as non-target organisms such as insects affected by the cultivation of an insecticide-producing GMO crop (for example, if they consume the pollen of such plants).[342] The panel adopted a similarly

[340] *Ibid.*, Annex A(1), paras. (a) to (d).
[341] *GMO*, [7.219]. [342] *Ibid.*

broad interpretation of the phrase 'risks arising from' – the terminology used in three of the four Annex A definitions of SPS measures. In the panel's view this phrase was 'broad and unqualified',[343] allowing its application to both actual and *potential* risks, as well as those risks 'that arise *indirectly* or *in the longer term* from pests, diseases, disease-carrying organisms or disease-causing organisms'.[344]

These findings suggested that the SPS Agreement is not confined simply to risk situations for which there are 'direct and immediate' links between a product and potential harms to human, animal or plant life or health associated with pests and diseases.[345] Instead, provided a plausible chain of causation can be demonstrated or hypothesised to connect a product with a given health or environmental risk, which is in some way connected to a pest or disease introduction, on the panel's analysis a trade-restrictive measure directed to mitigating that risk is an SPS measure. In the context of GMOs, this meant that both concerns related to their potential for direct adverse effects as 'pests' (for instance, the scenario where GMO crops escape and establish in other areas) *and* their possible 'pest effects' (for example, through out-crossing with other plants) were matters appropriately treated as SPS risks.

In the SPS Agreement the term 'pest' is undefined (other than a qualifying footnote that states that 'pests' include 'weeds');[346] however, the panel again developed a very broad interpretation of this term, relying on its ordinary meaning and context. According to the panel, the term 'pest' as used in the SPS Agreement connoted 'an animal or plant which is destructive, or causes harm to the health of other animals, plants or humans, or other harm, or a troublesome or annoying animal or plant'.[347] This interpretation not only departed from narrower definitions of pests under relevant international standards, such as those of the IPPC,[348] but also extended to a broad range of 'other harms'. According to the panel, a GMO was a pest if by growing where it is not wanted it 'may necessitate control or eradication efforts by a farmer (e.g., in the case of weeds) or diminish the economic value of the crop the farmer is seeking to grow (e.g., because his/her market is non-GMO with low or little tolerance for impurities)'.[349] At a later point

[343] *Ibid.*, [7.225]. [344] *Ibid.*, [7.226]. [345] *Ibid.*

[346] SPS Agreement, Annex A, fn. 4. [347] *GMO*, [7.240].

[348] The relevant definition used by the IPPC defined the term 'pest' as '[a]ny species, strain or biotype of plant, animal or pathogenic agent *injurious to plants or plant products*'.

[349] *GMO*, [7.244].

in its report, the panel suggested that the phrase 'other damage' used in the final category of the Annex A definition of SPS measures might have an even broader scope, extending to damage to property or infrastructure (such as water intake systems), economic damage (through lost sales), damage to non-biological components of the environment (such as soil nutrient cycles), or adverse effects on the dynamics of species in the broader, receiving environment.[350] By coupling this understanding of the term pest, with its earlier ruling as to the inclusion of indirect effects of GMOs within the scope of the SPS Agreement, the panel classified a wide range of potential environmental effects of GMOs as SPS matters.

In relation to health risks, the panel's approach was similarly broad and also driven by a close analysis of the text of relevant definitions of terms in the SPS Agreement. For example, relying on dictionary definitions of words such as 'additives', the panel concluded that genes could be considered 'substances added in the manufacture of the food plant'.[351] This interpretation represents an artificial understanding of the role of introduced genes such as antibiotic resistance marker genes in GMOs,[352] and again departed from relevant international practice (for instance, the term 'additives' used in Codex standards is restricted to substances added during food production processes).[353] Nonetheless, the panel persisted with a literal approach in its interpretation of other terms in Annex A(1)(b), such as the word 'food'. This led the panel to conclude that possible risks associated with consumption of the pollen or seeds of GMO plants by insects and wild fauna were properly regarded as food safety risks, despite the overtly environmental nature of these concerns.[354]

The outcome of the panel's analysis of the scope of the SPS Agreement in the *GMO* case was that the entire EC legislative scheme relating to the environmental release of GMO crops, as well as a substantial portion of its regulations dealing with novel food authorisations, were found to be SPS measures. The panel reached this conclusion notwithstanding the multiplicity and breadth of the risk concerns underlying the EC's

[350] *Ibid.*, [7.370]. See also [7.299].

[351] *Ibid.*

[352] Such genes are not so much 'added' as integrated into the genetic material of the GM plant. Moreover, it is not the gene itself, but rather the protein produced if the gene is expressed, that is the substance that may be linked to adverse health effects for consumers.

[353] *GMO*, [7.299]. [354] *Ibid.*, [7.292].

regulatory system for GMOs.[355] These interpretations by the panel – if they withstand scrutiny in future cases – effect a seismic shift in respect of the potential scope of operation of the SPS Agreement. As regards national regulatory frameworks for GMOs, even the broadest among them (such as the EC scheme) are, on the panel's approach, subject to the disciplines of the SPS Agreement. Thus although the *GMO* panel did not rule upon the 'safety' of GMOs or their 'likeness' to non-GMO products,[356] its interpretation of the SPS Agreement suggests that in any subsequent challenge to GMO regulation, SPS disciplines will be of primary relevance. Accordingly, matters of GMO safety would be assessed according to the scientific basis for any risk concerns and would stand or fall on the risk assessments produced in support of the potential for detrimental health or environmental impacts.

Beyond the GMO context, the panel's interpretations of the SPS Annex A definitions open up the potential for other domestic environmental laws concerned generally with issues of biodiversity to become subject to SPS scrutiny and challenge.[357] Conceivably this could mean that aspects of environmental regulation aiming to safeguard biodiversity and natural ecosystems from the indirect or long-term effects of pest introductions become SPS matters requiring a scientific justification.[358] If this were to occur, it would take SPS disputes into entirely new territory and expose a much greater range of domestic health and environmental regulations to potential SPS oversight, and with it the institutional rigours of the WTO regime.[359]

[355] Scott, *The WTO Agreement on Sanitary and Phytosanitary Measures*, p. 16. The only risk concern found to be potentially outside the SPS Agreement was one referenced by the novel food regulation directing labelling to prevent misleading consumers. Hence, while the EC procedures for approval of GMOs were held to be directed to risks of the type covered by the SPS Agreement, the panel found procedures for the approval of foods and food ingredients set out in Regulation 258/97 were 'in part' SPS measures: *GMO.*, [8.4].

[356] *GMO.*, [8.3].

[357] It is likely that the trade effects of the measures would need to be substantial in order to produce WTO challenges in the form of SPS disputes. Nonetheless, even the potential for SPS coverage may exert a chilling effect on domestic regulatory processes in the environmental sphere as regulators keep one eye to issues of international legal compatibility.

[358] See further Conrad, 'PPMs, the EC-Biotech Dispute and Applicability of the SPS Agreement'.

[359] See Scott, *The WTO Agreement on Sanitary and Phytosanitary Measures*, p. 17 commenting that an expansive reading of the scope of the SPS Agreement 'implies SPS imperialism of a kind which is by no means neutral from the point of view of regulating Member States'.

Evaluation of the scientific basis of safeguard measures

Although the panel determined that potential GMO risks examined under the EC legislation 'are the types of risks covered by the SPS Agreement',[360] its findings were not of critical importance for the EC regulatory scheme as such, given the complainants' decision not to challenge the scheme's WTO-consistency.[361] However, this was not the case for EC member states' safeguard measures, which the panel also found were addressed to various GMO risks that came within the ambit of the SPS Agreement.[362] The EC sought to argue that the safeguard measures – having been adopted on a provisional basis – should be analysed solely under Article 5.7 of the SPS Agreement. It further contended that it was the complainants, rather than the EC itself, which bore the burden of proving that Article 5.7 was not appropriately invoked by demonstrating that relevant scientific evidence was not insufficient.

The panel, in a seeming departure from previous case law on Article 5.7, agreed with the EC that the provision should be characterised as a 'right', and not as an 'exception', vis-à-vis both Articles 2.2 and 5.1.[363] However, the panel believed that it was inappropriate to begin its analysis with Article 5.7, finding instead that it should first determine if the member states' safeguard measures failed Article 5.1, before examining their consistency with Article 5.7.

At this point in its report, it might have been expected that the panel would have turned to the vast amount of scientific evidence and expert advice that had been compiled over the course of the hearings. Towards the end of its report, the panel observed that this evidence indicated:

> that many of the identified concerns are highly unlikely to occur in practice (e.g., the transfer of antibiotic resistance from marker genes used in the production of some biotech plants to bacteria in the human gut) [whereas] other identified concerns, such as those relating to the development of pesticide-resistance in target insects through exposure to pesticides (including those

[360] *GMO*, [8.4].

[361] Hence the panel did not determine 'whether the European Communities has a right to require the pre-marketing approval of biotech products' nor the question of the WTO-consistency of the EC's regulatory scheme: *ibid.*, [8.3].

[362] *Ibid.*, [8.9].

[363] *Ibid.*, [7.2969]. As the panel pointed out, this ruling would have had important implications for the application of the burden of proof to Article 5.7 claims: [7.2976].

incorporated into biotech plants) have indeed been documented to occur, including with respect to non-biotech crops.[364]

Nevertheless, the panel made no mention of the extensive expert advice it received at any stage in its analysis of the scientific basis of the member states' safeguard measures. This was particularly troubling given that the panel purported to assess the adequacy of various studies as risk assessments for the purposes of the SPS Agreement, and to determine the sufficiency or otherwise of relevant scientific evidence as a basis for provisional restrictions on the marketing of some GMOs. While other SPS decisions have certainly revealed the traps for panels of too heavy a reliance on scientific evidence as a measure of SPS risk, the panel's approach in the *GMO* dispute demonstrated the equal pitfalls of an analysis of risk regulatory measures employing the language of science but without an informed understanding of its inherent uncertainties and complexities in real world risk management contexts.[365]

The panel's initial examination of each of the safeguard measures under Article 5.1 focused upon whether any of the documentation put forward by the member states concerned, or relied upon by the EC, amounted to a risk assessment.[366] In light of the Appellate Body's findings in *Salmon* that the SPS Agreement requires different standards of risk assessment for different types of risks, the panel distinguished between risks related to the action of GMOs as pests or their pest effects, and risks relating to the presence of transgenes from GMOs in food. For the former, a risk assessment evaluating the likelihood or probability of harm was required, whereas for the latter an evaluation of the potential or possibility for adverse effects sufficed.

The panel determined whether these standards had been met by looking at the way scientific findings or conclusions about risk were expressed in the documents cited by the EC. For instance, in relation to one of the studies put forward in support of an Austrian safeguard measure, the panel noted that:

[364] *Ibid.*, [8.5].

[365] See also Alexia Herwig, 'Whither Science in WTO Dispute Settlement?', *Leiden Journal of International Law*, 21 (2008), 845.

[366] Howse and Horn, 'Measures Affecting the Approval and Marketing of Biotech Products', 76–7 have critiqued this approach as introducing a superfluous formalistic procedural constraint on a member's right to regulate, which is at odds with the very nature of scientific inquiry, where knowledge is advanced through a continuous process of scientists questioning and engaging with the previous work of their colleagues.

the document states that 'there are possibilities of direct risks which can be assessed within some limits according to the status of science and technology'. In addition, the study cites two analyses regarding environmental risk assessment of releasing GMOs. A quote from the first analysis indicates that 'the ecological impact of transgenic grasses *may* be pervasive' (emphasis added). The second analysis is said to demonstrate that 'the combination of natural gene pools through synthetic genes is incalculable in principle in predictive risk assessment'.[367]

The quotes chosen by the panel in this example suggested that the study concerned was attempting to highlight uncertainties in evaluating the ecological risks of GMOs and the difficulty of applying conventional risk assessment techniques in such circumstances. However, the panel took these statements to indicate 'a lack of evaluation of likelihood', meaning that the study 'does not meet the definition of a risk assessment as provided in Annex A(4), and therefore does not constitute a risk assessment within the meaning of Annex A(4) and Article 5.1'.[368]

The panel's application of other SPS risk assessment requirements developed by the Appellate Body – for example, the specificity requirement, and the insufficiency of 'some evaluation' of the likelihood of pest or disease risks – was equally mechanical in character. Thus the panel faulted a report put forward in support of a French measure, observing that although it did 'appear to provide some evaluation of the likelihood of entry, establishment or spread of one of the "pests" of concern' it did 'not provide any analysis of the associated potential biological and economic consequences of these hybrids, nor does it purport to evaluate the likelihood of entry, establishment or spread of these hybrids according to the SPS measures which might be applied'.[369] Likewise, a number of studies showing adverse effects from insecticide-producing GMO plants on non-target species were found by the panel not to constitute specific risk assessments in the absence of tests on the same variety of plant as was subject to the safeguard measure, or without evidence of the extent to which similar effects might arise under field conditions.[370]

According to the panel, however, the most significant failing of the various studies and reports cited in support of member states'

[367] *GMO*, [7.3044].

[368] *Ibid.*, [7.3046]. Similar findings were made in relation to other studies and documentation put forward in support of the safeguard measures: [7.3049], [7.3096], [7.3146], [7.3148], [7.3151], [7.3170], [7.3171].

[369] *Ibid.*, [7.3120]. [370] E.g., *ibid.*, [7.3148].

safeguard measures was their failure to evaluate purported risks. This failure was evidenced, in its eyes, by the description of environmental outcomes in terms of possibilities, rather than probabilities;[371] the lack of field-based assessments of adverse effects predicted in laboratory studies;[372] and the assertion of potential harms without a prior detailed assessment of 'relevant data and information'.[373] By contrast, what the panel seemed to be searching for was 'a complete, self-contained, scientific evaluation' of particular GMO risks.[374] This imposes high expectations on the quality of risk assessments performed for this purpose; expectations that may be unattainable in respect of many ecological and health risks, especially those associated with new technologies.

More problematic, though, was the panel's assumption that evidence of the production (or otherwise) of a complete, self-contained, scientific evaluation could be gleaned simply by looking to the language used in the relevant scientific documentation. Words such as 'may' or 'suggests' were treated by the panel as indications of an incomplete risk evaluation, without questioning whether these words have the same connotations for scientists as they do in legal or diplomatic discourse. Moreover, the panel made no enquiry as to the adequacy of current science on GMOs to provide the basis for comprehensive, specifically focused, case-by-case assessments of the likely health and environmental effects of varieties of GMO plants. Instead it simply assumed that suggestions of uncertainties, the lack of field data and the inability to reach definitive conclusions about risk were failings of the studies as risk assessments and not indications of the unsettled state of the underlying science.

Arguably the panel's findings that the various studies and reports cited by the EC in the *GMO* case were not risk assessments might have been simply a reflection of the insufficiency of this body of evidence as a basis for an adequate assessment of risks satisfying the requirements of Article 5.1. As discussed above, the Appellate Body in *Apples* found that Article 5.7 measures may still be adopted in such circumstances, provided the other requirements of the provision are also met. For many, GMOs are the quintessential case where insufficiencies exist in available scientific evidence about health and environmental risks, especially regarding the potential for adverse effects over the long term. The EC sought to argue that the way in which a particular regulatory

[371] E.g., *ibid.*, [7.3041]. [372] E.g., *ibid.*, [7.3098], [7.3099], [7.3148].
[373] E.g., *ibid.*, [7.3167], [7.3205]. [374] *Ibid.*, [7,3188].

authority reacts to such uncertainties is a function of its regulatory goals. Thus, if an authority is required to achieve a very high standard of protection or must satisfy the expectations of a society that is exceptionally sensitive regarding risks for which little information is to hand, it is likely to view relevant scientific evidence as insufficient for the purposes of a comprehensive risk assessment.[375] On other hand, in a society enthusiastic about the technology involved or unconcerned by risks for which data is currently poor, regulators might well take a different view of the sufficiency of the available evidence.[376]

The panel, however, rejected such a 'relational' view of the concept of insufficient scientific evidence.[377] Instead it found the relevant scientific evidence to be sufficient for risk assessment purposes given the fact that the EC-level scientific committees had been prepared to issue, and later reaffirm, favourable risk assessments of the products subject to Member State safeguard measures.[378] As the panel highlighted, the existence of favourable risk assessments issued by EC scientific committees invited questions as to 'how and why [member states] assessed the risks differently', particularly in cases where alternative assessments turned on 'possible uncertainties or constraints in the risk assessments in question'.[379] Nonetheless, in determining the sufficiency or otherwise of the scientific evidence for the purpose of SPS risk assessment, the panel undertook no examination of these issues (an approach that would now seem to be at odds with the Appellate Body's rulings in *Hormones II* on the evaluation of 'insufficient' scientific evidence).[380] Rather, the conclusion of a risk assessment by an EC scientific committee indicated, in the panel's view, the sufficiency of the underlying scientific evidence. The panel's approach made no allowance for the continually evolving state of scientific knowledge,[381] and seemed to assume that *any* risk assessment would be adequate for *all* purposes,

375 *Ibid.*, [7.3226]. 376 *Ibid.*, [7.3227]. 377 *Ibid.*, [7.3234].

378 *Ibid.*, [8.9]. As Howse and Horn, 'Measures Affecting the Approval and Marketing of Biotech Products', 81 note, the panel's approach suggested that to invoke Article 5.7 'a Member would have to show that no risk assessment has been attempted; once there is a risk assessment, no matter to what extent that very assessment exercise reveals the *limits* of the evidence as a basis for deciding *regulatory choices*, there is no right to proceed in a precautionary manner under article 5.7'.

379 *GMO*, [7.3085]. 380 See particularly *Hormones II*, [703].

381 Lang, 'Provisional Measures under Article 5.7 of the WTO's Agreement on Sanitary and Phytosanitary Measures', 1095. See also Mary E. Footer, 'Post-normal science in the multilateral trading system: social science expertise and the EC-Biotech panel', *World Trade Review*, 6(2) (2007), 294.

regardless of whether Member State authorities framed the risks of concern in the same way as the EC scientists, and notwithstanding the differing significance that might be attached by each to any uncertainties in the available scientific information.[382]

Underlying the panel's approach seemed to be a notion of risk assessment as an objective exercise that yields universally valid predictions about potential health and environmental harms. The panel distinguished between the role of scientists, whose task it was 'to assess objectively the existence and magnitude of a risk' and WTO members asked to make rational judgments about the need for risk management measures.[383] The panel acknowledged that factors such as limited data, which 'affect scientists' level of confidence in a risk assessment they have carried out', might in principle support a precautionary approach.[384] However, the suggestion was that this would only be the case where the risks and uncertainties to which precautionary measures were directed coincided with those evaluated by scientists in the course of risk assessment.

The turn to process review

While the panel in the *GMO* case made several potentially far-reaching findings regarding the scope of the SPS Agreement and its standards of risk assessment, these matters were ultimately not the focus of the decision. The major portion of the panel's report was concerned with aspects of the EC's regulatory *process*, in particular, the time taken by the EC in administering its regulations for GMO crops and food products. In part this reflected the way the case was argued by the complainants, who concentrated on the EC's de facto moratorium on GMO product approvals. The panel's election to take its reasoning down a similarly procedural path has been seen by some as a means of side-stepping divisive political issues over GMO safety, leaving these to another day and (possibly) another WTO dispute.[385] Alexia Herwig

[382] See also Herwig, 'Whither Science in WTO Dispute Settlement?', 834. In Annex K of its report the panel clarified its rulings, noting that it 'was ultimately not persuaded that in relation to the products subject to the member State safeguard measures challenged in this case, the scientific evidence available at the relevant time did not allow the performance of an assessment of the risk in human societies, or natural environments, as they actually exist, in accordance with the provisions of Article 5.1 and Annex A(4) of the SPS Agreement'.

[383] *GMO*, [7.3243]. [384] *Ibid.*, [7.3244].

[385] Gavin Goh and David Morgan, 'Political Considerations and Pragmatic Outcomes in WTO Dispute Rulings', *University of New South Wales Law Journal*, 30 (2007), 481–2.

speculates that the panel may have also 'realized that examining the risk of GMOs seriously would have brought to light significant uncertainties and unclear standards of rational decision-making'.[386]

The panel's focus on process and the approach that it took in assessing matters of procedural propriety potentially have significance beyond the confines of the *GMO* case. This is because of the emergence of an increasing number of reform proposals (discussed further in Chapter 7), which advocate that where international bodies are charged with determining the compliance of national risk measures with international standards, they should undertake a review of the process by which decisions on such measures are reached, rather than their technical accuracy. In this context, the results of the review undertaken by the *GMO* panel provide much food for thought regarding the utility of process-based approaches in international risk governance.

The basis of the complainants' claims of impropriety with regard to the EC's regulatory process for GMOs was that, having put in place detailed procedures for decision-making on GMO applications, the EC ought to have complied with these, rather than instituting a de facto, unilateral suspension of the approval process.[387] Its vigorous denial of the existence of any moratorium notwithstanding, the EC sought to put forward justifications for the length of time taken in processing particular product applications and the failure to issue a final decision on any product between June 1999 and August 2003. The EC's proffered reasons invoked the complexities of the scientific and regulatory context during the relevant period. Essentially they amounted to a call for the panel to respect the deliberative nature of the EC's regulatory process, which had involved a complex interplay between the international arena, Community institutions, national governments and citizens of the EC. On the regulatory front, for instance, the EC had been engaged in major revisions of its regulatory scheme, resulting in the introduction of a revised directive in 2001 and consideration of the need for new requirements on the traceability and labelling of GMO products. It had also been an active participant in international discussions regarding biotechnology regulation, both in Codex and under the auspices of the Convention on Biological Diversity.

The EC argued that debate over regulatory approaches at the domestic and international levels reflected the ongoing evolution of relevant

386 Herwig, 'Whither Science in WTO Dispute Settlement?', 833.

387 See also Scott, 'European Regulation of GMOs'.

science regarding the potential adverse effects of GMOs, highlighting novel risks and uncertainties.[388] New studies cited by the EC included research concerning the possible ecological impacts of GMO crops on non-target species, as well as the broader implications of GMOs for agricultural practices and organic farming.[389]

In analysing the reasons put forward by the EC to explain the time taken in administering its GMO regulatory process, the panel looked to the provisions of Annex C(1)(a) regarding the avoidance of 'undue delay'. The panel ruled that these provisions imposed an obligation on the EC to ensure that its authorisation and approval processes for GMO products were 'undertaken and completed' without an 'unjustifiable' (that is, undue) 'loss of time'.[390] While the panel appreciated that a determination of whether delay was unjustifiable would necessarily require a case-by-case assessment taking account of the relevant facts and circumstances, it appeared to view this test as one that would be unproblematic to apply in practice.[391] Indeed, from the panel's perspective, this might well have been so as it argued that its interpretation was informed by 'the object and purpose of the SPS Agreement', in particular, the goal of establishing 'a multilateral framework of rules and disciplines to guide the [...] enforcement of sanitary and phytosanitary measures in order to minimize their negative effects on trade'.[392] This implied that in assessing the justifiability of delays in SPS-related approval processes, it was the potential for detrimental trade effects, rather than the possible benefits for health and the environment (or for that matter, democratic legitimacy), which was the panel's primary consideration.

This underlying understanding of the purpose of the Annex C disciplines was also evident in the way the panel approached the scientific reasons put forward by the EC as a justification for delays. According to the EC, its regulatory approach was a 'prudent and precautionary' one, reflecting the 'scientific complexity and uncertainty' surrounding GMOs.[393] In its analysis of the EC's justification for approval delays in light of 'evolving science', the panel agreed that Annex C(1)(a) did not 'preclude the application of a prudent and precautionary approach to identifying, assessing and managing risks to human health and

[388] *GMO*, [7.1514].

[389] See the various studies cited by the EC in respect of member states' safeguard measures and analysed by the panel at *ibid*., [7.2560]ff.

[390] *Ibid*., [7.1495]. [391] *Ibid*., [7.1497].

[392] *Ibid*., [7.1499]. [393] *Ibid*., [7.1520]–[7.1521].

the environment arising from GMOs and GMO-derived products'.[394] Further, it stated that in some cases this approach might justify a member 'in requesting further information or clarification of an applicant in a situation where another Member considers that the information available is sufficient to carry out its assessment and reach a decision on an application'.[395]

Ultimately, however, the panel found that precaution must always be 'subject to reasonable limits, lest the precautionary approach swallow the discipline imposed by Annex C(1)(a), first clause'.[396] In the panel's view, the crucial factor for demarcating the line between 'genuine caution and prudence' and 'a pretext to delay the completion of an approval procedure' was the 'core obligation' under Annex C(1)(a) 'to come to a decision on an application'.[397] According to the panel, this core obligation held even 'in view of evolving science and a body of available scientific information and data that is still limited'; hence such a situation 'in and of itself would not warrant delays in the completion of approval procedures'.[398] This bald assertion served to undercut substantially the panel's support for a 'prudent and precautionary approach' expressed but a few paragraphs earlier in its report.

Delay, whether to consider new scientific information, to undertake further research or to gain a better understanding of different societal framings of risk, is generally considered to be inherent in a precautionary risk assessment process. The broadest versions of the precautionary principle risk regulatory paradigm would also contemplate the possibility of indefinite delays, since some uncertainties might be intractable yet still give rise to public concern, whereas others might only be resolvable after a generation or more of experience with the technology or product involved. However, the panel's vision of the permissible scope of a precautionary approach appeared to be more limited, essentially aligned with the ambit of Article 5.7 of the SPS Agreement. The panel ruled that if relevant scientific evidence is insufficient for SPS risk assessment, then a member's option is to adopt provisionally an SPS measure on the basis of available pertinent information.[399] Consequently, 'evolving science, scientific complexity and uncertainty, and limited available scientific information or data are not, in and of themselves, grounds for delaying substantive approval decisions' as even in a case of insufficient scientific evidence a substantive decision

[394] *Ibid.*, [7.1522]. [395] *Ibid.* [396] *Ibid*, [7.1523].
[397] *Ibid.*, [7.1522]–[7.1523]. [398] *Ibid.*, [7.1524]. [399] *Ibid.*, [7.1525].

(that is, application of provisional measures) is still envisaged by the SPS Agreement.[400]

Remarkably the panel's analysis in this part of its report did not make any reference to the extensive expert advice and submissions the panel received about the strengths and weaknesses of the available scientific data regarding the risks of GMOs. At the very least, a consideration of this information (including that put forward in *amicus* briefs submitted by various non-governmental organisations (NGOs) and social science experts)[401] might have enhanced the panel's understanding of the difficulties of definitively assessing the sufficiency or insufficiency of available scientific studies where information is evolving and the appropriate parameters of risk assessment remain a matter of scientific and social debate. The panel's appreciation of the kinds of uncertainties that affect risk assessment instead reflected a narrow, technical view of uncertainty as something arising in the course of (and containable within) scientific risk evaluation.[402]

Accordingly, the panel regarded uncertainty as a matter to be taken into account in reaching a substantive decision, rather than one that 'inherently affect[s] a Member's ability to reach substantive decisions on an application'.[403] This approach apparently reflected the panel's primary concern with trade impacts, even though it recognised that 'where science evolves and there is limited available scientific evidence, a deferral of substantive decisions might allow for better decisions at a later point in time'.[404] Time-limited approvals, approvals subject to conditions or approvals subject to automatic review were all regulatory options to be preferred, in the panel's view, to delay in reaching substantive decisions on GMO applications.[405] It ruled that Annex C(1)(a) first clause could thus not be construed 'to allow Members to go into a

[400] *Ibid.*, [7.1526].

[401] The panel, however, found no need to have regard to this information: *ibid.*, [7.11]. On the role of *amicus* briefs in the case see further Robyn Eckersley, 'A Green Public Sphere in the WTO? The Amicus Curiae Interventions in the Transaltantic Biotech Dispute', European Journal of International Relations, 13(3) (2007), 329; Caroline E. Foster, 'Social Science Experts and Amicus Curiae Briefs in International Courts and Tribunals: The WTO Biotech Case', *Netherlands International Law Review*, 52(3) (2005), 433.

[402] On 'science-based' versus 'precaution-based' understandings of uncertainty see Andreas Klinke and Ortwinn Renn, 'A New Approach to Risk Evaluation and Management: Risk-Based, Precaution-Based, and Discourse-Based Strategies', *Risk Analysis*, 22(6) (2002) 1071.

[403] *GMO*, [7.1526]. [404] *Ibid.*, [7.1527]. [405] *Ibid.*

sort of holding pattern while they or other entities undertake research with a view to obtaining additional scientific information and data'.[406]

Ultimately, the panel's analysis of the EC's justifications for regulatory delay and its conclusion that this was 'undue' quibbled little with Canada's submission that 'a delay in undertaking and completing an approval procedure must be considered "undue" if the delay is caused by a measure which is not based on scientific evidence'.[407] While the panel did not go so far as to equate procedural with substantive (scientific) propriety, the examples it provided of qualifying situations of justified delay are instructive as to its overall appreciation of what would be necessary in order to depart from a timely implementation of SPS regulatory processes. The panel remarked that a delay not supported by scientific evidence might be acceptable where 'caused by a temporary government shutdown in the wake of a natural disaster or civil unrest' or 'if a Member is confronting an unforeseeable and sharp increase in the number of products submitted for approval' producing 'a short delay in the processing of some or all pending applications, due to the need for that Member to reallocate existing resources, or to obtain additional resources, to deal with the new situation'.[408] The panel also suggested that in situations where 'new scientific evidence comes to light which conflicts with available scientific evidence and which is directly relevant to *all* biotech products subject to a pre-marketing approval requirement … it might, depending on the circumstances, be justifiable to suspend all final approvals pending an appropriate assessment of the new evidence'.[409]

Absent war, famine or extraordinary administrative burdens, it would seem – on the panel's analysis – that justifications for regulatory delay must be grounded in scientific *evidence* (not considerations of uncertainty, or divergent public risk concerns) which is *directly relevant* to *all* biotech products. Reflecting this understanding, the panel's assessment of the approval process undertaken for each of the products cited by the complainants was concerned not with the complexities or uncertainties of the scientific information feeding into the decision-making process and the challenges for adequate risk deliberation that this might have posed, but simply with the time taken by regulatory authorities to administer applications in each case.

[406] *Ibid.*, [7.1527]. [407] *Ibid.*, [7.1500].
[408] *Ibid.* [409] *Ibid.*, [7.1532].

The panel's approach to the assessment of undue delay in the *GMO* case suggests that the scope for 'intrusive' international review of national SPS regulation is not limited to situations where decision-makers 'entangle themselves in evaluations of science'.[410] Viewed through the lens of a trade liberalisation ethos, the panel appeared to regard almost any departure from standard regulatory processes as undue, paying very little heed to whether time was needed to promote adequate deliberation on complex risk questions, or the regulatory challenges posed by multiple levels of governance and an evolving knowledge base. The results of the panel's analysis indicate that limiting international risk governance to a function of reviewing regulatory *process* may not provide as ready a solution to the question of an appropriate role for international law and institutions in risk regulation as some might hope.[411]

Science as an arbiter of SPS risk

More than a decade on from the introduction of the SPS Agreement and its novel requirements for WTO review of the scientific underpinnings of member's risk regulatory measures, there is now a body of SPS jurisprudence that addresses the most important interpretative questions raised by the Agreement's science and risk assessment provisions. The interpretations of these provisions developed through the process of SPS dispute settlement reflect a preference for a science-based evaluation of SPS risks as a constraint on the circumstances in which members can adopt domestic SPS measures. Under the WTO SPS Agreement science has thus emerged as an arbiter both of the nature of the risks that can form the basis for national risk regulatory measures, and the circumstances of evidentiary insufficiency (or uncertainty) that justify precautionary responses in the form of provisional measures.

The fallacy of this approach is that, as Oren Perez stresses, '[s]cience is not interested in making decisions about risk'.[412] Science does not supply answers to the crucial questions that arise in the review of risk regulation, such as the point at which scientific evidence is sufficient (or insufficient) for risk evaluation, the requisite degree of specificity needed to treat a study as evidence of risk, or the extent to which

[410] Guzman, 'Food Fears', 38.
[411] This issue is taken up further in Chapter 7.
[412] Perez, 'Anomalies at the Precautionary Kingdom', 278.

considerations of uncertainty should influence decisions as to the reliability of the available evidence or the adequacy of a risk assessment. In order to produce findings on risk (rather than simply on the state of relevant scientific knowledge), scientific evidence must be filtered through a variety of social and policy processes that inevitably leave an imprint on that evidence of particular value concerns and assumptions about the institutional structures for risk management.[413]

As the SPS jurisprudence demonstrates, where the law demands of science something it cannot deliver, this can only be achieved by deferring to science 'in its legally reconstructed image'.[414] What has come to be regarded as scientific evidence and risk assessment in the SPS context are increasingly particular versions of these two concepts that lean towards technical and sound science perspectives on risk regulation. They place a heavy burden on advising scientists who are asked for ever more accurate and detailed answers that frequently extend beyond the scope of the scientific method.[415] Fidelity to text is often cited as a reason for the particular interpretations of scientific evidence and risk assessment that have been developed in the SPS jurisprudence, but this is belied by the open-ended language used that does not foreclose multiple ways of understanding science and risk assessment. Indeed legal reconstruction of science in a narrower form may potentially expose science itself to critique and opposition, undermining what most acknowledge as its still vital role in the SPS system of providing 'standards of empirical validity' that are 'helpful in practical decision-making'.[416]

In the *Hormones* litigation the Appellate Body has displayed an awareness of the limitations of science as an arbiter of SPS risk, with attempts to craft broader notions of SPS risk assessment (for instance, one that would accommodate the evaluation of divergent opinions and real world risks), a more nuanced concept of '(in)sufficient' scientific evidence and a less intrusive standard of review applicable in the scrutiny of members' risk assessments. In a much-cited statement in the first *Hormones* case it also indicated that the evidentiary requirements for SPS measures might be adjusted depending on the qualitative characteristics of the risk situation at hand, such as in cases 'where

[413] Winickoff *et al.*, 'Adjudicating the GM Food Wars', 97.
[414] Perez, *Ecological Sensitivity and Global Legal Pluralism*, p. 127.
[415] Herwig, 'Whither Science in WTO Dispute Settlement?', 836.
[416] *Ibid.*, 842.

risks of irreversible, e.g. life-terminating, damage to human health are concerned'.[417] Potentially this opens up space for differentiated levels of review of members' SPS measures to be applied depending on the nature of the risks concerned (for example, quarantine risks versus health risks), though difficult questions remain regarding the basis upon which such differentiation might be made.[418] Indeed, the lack of clear pointers as to the appropriate 'normative dimensions' of risk regulation for the purposes of the SPS Agreement would seem to have been an important factor in encouraging WTO dispute settlement bodies to turn to science rather than tackle the explicitly political and value-based elements of national risk regulation.[419]

Conclusion

With the conclusion of the SPS Agreement in the Uruguay trade round, the WTO was brought into 'the regulatory network which shapes the global response to risks', making it 'a potential addressee of risk claims'.[420] It may not have been the intention of negotiators of the SPS Agreement to make the WTO a site for addressing 'regulatory polarization' in the area of health and environmental risk.[421] Indeed, disputes such as the *GMO* case have arguably presented WTO adjudicators with an 'impossible task', placing a significant strain on the dispute settlement system.[422] In the future, WTO members might be well advised to avoid the referral of such cases to dispute settlement (a realisation that the *Hormones* litigants appear now to have reached, albeit very

417 *Hormones*, [124]. See Herwig, 'Whither Science in WTO Dispute Settlement?', 835 and Howse and Mavroidis, 'Europe's Evolving Regulatory Strategy for GMOs', 342.

418 The present author, as well as other commentators, are increasingly seeking to grapple with this issue, e.g. Herwig, 'Whither Science in WTO Dispute Settlement?'; Perez, 'Anomalies at the Precautionary Kingdom'; Jacqueline Peel, 'Risk Regulation under the WTO SPS Agreement: Science as an International Normative Standard?', Jean Monnet Working Paper No. 2/2004, (2004). Chapter 8 returns to the question of how the SPS Agreement might strike a better balance between the normative goals of risk regulation, empirical standards of scientific validity and trade liberalisation objectives.

419 Christian Joerges, 'Law, Science and the Management of Risks to Health at the National, European and International Level – Stories on Baby Dummies, Mad Cows and Hormones in Beef', Colum. J. Eur. L., 7 (2001), pp. 2–3.

420 Perez, *Ecological Sensitivity and Global Legal Pluralism*, pp. 120–1.

421 Footer, 'Post-normal science in the multilateral trading system', 281.

422 Howse and Horn, 'Measures Affecting the Approval and Marketing of Biotech Products', 82.

belatedly). In addition, the consensus-based, 'deliberative' processes of the SPS committee might be more extensively used by members to seek to ameliorate their regulatory differences, even if not overcoming them entirely.

The alternative course of pursuing science-based dispute resolution under the SPS Agreement has not yielded the clear standards for differentiating between protectionist and legitimate risk measures that were originally hoped for. Rather what has emerged from the SPS jurisprudence is a complex array of rulings on questions of sufficient scientific evidence and appropriate risk assessment that place substantial constraints on the regulatory autonomy of WTO members in the SPS field. While the notions of science and risk assessment that have developed under the SPS Agreement reflect interpretations of particular treaty text, they have the potential for broader application in international law, with 'spillover effects' already evident in related spheres of national and global risk regulation.

This raises important questions as to what might be alternative trajectories for international law in developing an appropriate role for science and risk assessment. Certainly, the SPS regime is not the only global sphere in which issues concerning the application of scientific evidence and the conduct of risk evaluation arise. Other important international fora in which these matters have been considered include environmental treaty regimes, such as the Biosafety Protocol and conventions regulating harmful chemicals, international standard-setting bodies such as Codex, and inter-governmental science-policy institutions such as the Inter-governmental Panel on Climate Change. Within the WTO itself, competing avenues for the evaluation of risk measures also exist, such as application of the non-discrimination norms of the GATT. These different approaches to risk regulation – which share a focus on the international level of governance (and all its associated challenges) – are the subject of discussion and evaluation in the next chapter.

6 Case studies of science and risk regulation in international law

Introduction

In recent years the interface between science and international risk governance has become an important topic of policy debate and scholarly analysis. This reflects the growing role of international requirements and global bodies in governing risk decision-making, displacing the once pre-eminent place of national authorities in this field.[1] The lion's share of academic discussion concerning science and international risk regulation has taken place in respect of the SPS Agreement, addressed in the previous chapter. However, there are a number of other areas of international law where the role of science and expertise in risk regulation and ensuring environmental safety poses ongoing, complex issues. Beyond the sphere of the SPS Agreement, the social scientific literature has been far in advance of the legal scholarship in examining these questions. This literature has yielded many important insights, particularly concerning science–policy configurations in which scientific evidence and expertise can play their most effective role.[2]

Despite this, there remains a pressing need for legal analysis in the area. Such analysis can add to wider understanding of the role of science in international risk regulation by providing an evaluation of the contribution of specific laws (such as treaty texts) and legal institutions

[1] Alexander Farrell, Jill Jäger and Stacy VanDeveer, 'Overview: Understanding Design Choices', in Alexander Farrell and Jill Jäger (eds.), *Assessments of Regional and Global Environmental Risks: Designing Processes for the Effective Use of Science in Decisionmaking* (Washington DC: Resources for the Future, 2006) p. 1. See also the discussion in Chapters 2 and 3.

[2] See particularly Farrell *et al.*, 'Overview: Understanding Design Choices'; William C. Clark *et al.*, *Global Environmental Assessments: Information and Influence* (Cambridge, MA: MIT Press, 2006); Steinar Andresen *et al.*, *Science and Politics in International Environmental Regimes: Between Integrity and Involvement* (Manchester University Press, 2000).

(such as the interpretative work of the dispute settlement system of the WTO). Indeed, as the SPS jurisprudence discussed in the previous chapter illustrates, law may well represent a privileged and central venue for the constitution of regimes of knowledge.[3] International law and lawyers can also offer useful expertise and input on questions relating to institutional design.[4] This becomes important when exploring the kinds of institutional features that might shape the use of science and expertise in global risk governance in desirable ways.

This chapter aims to contribute to the emerging body of legal analysis concerning the role of science and expertise in global risk governance by examining several case examples where the interface between international law, scientific expertise and risk assessment has been an important issue. The first case example explores the resolution of health and environmental disputes under the General Agreement on Tariffs and Trade (GATT).[5] Like the SPS Agreement, the GATT is a treaty under the auspices of the WTO and its compulsory dispute settlement system, but one that looks to a general norm of non-discrimination – instead of scientific principles and a requirement of risk assessment – in order to distinguish legitimate from protectionist risk regulatory measures. Rather than scrutinising the scientific adequacy of national risk regulations, prominent GATT cases, such as the *Shrimp/Turtle* and *Asbestos* decisions,[6] suggest alternative tests for reviewing the WTO-compliance of trade-restrictive measures that may permit WTO members more autonomy in reaching decisions about acceptable types and levels of risk.

The second case example turns to examine the use of science in the Codex Alimentarius Commission (Codex), a body whose standard-setting work has come to assume an important role in WTO processes, especially those under the SPS Agreement. In the often politically

[3] Andrew T. F. Lang, 'Legal Regimes and Regimes of Knowledge: Governing Global Services Trade', LSE Law, Society and Economy Working Paper Series, WPS 15-2009, July 2009, available at http://ssrn.com/abstract=1423538.

[4] Andrew T. F. Lang, 'Some Sociological Perspectives on International Institutions and the Trading System', in Colin B. Picker, Isabella D. Burn and Douglas W. Arner (eds.), *International Economic Law: The State and Future of the Discipline* (Portland: Hart Publishing, 2008), pp. 86–7.

[5] General Agreement on Tariffs and Trade, 15 April 1994, 55 UNTS 194, 1867 UNTS 187, in force 1 January 1995 (GATT).

[6] *United States – Importation Prohibition of Certain Shrimp and Shrimp Products*, Report of the WTO Appellate Body, WT/DS58/AB/R, 12 October 1998 (*Shrimp/Turtle*); *European Communities – Measures Affecting Asbestos and Asbestos-Containing Products*, Report of the WTO Appellate Body, WT/DS135/AB/R, 12 March 2001 (*Asbestos*).

charged area of regulating food contaminants and additives, Codex is assisted in its standard-setting role by an independent expert advisory body known as the Joint FAO/WHO Expert Committee on Food Additives (JEFCA). Once relatively obscure, Codex's standard-setting processes – and JEFCA's part in these processes – have assumed much greater significance since 1995 and have taken on a self-consciously 'science-based' orientation.[7] An important factor in this transformation was the WTO's endorsement of Codex standards as the benchmark for national food safety measures to enjoy a presumption of meeting the requirements of the SPS Agreement and the GATT.[8]

With regard to reliance on science and expertise, environmental risk issues present a very different prospect to health and food safety concerns, given the higher levels of uncertainty that generally characterise scientific knowledge about environmental matters.[9] In response, environmental treaty regimes have adopted a range of approaches to the use of science to inform policy and legal development, with three case examples considered here. The first focuses on the arrangements for risk assessment and the use of scientific evidence under a treaty regime that has received almost as much attention in the trade sphere as in the environmental arena, namely the Biosafety Protocol to the Convention on Biological Diversity (Biosafety Protocol).[10] This treaty purports to adopt a precautionary approach to regulation of the transboundary transfer, handling and use of 'living modified organisms' resulting from modern biotechnology.[11] However, ongoing issues surrounding the interrelationship between the Biosafety Protocol and the SPS Agreement (most recently analysed by a WTO panel in the *GMO* case)[12] have heavily shaped the particular manifestation of precautionary risk regulation found in the former treaty.

[7] David E. Winickoff and Douglas M. Bushey, 'Science and Power in Global Food Regulation: The Rise of the Codex Alimentarius', *Science, Technology and Human Values* (2009), OnLineFirst doi: 10.1177/0162243909334242.

[8] SPS Agreement, Article 3.2.

[9] Daniel Haag and Martin Kaupenjohann, 'Parameters, Prediction, Post-normal Science and the Precautionary Principle – a Roadmap for Modelling for Decision-Making', *Ecological Modelling*, 144 (2001), 45.

[10] Cartagena Protocol on Biosafety to the Convention on Biological Diversity, 29 January 2000, Montreal, 2226 UNTS 208, in force 11 September 2003 (Biosafety Protocol).

[11] Biosafety Protocol, Article 1.

[12] *European Communities – Measures Affecting the Approval and Marketing of Biotech Products*, Reports of the Panel, WTO Docs WT/DS291/R, WT/DS292/R, WT/DS293/R, 29 September 2006 (*GMO*), [7.73]–[7.75].

The next two cases studies look at different examples of scientific advisory bodies, which have become a mainstay of all contemporary multilateral environmental regimes. Some such bodies draw on existing epistemic communities, others are treaty-specific groups developing policy-targeted reports, and yet others consciously isolate themselves from policy processes in order to focus on producing high-quality scientific assessments of given environmental issues.[13] Under the Persistent Organic Pollutants Convention (POPs Convention), which is the focus of the fourth case example, its constituent scientific advisory body has assumed a central role as the source of expert risk assessments informing state parties' decisions on the international listing and regulation of POPs chemicals.[14] However, assessments of the POPs initially included under the treaty regime, as well as new chemicals proposed for listing, have carefully melded scientific reports with information that takes account of important policy dimensions of the management of such chemicals.

The final case example discusses perhaps the most sophisticated scientific advisory body that has developed so far to assess and advise on environmental risk issues. This is the Inter-governmental Panel on Climate Change (IPCC), tasked with supplying assessments of climate risk to state participants in the international climate change regime. This case example provides interesting comparisons with previous case studies dealing with health and food safety risks, where critiques have often focused on the discounting of minority scientific opinion and inadequate attention paid to non-scientific factors. By contrast, a major concern in the climate change context has been how to translate an increasing level of scientific *consensus* regarding the severe risks posed by global warming into robust policy measures and stringent legal prescriptions at the international level.

The case examples discussed in the chapter cover a range of different health and environmental risk issues, as well as a variety of different institutional settings (for example dispute settlement, treaty regimes, subsidiary expert bodies under treaties, inter-governmental organisations and hybrid public–private standard-setting bodies). This diversity gives a sense of the many and varied settings in international

[13] Noelle Eckley, 'Drawing Lessons About Science-Policy Institutions: Persistent Organic Pollutants (POPs) under the LRTAP Convention', ENRP Discussion Paper E-99-11, (Kennedy School of Government, Harvard University, 1999), p. 1.

[14] Convention on Persistent Organic Pollutants, 23 May 2001, Stockholm, (2001) 40 ILM 532, in force 17 May 2004 (POPs Convention).

law and legal institutions where issues around the role of science and risk assessment are presently discussed and determined. Importantly, it also helps to illustrate the part that both the nature of the risk situations at issue, and specific institutional constraints and context, play in shaping the use of science and expertise in different international risk regulatory fora. While international law and institutions thus face many challenges in common that may promote opportunities for cross-institutional learning,[15] a commitment to global solutions, in and of itself, should not be mistaken as 'a substitute for politics'.[16]

Review of risk regulation under the GATT

The GATT, and the jurisprudence it has generated in health and environmental trade disputes, may seem an odd place to look for approaches to risk decision-making that offer more flexible global configurations for the use of science and expertise. As we saw in Chapter 4, conventional 'trade insider' perspectives on the GATT hold that it was not designed to accommodate broader, 'non-trade' social concerns and values relating to matters of health and environmental protection. Early attempts by GATT panels to settle trade disputes raising health and environmental issues were often decided in accordance with this 'pro-trade' perspective, attracting strident critiques from environmentalists and broader civil society.[17]

However, more critical, and historically aware, analyses of the GATT and the associated global trade regime stress the fallacy of assuming that the GATT is 'single-minded about trade'.[18] Article XX of the GATT refers explicitly to policy objectives extending beyond a narrow understanding of trade liberalisation, such as the protection of human health and the conservation of natural resources.

[15] Laurence R. Helfer and Anne-Marie Slaughter, 'Toward a Theory of Effective Supranational Adjudication', *Yale Law Journal*, 107 (1997), 273.

[16] David Kennedy, 'A New World Order: Yesterday, Today and Tomorrow', *Transnational Law and Contemporary Problems*, 4 (1994), 375.

[17] See Steve Charnovitz, 'The Environment vs. Trade Rules: Defogging the Debate', Environmental Law, 23 (1993), 475; Daniel Esty, *Greening the GATT: Trade, Environment, and the Future* (Washington DC: Center for International Environmental Law, 1994); and John H. Knox, 'The Judicial Resolution of Conflicts between Trade and the Environment', Harv. Envtl. L. Rev., 28 (2004), 4–13 discussing the 1991 and 1994 *Tuna/Dolphin* rulings and the reaction of environmentalists.

[18] Steve Charnovitz, 'Linking Topics in Treaties', U. Pennsylvania J. Int'l Econ. L., 19 (1998), 344.

Moreover, as Steve Charnovitz has pointed out, '[f]rom its inception, the international trade regime has included goals besides trade liberalization.'[19] Robert Howse's examination of the historical foundations of the trading system shows that 'the postwar trading order addressed itself in its very conception and structure to "trade and ..."'.[20] In this sense, the post-Second World War trade regime constituted by the GATT concerned both 'trade' and 'non-trade' issues 'both because it proceeded from assumptions about how the two should interrelate, and because it saw its task as, in fundamental part, structuring the relationship between "trade and non-trade alternative measures"'.[21]

In seeking to deal with 'trade and' questions, the 'core dilemma' faced by the GATT is how to distinguish between legitimate and acceptable domestic policies on the one hand, and on the other, restrictive trade measures that represent 'cheating' on a government's trade liberalisation commitments in a way apt to undermine confidence in the trading system as a whole.[22] Under the SPS Agreement, as we saw in the previous chapter, science has been entrusted with a significant arbitral role as regards this question. By contrast, in the GATT a different path was taken. A non-discrimination norm was adopted in Article III to distinguish acceptable from unacceptable non-trade domestic policies, coupled with explicit exceptions for policies justified in terms of certain non-protectionist goals, including the above-mentioned exceptions for measures 'necessary to protect human, animal or plant life or health' (Article XX(b)) or 'relating to the conservation of exhaustible natural resources' (Article XX(g)). Adoption of trade-restrictive measures in accordance with the exceptions elaborated in Article XX is subject to the requirements of the introductory paragraph, or *chapeau*, of the article that specifies that such measures should not be:

> applied in a manner which would constitute a means of arbitrary or unjustifiable discrimination between countries where the same conditions prevail, or a disguised restriction on international trade.

[19] *Ibid.*, 332.

[20] Robert Howse, 'From Politics to Technocracy – and Back Again: The Fate of the Multilateral Trading Regime', *American Journal of International Law*, 96 (2002), 95.

[21] Andrew T. F. Lang, 'Reflecting on "Linkage": Cognitive and Institutional Change in the International Trading System', *Modern Law Review*, 70(4) (2007), 546 citing Howse, 'From Politics to Technocracy', 95.

[22] Howse, 'From Politics to Technocracy', 95–6.

This allowance for non-protectionist measures, which do not entail discrimination (at least in an unjustified or arbitrary sense), can be seen as a mechanism consistent both with allowing a wide scope for regulatory diversity and the disciplining of 'cheating', 'while minimizing the need for interference with the substance of domestic regulatory choices'.[23]

Nonetheless, prior to the advent of the WTO, panels in GATT disputes raising health and environmental matters appeared more concerned with safeguarding the coherence of the multilateral trading system, narrowly conceived. For instance, in the *Tuna/Dolphin* disputes, determined by dispute settlement panels in 1991 and 1994, unilateral measures adopted by one party as a means of influencing the health or conservation polices of other countries were seen as jeopardising the GATT as a 'multilateral framework for trade' among parties.[24] Commentators criticised the:

> tendency of panels to assume they understood the general purpose of a provision, and to give sense to it in light of that purpose, without regard to the individual words and phrases, almost always result[ing] in rulings tilted towards one particular value among the competing values at stake, namely that of liberal trade.[25]

However, steps taken by the Appellate Body in central cases, such as *Shrimp/Turtle*, to return to a close examination of the wording of the GATT treaty text and to reinvigorate the role of the Article XX *chapeau* in disciplining measures that amount to cheating on trade commitments are seen by a number of authors as crafting politically successful compromises in cases opposing trade to values of health or environmental protection.[26] Put another way, these cases offer a vision of an alternative means of conducting international judicial review of national decisions about the acceptability of particular risks that

[23] *Ibid.*, 97.

[24] *United States – Restrictions on Imports of Tuna*, GATT BISD, 39th Supp, 155, GATT Doc DS21/R, 3 September 1991 (*Tuna/Dolphin I* Panel Report), [5.27]; *United States – Restrictions on Imports of Tuna* GATT Doc DS29/R, 16 June 1994 (*Tuna/Dolphin II* Panel Report), [5.26]. This trend was continued in some early panel decisions after establishment of the WTO, although all were overturned on appeal to the Appellate Body.

[25] Robert Howse, *The WTO System: Law, Politics and Legitimacy* (London: Cameron May, 2007), p. 229.

[26] See, e.g., Howse, 'From Politics to Technocracy', 109.

does not call for intense scrutiny of the underlying science or expert evaluations of risk.

Shrimp/Turtle *and interpretation of the Article* XX chapeau

The case of *Shrimp/Turtle*, decided by the Appellate Body in 1998, with a reaffirmation of its findings in a compliance hearing in 2001,[27] is often regarded as a 'leitmotiv' for the clever resolution of tensions between the competing interests of trade and protection of the environment.[28] The dispute concerned a US measure that sought to condition access to the American shrimp market for exporting nations by ensuring that their shrimping fleets reduced by-catch of endangered turtle species through the installation of turtle excluder devices (TEDs) on the nets used. The US measure was originally applied by domestic environmental authorities in a restricted geographic area – principally the Western Atlantic and Caribbean – where efforts had been undertaken to transfer TED technology to local shrimpers and to engage the countries of the region in negotiations for a turtle conservation treaty. As a result of US court rulings on the proper interpretation of the authorising US legislation, American authorities were later forced to apply the ban on shrimp caught in a manner that harmed turtles to all nations exporting the product to the USA. This had a significant detrimental impact on shrimping industries in the South East Asian region where TEDs were not in widespread use. As a result, four WTO members in this region – India, Pakistan, Malaysia and Thailand – brought an action before the dispute settlement system alleging that the US measure was inconsistent with the GATT.[29]

While *Shrimp/Turtle* is often presented as a paradigmatic clash of trade and environmental concerns, the record of the panel hearing in the case reveals that the dispute was as much about the appropriate characterisation of the risks posed by human activities to endangered

27 *United States – Import Prohibition of Certain Shrimp and Shrimp Products*, WTO Doc WT/DS58/AB/RW, AB-2001-4, 22 October 2001 (Report of the Appellate Body).

28 Nathalie Bernasconi-Osterwalder and Maria Olivia, *EC-Biotech: Overview and Analysis of the Panel's Interim Report* (Washington DC: Center for International Environmental Law, 2006), p. 49.

29 For a more detailed overview of the facts in the dispute see Robert Howse, 'The Appellate Body Rulings in the Shrimp/Turtle Case: A New Legal Baseline for the Trade and Environment Debate', *Columbia Journal of Environmental Law*, 27 (2002), 491.

species of sea turtles.[30] The complainant countries led evidence of their respective turtle conservation policies that largely focused on protecting turtle eggs in nests and hatchlings in order to increase the number of young turtles making it from nesting areas back to the sea. The USA, however, saw risks to mature turtles of reproductive capacity as posing a more significant threat to the long-term survival of the species. It argued that, for this population, the greatest cause of mortality was drowning after entanglement in shrimping nets. Accordingly, the US policy of requiring TEDs on shrimping nets was designed to decrease adult turtle mortality, so improving the capacity for regeneration of the species.

Before the panel, a great deal of scientific expert evidence was presented regarding the risks faced by species of sea turtles and the effectiveness of different turtle conservation policies in addressing such risks.[31] (It is worth noting that presentation of extensive scientific evidence in GATT disputes raising health and environmental concerns is not unusual as the genuineness of these concerns and the need for trade measures to address them is generally seen to demand verification via objective scientific evidence.)[32] Consequently, one course open to the Appellate Body would have been to review the adequacy of the scientific evidence proffered by the USA in support of its TED requirement. However, the Appellate Body's decision in the case made no reference to the scientific material and expert evidence, either in evaluating the US claim that shrimping poses a genuine risk to the conservation of an 'exhaustible natural resource' (namely, endangered species of sea turtles) or in examining the appropriateness of the risk regulatory response instituted by US authorities. Instead, the Appellate Body adopted a two-pronged interpretative approach focusing first on the plain meaning of the text of GATT Article XX and the exception in paragraph (g), and secondly looking beyond the trade text – where necessary – to find 'points of political agreement in non-trade contexts, including international environmental instruments'.[33]

[30] *United States – Importation Prohibition of Certain Shrimp and Shrimp Products*, Report of the Panel, WT/DS58/R, 15 May 1998 (*Shrimp/Turtle* Panel Report).

[31] *Shrimp/Turtle* Panel Report, [5.10]–[5.432], Annex IV.

[32] Robert Hudec, 'Science and "Post-Discriminatory" WTO Law', B.C. Int'l & Comp. L. Rev., 26 (2003), 185.

[33] Knox, 'The Judicial Resolution of Conflicts between Trade and the Environment', 48.

Returning to the text of GATT Article XX

A notable feature of the Appellate Body's jurisprudence under the GATT has been its emphasis upon the ordinary meaning of the legal text as a starting point for interpreting the requirements of the Agreement.[34] By contrast, the panel in the *Shrimp/Turtle* case applied an overtly teleological understanding of GATT Article XX. This understanding was consistent with 'trade insider' perceptions of the objectives of the treaty, which emphasise its goals of 'the promotion of economic development through trade', 'liberalization of access to markets on a non-discriminatory basis' and 'a multilateral approach to trade issues'.[35] Ignoring the specific exceptions in favour of an analysis of the purpose of the Article XX *chapeau*, the panel determined that this paragraph 'only allows Members to derogate from GATT provisions so long as, in doing so, they do not undermine the WTO multilateral trading system, thus also abusing the exceptions contained in Article XX'.[36] According to the panel, trade measures that conditioned access to a domestic market on the basis of a unilaterally prescribed policy threatened the coherence of the multilateral trading system and could not be justified under the Article XX *chapeau*.[37]

This interpretation of the *chapeau* was roundly criticised by the Appellate Body as finding no basis in the text of GATT Article XX.[38] It emphasised that maintaining the multilateral trading system – while 'necessarily a fundamental and pervasive premise underlying the WTO Agreement' – was 'not a right or an obligation, nor is it an interpretative rule which can be employed in the appraisal of a given measure under the chapeau of Article XX'.[39] The Appellate Body also rejected the panel's reasoning that the acceptance of unilateral trade measures adopted for non-trade ends would threaten the collapse of the multilateral trading system. Adopting this understanding, it commented, could render Article XX 'inutile' since 'conditioning access to a Member's domestic market on whether exporting Members comply with, or adopt, a policy or policies unilaterally prescribed by the importing Member may, to some degree, be a common aspect of

[34] This was signalled in its first decision in *United States – Standards for Reformulated and Conventional Gasoline*, Report of the WTO Appellate Body, WT/DS2/AB/R, 29 April 1996 (*Gasoline*).

[35] *Shrimp/Turtle* Panel Report, [7.42–7.43].

[36] *Ibid.*, [7.44]. [37] *Ibid.* [7.45].

[38] *Shrimp/Turtle*, [121]. [39] *Ibid.*, [116].

measures falling within the scope of one or another the exceptions (a) to (j) of Article XX'.[40]

This decision paved the way for the Appellate Body to consider the circumstances in which members might be permitted to adopt trade-restrictive measures in pursuance of one of the non-protectionist goals specified in Article XX. Sensibly, the Appellate Body rejected the panel's top-down, *chapeau*-first approach, instead looking to see if the US measure could be first be provisionally justified in accordance with Article XX(g) before turning to the requirements of the article's introductory paragraph. As discussed further below, the Appellate Body reached the conclusion that the US measure did meet the requirements of paragraph (g) as a genuine environmental measure addressed to the conservation of exhaustible natural resources, namely, the endangered species of sea turtles at issue in the case.

The Appellate Body then turned to the introductory *chapeau* to evaluate whether the US measure – albeit adopted to address a legitimate environmental risk – was *applied* in an unjustifiable or arbitrarily discriminatory manner. According to the Appellate Body, this was where the problems with the US measure lay, as there were several aspects of the implementation of the measure that created arbitrary or unjustifiable discrimination. In summary, these were the failure to engage all affected countries 'in serious, across-the-board negotiations with the objective of concluding bilateral or multilateral agreements for the protection and conservation of sea turtles' prior to adoption of the measure;[41] differential levels of effort made to transfer TEDs to some of the US' trading partners but not others;[42] the 'rigidity and inflexibility' of the US programme in demanding other members 'adopt a comprehensive regulatory program that is essentially the same as the United States' program, without inquiring into the appropriateness of that program for the conditions prevailing in the exporting countries';[43] and the institution of a regulatory process for applying the US requirements that failed to meet 'minimum standards for transparency and procedural fairness'.[44]

Despite its strictures issued to the panel regarding the need to ground analysis in the GATT text, the Appellate Body's own interpretation of the Article XX *chapeau* did not have a clear textual basis. As John Knox notes, the Appellate Body made little effort to discern the ordinary meaning of the terms of the *chapeau*, such as 'arbitrary or unjustifiable

[40] *Ibid.*, [121]. [41] *Ibid.*, [166]. [42] *Ibid.*, [175].
[43] *Ibid.*, [177]. [44] *Ibid.*, [183].

discrimination' and 'a disguised restriction on international trade'.[45] Indeed, the interpretation of the *chapeau* put forward by the Appellate Body appeared to be inspired as much by a teleological approach as that of the panel, albeit casting the purpose of the Article XX *chapeau* in different terms. The Appellate Body found that the *chapeau*:

> embodies the recognition on the part of WTO Members of the need to maintain a balance of rights and obligations between the right of a Member to invoke one or another of the exceptions of Article XX, specified in paragraphs (a) to (j), on the one hand, and the substantive rights of the other Members under the GATT 1994, on the other hand.[46]

Hence, the task of construing the *chapeau* was

> essentially the delicate one of *locating and marking out a line of equilibrium* between the right of a Member to invoke an exception under Article XX and the rights of the other Members under varying substantive provisions … of the GATT 1994, so that neither of the competing rights will cancel out the other and thereby distort and nullify or impair the balance of rights and obligations constructed by the Members themselves in that Agreement. The location of the line of equilibrium, as expressed in the chapeau, is not fixed and unchanging; the line moves as the kind and the shape of the measures at stake vary and as the facts making up specific cases differ.[47]

Looking beyond the text to find political agreement

In essence, by 'converting the chapeau into a mechanism to balance opposing teleological interests', the Appellate Body opened the way to the incorporation of criteria beyond those specified in the GATT text for the purpose of judging the legitimacy of environmental risks put forward as the basis for national trade-restrictive measures.[48] Rather than being scientific in nature, these factors were largely based on procedural and transparency criteria, which commentators such as Joanne Scott have argued could be considered as 'bolstering rather than undermining democracy by virtue of their capacity to enhance the external accountability of states'.[49] In addition, the Appellate Body placed

45 Knox, 'The Judicial Resolution of Conflicts between Trade and the Environment', 56.
46 *Shrimp/Turtle*, [156].
47 *Ibid.*, [159], (emphasis added).
48 Knox, 'The Judicial Resolution of Conflicts between Trade and the Environment', 56–57.
49 Joanne Scott, 'European Regulation of GMOs: Thinking About "Judicial Review" in the WTO', in Michelle Everson and Ellen Vos (eds.), *Uncertain Risks Regulated* (Milton Park: Routledge-Cavendish, 2009) p. 295.

significant weight on multilateralism in international trade relations as a principle that attracts widespread political support. In contrast to the panel, which objected to unilateral environmental trade measures on the basis that their widespread adoption might undermine the multilateral trading system, the Appellate Body focused on 'a more deeply felt political objection to unilateral trade restrictions aimed at protecting a common resource: their disregard of the conditions in and views of less powerful states with respect to the resource'.[50]

Multilateralism and interstate cooperation to address shared problems are values that equally find support in the broader international sphere. As the Appellate Body noted, many international environmental instruments, including foundational documents in the field such as the Rio Declaration on the Environment and Development,[51] stress the importance of a cooperative approach to managing global environmental issues and shared resources. The Appellate Body relied on such instruments in contending for the desirability of serious efforts to reach multilateral agreement on issues of turtle conservation before resorting to unilateral measures.[52]

It also looked to widely ratified multilateral environmental treaties in interpreting ambiguous phrases in the GATT, such as Article XX(g)'s reference to 'exhaustible natural resources'. On this basis, the Appellate Body rejected a narrow understanding of exhaustible natural resources as limited to non-living, finite resources such as fossil fuels and ruled that Article XX(g) extended to measures designed to conserve biological resources, such as animal populations, which were 'exhaustible' in the sense of being exposed to threats that could reduce the capacity for species' survival.[53] Further, the Appellate Body treated the question of the legitimacy of the US' environmental purpose as one that should be viewed 'in the light of contemporary concerns of the community of nations about the protection and conservation of the environment'.[54] It ascertained these concerns both from the endorsement of 'the objective of sustainable development' in the preamble to the WTO Agreement, as well as from environmental treaties

[50] Knox, 'The Judicial Resolution of Conflicts between Trade and the Environment', 57. One of the major objections to unilateral trade measures is the fact that only powerful trading blocs, such as USA and the EC, have capacity to employ them for health or environmental ends.

[51] Rio Declaration on Environment and Development, A/CONF.151/26 (Vol. I) (1992), Principle 12.

[52] *Shrimp/Turtle*, [168]. [53] *Ibid.*, [127]–[134]. [54] *Ibid.*, [129].

specifically dealing with the endangered and migratory nature of sea turtles.[55]

The *Shrimp/Turtle* case can thus be read as an endorsement of international practices for discerning legitimate environmental risk concerns that eschew 'objective' scientific criteria in favour of reliance on shared political values (multilateralism, fairness, transparency). In this context, the Appellate Body's resort to multilateral instruments beyond the trade sphere (such as widely ratified treaties recognising the existence of a particular environmental threat or general international legal principles endorsing cooperation in managing shared resources) served to ameliorate the potential for arbitrariness inherent in the balancing test adopted in its interpretation of the Article XX *chapeau*. This was all the more necessary given that WTO members themselves have not provided any substantive indication of their normative objectives with respect to the relationship between trade and the environment,[56] beyond endorsing the overall goal of 'mutual supportiveness' between trade and environmental interests.[57]

Evaluating the necessity of measures: weighing and balancing test

In *Shrimp/Turtle* an important part of the Appellate Body's approach to evaluating Article XX was its appreciation of the need to maintain a balance of rights and obligations under the GATT in marking out the line of equilibrium between them in any particular dispute. A balancing test was also a feature of the *Asbestos* case; another Appellate Body GATT decision frequently put forward as evidence of the capacity of WTO review to navigate 'the delicate interrelationship of values and interests' that arise in trade disputes over health and environmental measures.[58]

Whereas the *Shrimp/Turtle* case focused on the provisions of Article XX(g), the decision in *Asbestos* was concerned with a measure adopted on health grounds for which justification was sought under Article XX(b). This exception was the precursor of the SPS Agreement and,

[55] *Ibid.*, [129]–[132] referring to the Convention on Trade in Endangered Species of Wild Fauna and Flora, and the Bonn Convention, both of which list sea turtles as endangered species.

[56] *Shrimp/Turtle*, [154].

[57] *Ibid.*, referring to the preamble of the Ministerial Decision on Trade and Environment issued following the conclusion of the Uruguay trade round.

[58] Howse, 'From Politics to Technocracy', 109.

like Article 2.2 of that Agreement, requires a determination of the extent to which a measure is '*necessary* to protect human, animal or plant life or health' (emphasis added). Long-standing GATT jurisprudence, endorsed by the Appellate Body, interprets the necessity test in Article XX(b) as a requirement for the member adopting a health measure to choose the least trade-restrictive option.[59] Strictly applied – and especially if strong scientific evidence is demanded to demonstrate the existence of a health risk or the appropriateness of a particular regulatory approach – the necessity test in Article XX(b) could serve as an avenue for stringent scientific review of members' measures and the institution of sound science requirements.

However, in the *Asbestos* case the approach taken to evaluating the necessity for health measures under Article XX(b) of the GATT – known as the 'weighing and balancing' test – involved the application of an apparently more deferential review standard. The Appellate Body first applied the weighing and balancing test in the case of *Korea Beef* in evaluating the concept of necessity under Article XX(d), which speaks of measures 'necessary to secure compliance with laws or regulations'.[60] In its *Korea Beef* decision, the Appellate Body explained the necessity test as requiring:

> a process of weighing and balancing a series of factors which prominently include the contribution made by the compliance measure to the enforcement of the law or regulation at issue, *the importance of the common interests or values protected by that law or regulation*, and the accompanying impact of the law or regulation on imports or exports.[61]

It is the reference to the importance of the shared interests or values underlying a particular regulation as a factor competing with trade concerns in the evaluative process that potentially opens up scope for domestic regulatory choices to be given greater leeway in WTO review under the GATT.[62]

59 See generally, Catherine Button, *The Power to Protect: Trade, Health and Uncertainty in the WTO* (Oxford, Hart Publishing, 2004), pp. 29–37.

60 The paragraph indicates that its particular sphere of operation is with respect to laws or regulations relating to customs enforcement, the enforcement of certain types of monopolies, the protection of patents, trade marks and copyright, and the prevention of deceptive practices.

61 *Korea – Measures Affecting Imports of Fresh, Chilled and Frozen Beef*, Report of the WTO Appellate Body, WT/DS169/AB/R, 11 December 2000 (*Korea Beef*), [164] (emphasis added).

62 In the context of Article XIV of the General Agreement on Trade in Services, which also includes a necessity test, the Appellate Body indicated in *United States – Measures*

The weighing and balancing test in *Asbestos*

In the case of *Asbestos*, the WTO dispute settlement system was called upon to determine the GATT legality of a French trade ban on asbestos and asbestos-containing building products, challenged by Canada. The health risks posed by asbestos fibres are the subject of a substantial body of scientific evidence, which was clearly regarded as authoritative by both the panel deciding the case and the Appellate Body.[63] The Appellate Body also gave great weight to the health objective pursued by the French measure, observing:

> In this case, the objective pursued by the measure is the preservation of human life and health through the elimination, or reduction, of the well-known, and life-threatening health risks posed by asbestos fibres. The value pursued is both vital and important in the highest degree.[64]

The panel hearing the case determined that the French measure was discriminatory, violating Article III:4 of the GATT,[65] but found that it could be justified under Article XX(b).[66] In assessing the health risks posed by asbestos products and the necessity of the French ban in this regard, the panel contended that it was 'not its function to settle a scientific debate, not being composed of experts in the field of the possible human health risks posed by asbestos' and hence that it did 'not intend to set itself up as an arbiter of the opinions expressed by the scientific community'.[67] Instead it saw its task as being akin to that performed by panels in SPS disputes, namely 'to determine whether there is sufficient scientific evidence to conclude that there exists a risk for

Affecting the Cross-Border Supply of Gambling and Betting Services, Report of the WTO Appellate Body, WT/DS285/AB/R, 7 April 2005 that the weighing and balancing process inherent in the necessity analysis 'begins with an assessment of the "relative importance" of the interests or values furthered by the challenged measure': [306].

63 This was evident from the Appellate Body's references to the expert advice compiled by the Panel: *Asbestos*, [135], [162], [166].

64 *Ibid.*, [172].

65 *European Communities – Measures Affecting Asbestos and Asbestos-Containing Products*, Report of the Panel, WT/DS153/R, 18 September 2000 (*Asbestos* Panel Report), [8.158]. The Panel rejected the EC's argument that the health risks posed by asbestos were a relevant factor to consider in evaluating whether asbestos-containing and non-asbestos-containing products were 'like' in the sense required by Article III:4: [8.130]–[8.132]. The Appellate Body, on the other hand, was 'very much of the view that evidence relating to the health risks associated with a product may be pertinent in an examination of "likeness" under Article III:4 of the GATT 1994': [113]. See also [114]–[115].

66 *Asbestos* Panel Report, [8.241].

67 *Ibid.*, [8.181].

human life or health and that the measures taken by France are necessary in relation to the objectives pursued', basing its conclusions 'on the scientific evidence put forward by the parties and the comments of the experts consulted'.[68]

On appeal the Appellate Body saw nothing amiss with the panel's evaluation of the credibility and weight to be attached to the scientific evidence that had been before it. As a general matter, it indicated that it would only be prepared to interfere with the panel's appreciation of the scientific evidence where it was satisfied that the panel had 'exceeded the bounds of its discretion, as the trier of facts, in its appreciation of the evidence', which it held was not the case.[69] On the contrary, the Appellate Body noted the general expert consensus – shared by the scientists consulted by the panel, as well as international bodies such as the International Agency for Research on Cancer and the World Health Organization – concerning the human health risks posed by asbestos fibres.

This left the question of the necessity of the French measure, which had been designed with the goal of bringing a 'halt' to the spread of asbestos-related health risks in the importing country.[70] In demonstrating the need for a trade ban, the Appellate Body held that there was no requirement for France to quantify the risk to human life or health posed by asbestos fibres, which might be evaluated either in 'quantitative or qualitative terms'.[71] Instead, in assessing whether the French ban was necessary for achieving the country's chosen level of protection against health risks, the Appellate Body found that a similar process of 'weighing and balancing' to that discussed in its *Korea Beef* ruling was appropriate.[72] Given that the health value underlying the ban was considered to be 'both vital and important in the highest degree',[73] the Appellate Body came to the conclusion that a ban was 'necessary' as the most appropriate regulatory measure. In particular, it held that France could not have been reasonably expected to employ any alternative measure to a ban because this would have involved a continuation of the very risk that France was seeking to halt, thus effectively preventing France from achieving its chosen level of health protection.[74]

[68] *Ibid.*, [8.182]. [69] *Asbestos*, [162]. [70] *Ibid.*, [168].
[71] *Ibid.*, [167]. Arguably, given the existence of a well-established health risk, it was a scenario in which quantitative risk assessment was possible.
[72] *Ibid.*, [171], [172]. [73] *Ibid.*, [172]. [74] *Ibid.*, [174].

Deference to national regulatory judgments about risk?

The Appellate Body's *Asbestos* decision was widely celebrated by health advocates as a sign that the WTO dispute settlement system – at least in the context of the GATT – will afford significant regulatory autonomy to countries attempting to manage (health) risks via trade measures.[75] M. Gregg Bloche and Elizabeth Jungman, for example, argue that the Appellate Body's decision heralds the 'treatment of protection for health as an interpretative principle, calling for less onerous standards of proof and review for trade restraints when health is at stake'.[76] Some further evidence in support of this claim is supplied by subsequent GATT cases determined by the Appellate Body, such as the *Retreaded Tyres* dispute decided in December 2007.[77]

In this case the EC challenged a Brazilian ban on imports of retreaded tyres, which Brazil argued was necessary to protect human, animal and plant life and health given the human health and biodiversity risks posed by the accumulation of waste tyres in its territory.[78] The Appellate Body evaluated the panel's application of the 'weighing and balancing' test in respect of the Brazilian measure, which included findings that risks of dengue fever and malaria arising from the accumulation of waste tyres and the objective of protecting human life and health against such diseases were 'vital and important in the highest degree'.[79] The panel had also noted that the objectives of the Brazilian

[75] See, eg, Marie-Claire Cordonier Segger and Markus Gehring, 'The WTO and Precaution: Sustainable Development Implications of the WTO Asbestos Dispute', *Journal of Environmental Law*, 15 (2003), 289; Laura Lavitz, 'The World Trade Organisation Appellate Body Report, *European Communities – Measures Affecting Asbestos and Asbestos-Containing Products*, March 12 2001, WT/DS135/AB/R', Minnesota Journal of Global Trade, 11 (2002), 43.

[76] M. Gregg Bloche and Elizabeth Jungman, 'Health Policy and the WTO', *Journal of Law, Medicine and Ethics*, 31 (2003), 532.

[77] Brazil – Measures Affecting Imports of Retreaded Tyres, Report of the Appellate Body, WT/DS332/AB/R, 3 December 2007 (*Retreaded Tyres*).

[78] Particular risks of concern included those from mosquitoes which use tyres as breeding grounds and the exposure of human beings to toxic emissions caused by tyre fires which may cause loss of short-term memory, learning disabilities, immune system suppression, cardiovascular problems, cancer, premature mortality, reduced lung function, suppression of the immune system, respiratory effects, heart and chest problems. Further risks to animal and plant life and health include the exposure of animals and plants to toxic emissions caused by tyre fires and the transmission of mosquito-borne disease (e.g. dengue) to animals.

[79] *Brazil – Measures Affecting Imports of Retreaded Tyres*, WT/DS332/R, 12 June 2007 (Report of the Panel) [7.210].

import ban encompassed environmental protection – a value which both it and the Appellate Body considered to be important.[80] In the Appellate Body's view, there was nothing erroneous in the panel's reasoning that, in light of the importance of the interests protected by the import ban, the contribution of the ban to the achievement of its objective outweighed its trade-restrictiveness.[81]

Moreover, even though the Appellate Body recognised that a ban is a severe form of trade restriction – indicating that the measure should be 'apt to make a material contribution to the achievement of its objective'[82] – its assessment of the link between the health and environmental goals of the Brazilian measure and the regulatory approach taken was cognisant of the challenges involved in dealing with complex health and environmental problems. It observed that:

> certain complex public health or environmental problems may be tackled only with a comprehensive policy comprising a multiplicity of interacting measures. In the short-term, it may prove difficult to isolate the contribution to public health or environmental objectives of one specific measure from those attributable to the other measures that are part of the same comprehensive policy. Moreover, the results obtained from certain actions – for instance, measures adopted in order to attenuate global warming and climate change, or certain preventive actions to reduce the incidence of diseases that may manifest themselves only after a certain period of time – can only be evaluated with the benefit of time.[83]

In this context, the Appellate Body indicated that the evidence or data relied upon by the panel might pertain 'to the past or the present', or might consist of 'quantitative projections in the future, or qualitative reasoning based on a set of hypotheses that are tested and supported by sufficient evidence'.[84]

These findings notwithstanding, the weighing and balancing test employed by the Appellate Body in cases considering Article XX(b) leaves open a number of important questions pertinent in evaluating the extent to which it supports regulatory diversity in protecting against risks to human health and the environment. At one level the test paves the way for an explicit consideration of the *value* of particular health and environmental objectives in determining whether risk regulatory action is necessary despite potential trade impacts. On the other hand, this immediately raises problems as to how the international

[80] *Ibid.*, [7.112]; *Retreaded Tyres*, [179]. [81] *Retreaded Tyres*, [179].
[82] *Ibid.*, [150]. [83] *Ibid.*, [151]. [84] *Ibid.*

decision-makers involved in WTO dispute settlement might legitimately go about evaluating the importance of the common interests or values implicated in any risk dispute. As Timothy Reif and Julie Eckert point out, there are few established guidelines for determining the relative substantive importance of national interests or values in the health or environmental sphere.[85] In this respect, the Appellate Body's alternative approach – evident in *Shrimp/Turtle* – of looking to procedural and transparency criteria in evaluating the genuineness of risk concerns may offer more promise.

It also remains unclear whether the weighing and balancing test would favour a deferential approach to domestic regulatory autonomy where the risks at issue extend beyond situations of well-founded harms. There is a strong suggestion in the *Asbestos* case that the willingness of WTO dispute settlement decision-makers to find that the values at stake were vital was predicated on a finding of proven (or at least scientifically well-supported) health risk.[86] In other words, it appeared to be the strong scientific consensus surrounding the existence of asbestos-related health risks that founded the Appellate Body's ruling that there was also substantial social consensus on the importance of the interests at stake. As yet, WTO panels and the Appellate Body have not encountered a case under the GATT where the risk claims at issue are the subject of significant international controversy.[87] If faced with a situation where disputes remain over the relative importance of different health or environmental values – most likely heightened by issues of scientific uncertainty – WTO dispute settlement decision-makers might well return to seeming less controversial factors in weighing and balancing the necessity for particular trade measures. These factors might focus on the effectiveness of measures in achieving their regulatory ends (which could be expected to draw on scientific knowledge) and their efficiency (relying on estimates of economic costs).

[85] Timothy Reif and Julie Eckert, 'Courage You Can't Understand: How to Achieve the Right Balance Between Shaping and Policing Commerce in Disputes Before the World Trade Organization', *Columbia Journal of Transnational Law*, 42 (2004), 689.

[86] Button, *The Power to Protect*, p. 4. See also, Mary Footer and Saman Zia-Zarifi, 'European Communities – Measures Affecting Asbestos and Asbestos-Containing Products: The World Trade Organization on Trial for its Handling of Occupational Health and Safety Issues', *Melbourne Journal of International Law*, 3 (2002), 120, 142. Equally in *Retreaded Tyres*, risks such as malaria transmission and the toxicity of tyre fire fumes were well established.

[87] Claims under the GATT were raised by Canada and Argentina but not addressed by the panel in the *GMO* case.

Codex standard-setting and the role of JEFCA

As we saw in the previous chapter, an important element of the attempt made in the SPS Agreement to strengthen GATT disciplines for separating legitimate from protectionist SPS measures was the reliance placed on the epistemic authority of international, science-based standard-setting bodies. The three 'relevant international organizations' singled out by the SPS Agreement were the International Office of Epizootics, the International Plant Protection Convention and – in the area of food safety and human health protection – the Codex Alimentarius Commission. In line with the provisions of Article 3 of the SPS Agreement (as they were interpreted by the Appellate Body in the *Hormones* case) national SPS measures which conform to 'international standards, guidelines or recommendations' issued by the three organisations are presumptively valid under the SPS Agreement and the GATT, whereas those that result in a higher level of SPS protection than achieved by measures based on relevant international standards, guidelines or recommendations require a justificatory risk assessment meeting the requirements of Article 5 of the SPS Agreement.[88]

The authority accorded by negotiators of the SPS Agreement to the standards of international organisations, such as Codex, drew on a perception of these bodies as 'supposedly devoted' to 'a universalist framework of epistemic warrant, namely risk analysis'.[89] In reality, before the spotlight of the SPS negotiations cast them into stark relief, even the risk assessment procedures of Codex – which were the most advanced of the three organisations – were not applied consistently as part of a formalised, scientific framework for global food safety regulation. Instead, Codex's risk analysis approach, and the involvement of scientific advisory bodies in that process, evolved significantly in response to the new pressures created by the SPS Agreement.

As David Winickoff and Douglas Bushey have argued, this meant that 'the SPS negotiators and the trading regime had to produce the very science-based agency [they] had identified as its foundation'.[90] At the same time, legitimation of its authority by the WTO has allowed Codex to broach difficult questions regarding the role of science in international

[88] SPS Agreement, Articles 3.2, 3.3. *European Communities – Measures Concerning Meat and Meat Products (Hormones)*, WT/DS26/AB/R, AB-1997–4 (Report of the Appellate Body) [75].

[89] Winickoff and Bushey, 'Science and Power in Global Food Regulation', 5.

[90] *Ibid*, 7.

risk regulation. Its resulting processes for the use of science and the involvement of expert bodies such as JEFCA place a strong emphasis on 'the principle of sound scientific analysis and evidence' in line with the prerogatives of the SPS Agreement.[91] However, the incorporation of procedural requirements that promote consensus-based decision-making and involve international NGOs in the formulation of standards also represent a bona fide, if not always successful, attempt to mix 'technocratic and democratic elements' in order to accommodate the particular role played by values and cultural concerns in the food safety area.[92]

Standard-setting process in Codex

Codex, established in 1963, is an inter-governmental body with a membership that currently stands at 181 states.[93] It operates within the framework of the Joint FAO/WHO Food Standards Programme with the purpose of protecting the health of consumers and ensuring fair practices in the food trade.[94] The majority of Codex's work is undertaken by committees consisting of member state delegates (for example the Committee on Food Additives), with administrative assistance provided by a secretariat and the input of independent scientific advice from joint FAO/WHO expert bodies (for example JEFCA). The principal product of Codex's work is the Codex Alimentarius (or food code), containing internationally adopted standards, guidelines, codes of practice and other recommendations regarding food safety.[95] Before they achieved a more authoritative status with the conclusion of the SPS Agreement, adoption of Codex standards by countries to guide their own national food safety measures occurred on a purely voluntary

91 See 'General Principles of the Codex Alimentarius': Codex Alimentarius Commission, *Procedural Manual*, 18th edn (Rome: Food and Agriculture Organisation of the United Nations, 2008) (Codex Procedural Manual), p. 16, principle 1.

92 Winickoff and Bushey, 'Science and Power in Global Food Regulation', 10.

93 Membership of the Codex Alimentarius Commission is open to all member nations and associate members of the FAO and/or WHO.

94 Statutes of the Codex Alimentarius Commission, Codex Procedural Manual, Article 1. The organisation's consumer protection objectives extend beyond the narrow focus of the SPS Agreement on food safety as a health issue.

95 Standards usually relate to product characteristics, e.g. maximum limits for residues of veterinary drugs in certain foods. Codes of practice define production, processing, manufacture, transport and storage for foods or food groups to ensure food safety. Guidelines are either principles setting out the policy applicable in a particular area, e.g. principles for risk analysis of foods derived from modern biotechnology, or guidelines interpreting principles. See WHO/FAO, *Understanding the Codex Alimentarius*, 3rd edn, (Rome, 2006), pp. 10–11.

basis. Codex nevertheless served an important function as a forum for the discussion of food policy and the dissemination of pertinent scientific information and best practice regulation.[96]

The procedures of Codex for agreeing on food safety standards are well defined and, in comparison to many international bodies, relatively transparent and participative. The general process for establishing a new standard involves eight steps, including the consideration of expert advice and two rounds of consultation with member governments (see Figure 6.1).[97] International NGOs, such as industry groups or organisations representing consumer interests, which have meet Codex requirements for the conferral of observer status, are able to speak at meetings and are entitled to receive copies of relevant documents such as draft standards.[98] With Codex's increased international profile, many new NGO actors such as consumer protection organisations have sought to become involved in its work, such that it is acknowledged to have 'opened up from being a backwater where industry dominated'.[99] This openness has led some to present Codex 'as an example of good practice in terms of its relations with NGOs and its willingness to accept their input into its work'.[100] The addition of new voices in the standard-setting process certainly means that – where agreement can be reached – its results are more likely to take account of broader constituencies of both countries and interest groups.[101]

Nevertheless, obstacles to a truly 'deliberative' process remain, which have been exacerbated by the heightened global profile of the organisation. For instance, Michael Livermore identifies as key inequities hindering 'true deliberation': lower levels of participation in the organisation by developing countries compared with developed countries; a lack of voice for consumer concerns not aligned with governments'

[96] Michael Livermore, 'Authority and Legitimacy in Global Governance: Deliberation, Institutional Differentiation, and the Codex Alimentarius', N.Y.U. Law Rev., 81 (2006), 774.

[97] There is also provision for an accelerated procedure (with Stages 6 and 7 removed) invoked on the basis of a two-thirds majority of the votes cast by members.

[98] See 'Principles Concerning the Participation of International Non-Governmental Organizations in the Work of the Codex Alimentarius Commission', *Codex Procedural Manual*, p. 23.

[99] Mark A. Pollack and Gregory C. Shaffer, *When Cooperation Fails: The International Law and Politics of Genetically Modified Foods* Oxford University Press, 2009), p. 172 quoting an interview with Michael Hansen of Consumer's Union.

[100] FAO/WHO, Report of the Evaluation of the Codex Alimentarius and other FAO and WHO Food Standards Work, 15 November 2002, [146].

[101] Pollack and Shaffer, *When Cooperation Fails*, p. 172.

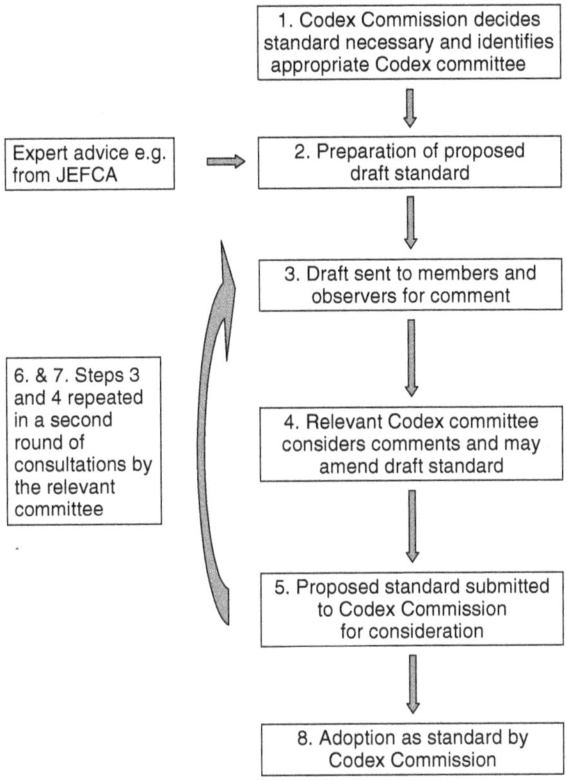

Figure 6.1: Standard-setting process in Codex

interests given the absence of voting rights for NGOs; and the still prevalent 'bias' in NGO participation towards industry groups over consumer organisations.[102]

Another aspect of the standard-setting process in Codex that has come under strain with the increase in the international authority accorded to Codex's work is its customary adherence to the principle of consensus decision-making. Although provision exists for Codex standards to be adopted on the basis of a simple majority of the votes cast by those members present at a particular meeting, the organisation's Procedural Manual urges the Commission to:

[102] Livermore, 'Authority and Legitimacy in Global Governance', 783–6. See also Gráinne de Búrca, 'Developing Democracy Beyond the State', Colum. J. Transnat'l L., 46 (2008), 234; Andrew Morriss and Roger Meiners, 'Borders and the Environment', *Environmental Law*, 39 (2009), 145.

make every effort to reach agreement on the adoption or amendment of standards by consensus. Decisions to adopt standards are taken by voting only if such efforts to reach consensus have failed.[103]

This commitment to consensus decision-making was seriously tested in the aftermath of the coming into force of the SPS Agreement as states realised the stakes of consenting (or at least abstaining from objecting) to Codex standards with which they did not agree. During 1995, for instance, a new standard on hormone residues in meat did not achieve consensus within Codex as a result of European opposition. The USA initiated a secret vote on the standard (the first such vote in Codex's history), which was approved by a narrow majority of the members present.[104] In the first *Hormones* dispute the failure of the EC to base its hormone measures on the relevant Codex standard was found to require the EC to produce a justificatory risk assessment for its divergent approach, notwithstanding the EC's opposition to the setting of an international standard for hormone residues within Codex.

Several other instances of voting on standards in subsequent years led some commentators to predict that there would be growing resort to non-consensus decision-making processes in Codex as a result of the influence of the SPS Agreement.[105] That this has not come to pass is testament to the diligent work of Codex since 2000 to ensure widespread support for its standards. Consensus decision-making has been restored as the normal practice of the institution via adherence to certain 'measures to facilitate consensus' agreed in 2003 and now set out in Codex's Procedural Manual.[106] These measures are phrased as recommendations for the Commission to:

- Refrain from submitting proposals for new standards 'where the scientific basis is not well-established on current data and, where necessary, [to] carry out further studies in order to clarify controversial issues';

[103] Codex Procedural Manual, p. 14.

[104] Doaa Motaal, 'The "Multilateral Scientific Consensus" and the World Trade Organization', J. World Trade, 38(5) (2004), 866. Votes in favour numbered 33, with 29 against and 7 abstentions.

[105] Terence Stewart and David Johanson, 'The SPS Agreement of the World Trade Organization and International Organizations: The Roles of the Codex Alimentarius Commission, the International Plant Protection Convention, and the International Office of Epizootics', Syracuse J. Int'l L. & Commerce, 26 (1998), 45.

[106] Codex Procedural Manual, p. 16 (Rule XII(2)).

- Provide for 'thorough discussions and documentation of the issues' at committee meetings;
- Organise informal meetings of the parties where disagreements arise provided that the objectives of any such meetings are clearly defined and participation is open to all interested members and observers in order to preserve transparency;
- Redefine, where possible, the scope of subject matter being considered for the elaboration of standards to cut out issues on which consensus could not be reached;
- Ensure matters are not progressed further in the standard-setting process 'until all relevant concerns are taken into account and adequate compromises worked out';
- Emphasise to committees and their chairpersons that matters should not proceed to the Commission for decision 'until such time as consensus has been achieved at the technical level'; and
- Facilitate the increased involvement and participation of developing countries.

These measures envisage broad participation and transparency as key planks for building consensus on international food safety standards. As Winickoff and Bushey point out, the measures also promote the view that technical consensus is central to political agreement, which is consistent with the perception (of importance both for Codex and the WTO) that Codex standards are 'scientifically sound'.[107]

Nevertheless, the pursuit of consensus decision-making in Codex places some important limits on its capacity as a global governance body in the area of food safety regulation. The organisation has been most successful in establishing health-related standards where there exists a clear scientific basis. By contrast, regulation of food safety risks in areas of significant scientific controversy and/or where strong divergences arise over health or environmental values has proved more difficult for Codex. The arduous process that has accompanied the organisation's attempt to develop risk management policies governing transgenic foods is one such example.[108] This suggests that ultimately Codex may be best to eschew regulatory activity in respect of high-profile, politically divisive food safety concerns in favour of

[107] Winickoff and Bushey, 'Science and Power in Global Food Regulation', 14.

[108] See Pollack and Shaffer, *When Cooperation Fails*, pp. 166–71, although the authors note that Codex has achieved more success on technical aspects of this issue such as the elaboration of risk assessment guidelines for foods derived from biotechnology.

focusing on less contentious risks that are amenable to technical agreement.[109]

Codex's risk analysis framework

An important element of Codex's claim to generate 'scientifically sound' international standards regulating food safety is its processes for conducting science-based risk assessments. Prior to the 1990s Codex's principles and processes for evaluating information on food safety risks and devising harmonised standards were relatively informal and subject to different interpretation and application from meeting to meeting.[110] However, as the SPS negotiations took shape in the late 1980s and early 1990s, it became clear that Codex would need to revisit and modify its procedures to take account of the new recognition for Codex standards that was being proposed within the WTO framework. Discussions on this question began within Codex as early as 1991 and included a recommendation for 'an early review programme to examine all Codex standards as to their current relevance and sound scientific basis, with a view to facilitating international trade'.[111]

In 1997 Codex initiated a process for drafting uniform 'risk analysis principles and guidelines' that were originally intended for application both by the standard-setting body and by its member countries. Agreement proved difficult, resulting in lengthy negotiations and splitting of the exercise into two separate projects: one focused on developing a framework for application by Codex and its expert advisory bodies, and the other devising principles applicable by member governments. The Working Principles for Risk Analysis for Application in the Framework of the Codex Alimentarius (Codex Risk Analysis Principles) were eventually adopted in 2003,[112] with Working Principles for Risk Analysis for Food Safety for Application by Governments following some time later in 2007.[113]

Both sets of principles adopt a 'structured approach' to risk analysis comprised of three distinct, albeit 'closely linked', components of

[109] *Ibid.*, pp. 172–3.

[110] David Jukes, 'The Role of Science in International Food Standards', *Food Control*, 11(3) (2000), 182.

[111] See Report of the Nineteenth Session of the Joint FAO/WHO Codex Alimentarius Commission, Rome, 1–10 July 1991, Appendix 4, [10(i)].

[112] Codex Procedural Manual, p. 101 (Codex Risk Analysis Principles).

[113] Codex Alimentarius, Working Principles for Risk Analysis for Food Safety for Application by Governments, 1st edn, (FAO and WHO: 2007).

risk assessment, risk management and risk communication. The risk assessment component is envisaged as a scientific domain in the sense that the process is entrusted to experts and must be based on 'all available scientific data'.[114] Risk management decisions taken by the Codex Commission, on the other hand, are concerned with weighing policy options for the prevention and mitigation of identified risks. The Codex Risk Analysis Principles prescribe the need for 'a functional separation of risk assessment and risk management, in order to ensure the scientific integrity of the risk assessment'.[115] As discussed in Chapter 3, a stringent distinction between risk assessment and risk management is virtually impossible to maintain in practice given the extent to which risk assessment is shaped by value judgments. Perhaps in cognisance of this limitation, the Codex Risk Analysis Principles go on to recognise 'that risk analysis is an iterative process, and interaction between risk managers and risk assessors is essential for practical application'.[116] Winickoff and Bushey regard this statement as an acknowledgement of the value considerations that generally underpin risk assessment and the need for these considerations to be developed jointly by technical and policy people.[117]

There are further elements of the Codex Risk Analysis Principles that serve to ameliorate what might otherwise be a fairly strict, technically oriented notion of risk assessment. For example, risk assessments should 'seek and incorporate relevant data from different parts of the world, including that from developing countries' and should 'be based on realistic exposure scenarios'.[118] This suggests that in determining the scope of the risk assessment, alternative perspectives (for example of developing countries) and real world risk considerations (for example the potential for cumulative impacts or effects on high-risk populations) are relevant. In addition, attention is paid to questions of scientific uncertainty, albeit largely in a manner consistent with a technical risk perspective. Thus 'precaution' is recognised as 'an inherent element of risk analysis', meaning that the 'degree of uncertainty and variability in the available scientific information should be explicitly

[114] Codex Risk Analysis Principles, Principles 18 and 20.
[115] *Ibid.*, Principle 9.
[116] Codex Procedural Manual, p. 102.
[117] Winickoff and Bushey, 'Science and Power in Global Food Regulation', 8, although they also note that the risk assessment/risk management distinction remains problematic to the extent that it may render particular value choices more opaque.
[118] Codex Procedural Manual, p. 103.

considered in the risk analysis'[119] and '[c]onstraints, uncertainties and assumptions having an impact on the risk assessment should be explicitly considered at each step in the risk assessment and documented in a transparent manner'.[120] Moreover, when 'there is evidence that a risk to human health exists but scientific data are insufficient or incomplete', Codex is discouraged from proceeding to elaborate a food safety standard in favour of considering a more informal option, such as a code of practice, provided that such a text 'would be supported by the available scientific data'.[121]

Nonetheless, the Codex Risk Analysis Principles stop short of endorsing the precautionary principle as a valid approach to risk regulation in circumstances of scientific uncertainty. Instead the Principles speak of 'resolving the impact of uncertainty on the risk management decision',[122] a responsibility that is held to lie with risk managers not risk assessors, although it is clear that risk assessors routinely adopt various uncertainty management techniques, such as reliance on qualitative information and safety factors (as discussed below). Further, where there is judged to be 'sufficient scientific evidence', it is contemplated that Codex may proceed to elaborate a standard or related text despite some areas of uncertainty remaining, provided 'the assumptions used for the risk assessment and the risk management options selected ... reflect the degree of uncertainty and the characteristics of the hazard'.[123]

The risk analysis framework within Codex operates in conjunction with other policy dictates, such as the 'Statements of Principle Concerning the Role of Science in the Codex Decision-Making Process and the Extent to Which Other Factors are Taken into Account'.[124] The latter Statements of Principle reaffirm 'the principle of sound scientific analysis and evidence, involving a thorough review of all relevant information' as the foundation of Codex food safety standards.[125] However, they also require Codex to have regard 'where appropriate, to other legitimate factors relevant for the health protection of consumers and for the promotion of fair practices in the food trade'.[126] The criteria for the consideration of 'other legitimate factors' make clear that their

119 Codex Risk Analysis Principles, Principle 11.
120 *Ibid.*, Principle 23. 121 *Ibid.*, Principle 10.
122 *Ibid.*, Principle 25. 123 *Ibid.*, Principle 11.
124 Decision of the 21st session of the Commission, 1995 (Statements of Principle).
125 Statements of Principle, Principle 1.
126 *Ibid.*, Principle 2.

relevance lies at the stage of risk management and that consideration of such factors 'should not affect the scientific basis of risk analysis'. This suggests a limited role for non-scientific factors in the risk assessment process; for instance, such considerations could seemingly not be considered in determining the scope of the risk assessment and the hazards to be evaluated.

What exactly 'other legitimate factors' comprise was a matter of heated debate during the formulation of the Statements of Principle document, which occurred against the backdrop of controversies over hormone residues in beef and milk. Possible factors that have been identified by the Codex Secretariat as having been applied in previous Codex work include fairly standard risk management principles, such as the economic sustainability and technical feasibility of recommended measures and the application of safety factors in deriving permissible exposure levels from experimental, toxicological data. Considerable uncertainty surrounds the relevance for Codex risk management of other factors, such as the level of technological need for a particular food additive, environmental or consumer concerns and animal welfare issues.[127] Winickoff and Bushey note that, in any event, the debate over other legitimate factors in Codex has begun to fade in the last half decade as standardisation of the organisation's risk analysis process has advanced. Overall, these authors argue, the effect of development of the Codex Risk Analysis Principles has been 'to supplant other potential frameworks' (such as one based on the precautionary principle) and 'marginaliz[e] environmental, economic and other potential factors in food safety regulation'.[128]

Role of JEFCA in developing international food safety standards

Consistent with the 'scientifically sound' representation of Codex standard-setting is the long-standing role played by expert advisory bodies in the process. The oldest such body, JEFCA, has been in existence since 1956 as an expert committee focusing on evaluation of the safety of food additives, contaminants and veterinary drug residues.[129] While formally separate from the Codex as a joint FAO/WHO administered and financed body, JEFCA's primary role is to provide scientific advice to three Codex committees: the Committee on Food Additives,

127 Jukes, 'The Role of Science in International Food Standards', 190–2.

128 Winickoff and Bushey, 'Science and Power in Global Food Regulation', 9.

129 Contaminants and naturally occurring toxicants were added to JEFCA's mandate in 1972 and veterinary drug residues in 1987.

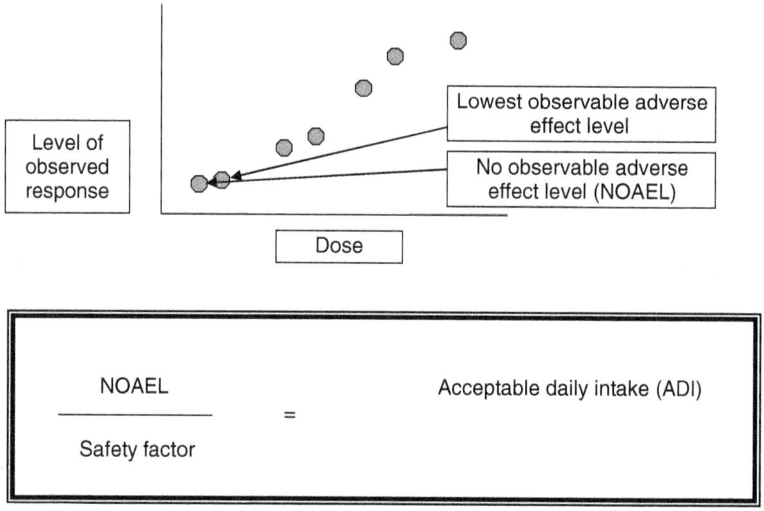

Figure 6.2: Deriving an ADI from animal studies

the Committee on Contaminants in Foods and the Committee on Residues of Veterinary Drugs in Food. For additives and food contaminants, JEFCA's advice takes the form of proposals for an acceptable daily intake (ADI) of the chemical concerned (Figure 6.2). For veterinary drug residues found in meat derived from animals administered a particular drug, its recommendations are expressed as maximum residue limits (MRLs).[130]

Both in its composition and functioning, JEFCA is an overtly scientific body. Its members are drawn from rosters compiled by the FAO and WHO of scientists who meet the organisations' requirements for independence, coupled with a certain level of expertise and experience. Its primary responsibility is to perform 'science-based, quantitative' risk assessments upon which the relevant Codex committees, and ultimately the Commission, base their risk management decisions.[131] The scientific nature of JEFCA's evaluations is emphasised by requirements

[130] ADI is a limit set to protect human health, based on toxicological concerns. MRLs are maximum levels of residue that are acceptable at the point of supply, based on good agricultural practice. Accordingly, MRLs are typically set much lower than ADIs. See Australian Pesticides and Veterinary Medicines Authority, 'Questions and Answer About Chemical Residues' at www.apvma.gov.au/residues/residues_faq.php.

[131] Risk Analysis Principles Applied by the Codex Committee on Food Additives and the Codex Committee on Contaminants in Food, Codex Procedural Manual, 111 (Food Additives Risk Principles).

for the body to limit itself 'to presenting its deliberations and conclusions ... in a complete and transparent manner' that 'should not include the consequences of its analyses on trade or other non-public health consequence [sic]'.[132] It is also subject to prescriptions to communicate to risk managers the 'magnitude and source of uncertainties' in its risk assessments, the procedures by which any uncertainties are estimated and the basis for default assumptions used to account for uncertainties.[133] While such requirements help to maintain a degree of separation between JEFCA's realm of scientific risk assessment and the committees' arena of risk management policy, they cannot eliminate entirely the overlap between risk assessment and risk management activities. JEFCA's risk assessment procedures will inevitably retain an element of science policy inherent in tasks such as identifying the hazards of concern to be evaluated (that is, the scope of the risk assessment), applying safety factors where ADIs or MRLs are derived from animal feeding studies, and determining the most probable exposure scenarios.

JEFCA's working procedures, as well as its criteria for selecting participating experts, are designed to emphasise the objectivity and universality of the scientific body's risk assessment conclusions. However, they are set against 'inclusiveness' requirements for geographically equitable representation of experts and the participation of developing country members that are in seeming tension with Codex's concern with scientific credibility and credentials.[134] For instance, the Codex Procedural Manual states that JEFCA's scientific experts should be selected 'taking into account geographic representation to ensure all regions are represented'.[135] In addition, a recent meeting on the topic of 'Enhancing Developing Country Participation in Scientific Advice Activities' recommended that 'due consideration' be given in the selection of participants to 'geographical and socioeconomic balance', although this should not be 'to the extent that it compromises scientific integrity'.[136] A similar concern with the 'representativeness' of JEFCA's risk assessment process is evident in directions in the Codex Procedural Manual for JEFCA to 'strive to base its risk assessments on

[132] *Ibid.*, [36]. [133] *Ibid.*, [34] and [35].

[134] Winickoff and Bushey, 'Science and Power in Global Food Regulation', 17.

[135] Food Additives Risk Analysis Principles, [25].

[136] FAO and WHO, FAO/WHO Framework for the Provision of Scientific Advice on Food Safety and Nutrition (Rome: FAO/WHO, 2007), p. 11.

global data, including data from developing countries' and to 'take into account regional differences in food consumption patterns'.[137]

In domestic risk assessment settings, attempts to expand the political or social interests represented on expert committees have often been seen to compromise their scientific authority.[138] However, Winickoff and Bushey argue that JEFCA and other Codex expert advisory bodies been able to achieve greater inclusiveness without undermining their epistemic authority by framing requirements for representativeness of the experts involved as contributing to building technical capacity in the food safety area in developing countries. Likewise, requirements for a range of 'global' data to be considered are accepted as necessary for building the credibility of JEFCA assessments by overcoming any perception of bias towards 'Northern' information. Accordingly, 'discourses of representation and sound science are made to converge rather than conflict, achieving the reconstruction of science advisory committees as hybrid zones of knowledge making and political negotiation.'[139]

Of course, a diverse geographic representation within JEFCA and the risk information it considers might generate its own problems, such as heightened potential for conflicting views on the food safety risks assessed, undermining the achievement of an international technical consensus. Prior to 1995 divergent scientific views within JEFCA were not considered to be problematic with the practice of the body being simply to record any areas of disagreement in the material presented to the Codex Commission. However, with the coming into force of the SPS Agreement and initiation of the first dispute of *Hormones*, presenting a single view in expert risk assessment was seen to be more important. JEFCA, like the Commission itself, has thus made a concerted effort to reach an agreed position on all questions put before the body. At least since 1997 these efforts would appear to have been highly successful with unanimity achieved in all instances. One of the advising experts in the *Hormones II* case, who was a member of JEFCA over the relevant period, described the process within the committee as follows:

> Generally what happens is that there is a discussion, there may be varying interpretations of a dataset, the experts get together over the period of a

[137] Food Additives Risk Analysis Principles, [30], [32].

[138] See the discussion of the operation of such committees in the US risk regulatory system in Sheila Jasanoff, *The Fifth Branch: Science Advisors as Policymakers* (Cambridge, MA: Harvard University Press, 1990).

[139] Winickoff and Bushey, 'Science and Power in Global Food Regulation', 18.

meeting and explore the various possibilities, bringing new information, or new insights and reach a common position ...[140]

The emphasis on reaching a consensus position within JEFCA 'introduces an important element of *democratic process* to what is ostensibly legitimated as an expert activity.[141] At the same time it creates a perception of technical consensus that has been useful in generating greater political consensus around the setting of international food safety standards within the Codex Commission.

Biosafety Protocol and precautionary risk regulation

The reverberations of the science and risk assessment requirements of the SPS Agreement have been felt not just in international standard-setting bodies, such as Codex, which have been brought into a close relationship with the WTO, but also farther afield in other, apparently unrelated areas of international law. The Biosafety Protocol, concluded in January 2000, is a prominent illustration of this phenomenon.[142] On the one hand, the Protocol has been hailed as 'one of the most explicit examples of operationalization of the precautionary principle/approach in any multilateral environmental agreement',[143] which would seemingly put it at the opposite end of the risk regulatory spectrum from the SPS Agreement.[144] On the other hand, the issue of the compatibility between international controls on the transboundary movement of LMOs and the trade agreements of the WTO was a major element of the negotiations for the Biosafety Protocol,[145] and would appear to have

[140] United States - Continued Suspension of Obligations in the EC-Hormones Dispute, Addendum to the Report of the Panel, WT/DS320/R/Add.7, (Annex G), [511].

[141] Winickoff and Bushey, 'Science and Power in Global Food Regulation', 16 (emphasis in original).

[142] The Protocol now has 157 parties, although most represent 'the importer rather than the exporter perspective on GMO trade': Robert Falkner and Aarti Gupta, *Implementing the Biosafety Protocol: Key Challenges* (London: Chatham House, 2004), p. 2.

[143] Ruth Mackenzie *et al.*, *An Explanatory Guide to the Cartagena Protocol on Biosafety*, IUCN Environmental Policy and Law Paper No. 46, (Gland: IUCN, 2003), p. 14.

[144] In *Hormones* the Appellate Body stated that the precautionary principle 'has not been written into the SPS Agreement as a ground for justifying SPS measures that are otherwise inconsistent with the obligations of Members set out in particular provisions of that Agreement': [124].

[145] The question of the interrelationship between the Protocol and WTO rules has been the focus of much scholarship on the Biosafety Protocol: see Gretchen Gaston and Randall Abate, 'The Biosafety Protocol and the World Trade Organization: Can

played an influential role in shaping the version of precautionary risk assessment eventually included in the treaty.

Flowing from this institutional setting, the Biosafety Protocol employs particular ways of 'framing' the issue of biosafety risk that 'serve to mobilize different interests and actor coalitions, legitimize distinct kinds of knowledges and expertise, and validate certain solutions while excluding others'.[146] While the Protocol is now nearly a decade old, it only took effect on 11 September 2003, with regular meetings of the parties commencing in February 2004. At this early stage, framings of the biosafety issue that emerged during negotiations for the Protocol continue to shape its evolution.[147] Moreover, despite subsequent meetings of the parties achieving some progress in elaborating operational aspects of the Biosafety Protocol,[148] a 'stark divide' remains between countries over the risk regulatory approach applicable within the international regime and the extent to which this should be based upon the precautionary principle.[149]

the Two Coexist?', *Pace International Law Review*, 12 (2000), 107; Brett Grosko, 'Genetic Engineering and International Law: Conflict or Harmony? An Analysis of the Biosafety Protocol, GATT and the WTO Sanitary and Phytosanitary Agreement', *Virginia Environmental Law Journal*, 20 (2001), 295; Maria Julia Oliva, 'The Cartagena Protocol on Biosafety and the Agreement on Sanitary and Phytosanitary Measures: What will Decisions Regarding GMOs have to be Based On?', *International Legal Perspectives*, 13 (2002), 22; Olivette Rivera-Torres, 'The Biosafety Protocol and the WTO', B.C. Int'l & Comp. L. Rev., 26 (2003), 263; Sabrina Safrin, 'Treaties in Collision? The Biosafety Protocol and the World Trade Organization Agreements', Am. J. Int'l L., 96 (2002), 606; Terence Stewart and David Johanson, 'A Nexus of Trade and the Environment: The Relationship between the Cartagena Protocol on Biosafety and the SPS Agreement of the World Trade Organization', Colorado J. Int'l Envt'l Law & Policy, 14 (2003), 1; Samuel Blaustein, 'Splitting Genes: The Future of the Genetically Modified Organisms in the Wake of the WTO/Cartagena Standoff', *Pennsylvania State Environmental Law Review*, 16 (2008), 367; Abdul Haseeb Ansari, 'Biosafety Protocol, SPS Agreement and Export and Import Control of LMOs/GMOs', *Journal of International Trade Law and Policy*, 7(3) (2008), 139.

[146] Aarti Gupta, 'Problem Framing in Assessment Processes: The Case of Biosafety', in Ronald B. Mitchell *et al.* (eds.), *Global Environmental Assessments: Information and Influence* (Cambridge, MA: MIT Press, 2006), p. 57.

[147] *Ibid.*, p. 58.

[148] For instance, the parties have established an Ad Hoc Technical Expert Group on Risk Assessment and Risk Management to develop a 'roadmap' on the necessary steps to conduct risk assessment in accordance with the Biosafety Protocol. The Group held its first meeting in April 2009. See Report of the First Meeting of the Ad Hoc Technical Expert Group on Risk Assessment and Risk Management under the Cartagena Protocol on Biosafety, UNEP/CBD/BS/AHTEG-RA&RM/1/3, 30 April 2009.

[149] Pollack and Shaffer, *When Cooperation Fails*, p. 155.

Framing of biosafety risk under the Biosafety Protocol

The Biosafety Protocol was negotiated between 1996 and 2000 under the auspices of Article 19(3) of the Convention on Biological Diversity (CBD), which directs state parties to:

> consider the need for and modalities of a protocol setting out appropriate procedures, including, in particular, advance informed agreement, in the field of the safe transfer, handling and use of any living modified organism resulting from biotechnology that may have adverse effect on the conservation and sustainable use of biological diversity.[150]

Out of the broad array of ecological, health-related, socio-economic and ethical concerns surrounding GMOs (designated 'living modified organisms' or LMOs in the Protocol),[151] the CBD thus includes mention of only a subset of risks pertaining to the potential adverse effects of GMOs on biodiversity.[152] This initial framing of the biosafety issue at the global level has had important consequences for the scope and nature of the Protocol by supporting distinctions made between different categories of LMOs.

Whereas LMOs with potential, direct biodiversity impacts – such as those intended for deliberate release into the environment of an importing country (for example as seeds) – were widely accepted as being a legitimate concern of the Protocol, the inclusion of other LMOs posing more indirect risks to biodiversity via their potential adverse effects on human health – such as LMOs intended for food, feed or processing (LMO-FFPs, for example crops) – was heavily contested. In addition, the focus on adverse ecological effects raised questions over whether broader, possible socio-economic impacts of GMO agriculture (for example the implications for traditional agriculture in developing countries, as well as the institutional capacity of such countries to

150 Convention on Biological Diversity, opened for signature 5 June 1992, 1760 UNTS 79 (entered into force 29 December 1993).

151 The USA pressed for this language in the Convention in order to deflect attention away from *genetic* modification as the focus of global regulatory attention: Rivera-Torres, 'The Biosafety Protocol and the WTO', 271.

152 For overviews of the social and ethical issues surrounding GMOs, as well as socio-economic dimensions of particular concern to developing countries, see Nuffield Council on Bioethics, *Genetically Modified Crops: The Social and Ethical Issues* (London: Nuffield Council on Bioethics, 1999); Nuffield Council on Bioethics, *The Use of Genetically Modified Crops in Developing Countries* (London: Nuffield Council on Bioethics, 2003).

manage risks associated with GMOs) could validly be dealt with by the Protocol. A further complicating factor was the high degree of uncertainty surrounding not just the risks but also the potential benefits of GMO agriculture. This raised the relevance of the precautionary principle for the assessment and management of GMOs. As Aarti Gupta has remarked, biosafety thus presented 'an anticipatory governance challenge'; one where the shape of the problem to be managed and the desired outcomes were unclear.[153]

The provisions of the Biosafety Protocol pertaining to requirements for scientific evidence and risk assessment reflect these constraints on the global framing of the problem of biosafety. Science and risk assessment play an important part in the decision-making processes mandated under the Protocol given the call for biosafety risk assessments 'carried out in a scientifically sound manner', which take into account 'relevant technical and scientific details' regarding the LMO concerned.[154] At the same time, the Protocol text declares states' awareness of 'the rapid expansion of modern biotechnology and the growing public concern over its potential adverse effects'.[155] Consequently, the Protocol gives a prominent place to the precautionary principle, mentioning 'the precautionary approach contained in Principle 15 of the Rio Declaration' in both its preamble and objective. The latter provision, moreover, extends the concern of the Protocol from purely that of 'contribut[ing] to ensuring an adequate level of protection in the field of the safe transfer, handling and use of living modified organisms resulting from modern biotechnology that may have adverse effects on the conservation and sustainable use of biological diversity', so as also to take into account risks to human health.[156]

The centrepiece of the Biosafety Protocol is its procedure for 'advance informed agreement' that requires exporters of certain LMOs to notify an importing party of a proposed import in order to allow the latter party to make an informed decision on whether it will permit the import to proceed.[157] Decisions taken by the importing party are to be made in light of a risk assessment, but the importing party can

[153] Gupta, 'Problem Framing in Assessment Processes', p. 61.
[154] Biosafety Protocol, Article 15.1, Annex III, paras. 3 and 9.
[155] *Ibid.*, preamble.
[156] *Ibid.*, Article 1. See also Article 4 dealing with the scope of the Biosafety Protocol.
[157] *Ibid.*, Articles 8 and 10.

require the exporter to carry out this assessment and bear its cost.[158] Importantly, the importing party is also granted rights to refuse the proposed import on a precautionary basis by way of a provision in the Biosafety Protocol that states:

> Lack of scientific certainty due to insufficient relevant scientific information and knowledge regarding the extent of the potential adverse effects of a living modified organism on the conservation and sustainable use of biological diversity in the Party of import, taking also into account risks to human health, shall not prevent that Party from taking a decision, as appropriate, with regard to the import of the living modified organism in question … in order to avoid or minimize such potential adverse effects.[159]

Reflecting the framing of biosafety in the CBD context as primarily an ecological concern with biodiversity protection, this seemingly broad provision for precautionary risk regulation is circumscribed in various ways. For a start, the fully-fledged advanced informed agreement procedure only applies prior to the first intentional transboundary movement of LMOs for deliberate release into the environment of an importing party (for example GMO seeds).[160] A different procedure applies in respect of LMO-FFPs on the basis that these LMOs do not pose direct ecological risks to the environment of an importing country.

Under the procedure applicable to LMO-FFP imports, exporting countries are required to notify a centralised Biosafety Clearing House within fifteen days of granting domestic approval for a new LMO variety, providing a risk assessment on which the domestic approval was based.[161] An importing country may still take a decision to ban import of the LMO concerned into its territory relying on similar precautionary grounds as are available under the advanced informed agreement procedure.[162] However, in the case of LMO-FFPs the onus lies with the importing country to seek out the relevant information regarding the risks potentially posed by a LMO even though countries may not know whether or when such commodities might be exported to their countries. This shift has important consequences for developing countries, in particular, which may not have the technological capacity to monitor information provided through the internet-based Biosafety Clearing

[158] *Ibid.*, Article 15. This differs from the SPS Agreement where the onus for producing a justificatory risk assessment lies with the importing country, although it can rely on a risk assessment performed by other countries or international organisations.

[159] Biosafety Protocol, Article 10(6).

[160] *Ibid.*, Article 7. [161] *Ibid.*, A11. [162] *Ibid.*, Article 11(8).

House, nor the institutional capacity to assess whether a given LMO is likely to be of concern within their particular national context. Such informational asymmetries are only exacerbated in conditions of scientific uncertainty, which make it more difficult for importing countries to determine the risks posed by import of a particular LMO-FFP.

Influence of WTO considerations in the Protocol negotiations

Commentators examining the negotiations for the Biosafety Protocol have uniformly pointed to issues over the Protocol's potential interrelationship with WTO rules – particularly the science and risk assessment requirements of the SPS Agreement – as an important factor influencing states' negotiating positions.[163] During the course of the Biosafety Protocol negotiations, three WTO rulings under the SPS Agreement were issued by the Appellate Body (*Hormones*, *Salmon* and *Varietals*). Moreover, discussions concerning international controls on the import of LMOs under the Protocol took place in the shadow of a transatlantic dispute over GMO risk regulation, which eventually culminated in the WTO case brought by the USA, Canada and Argentina against the EC in 2003. The perception of many US negotiators at the time was that the Biosafety Protocol was 'not a real environmental treaty aimed at alleviating environmental problems, but one to provide protection to the EU in any WTO litigation over GMOs'.[164] Whether this was the case or not, simmering tensions over the synergies or conflicts between the Protocol and WTO treaties imposed important constraints on negotiations for the former. In particular, this led to attempts to align the decision-making processes in the global biosafety regime with the 'technicalized risk-based understanding of safety within the WTO'.[165]

The central part played by trade concerns in the Biosafety Protocol negotiations worked to give prominence to the views of certain countries and negotiating blocs over others, resulting in the incorporation of particular representations of precautionary risk assessment. Minimising the level of conflict with applicable rules of the WTO and reducing the impacts of the Protocol on the agricultural commodity trade were the

[163] Aarti Gupta, 'Governing Trade in Genetically Modified Organisms: The Cartagena Protocol on Biosafety', *Environment*, 42(4) (2000), 24; Gupta, above 'Problem Framing in Assessment Processes', p. 60; Gaston and Abate, 'The Biosafety Protocol and the World Trade Organization', 112; Holly Saigo, 'Agricultural Biotechnology and the Negotiation of the Biosafety Protocol', *Georgetown International Environmental Law Review*, 12 (2000), 811.

[164] Pollack and Shaffer, *When Cooperation Fails*, p. 153.

[165] Gupta, 'Problem Framing in Assessment Processes', p. 62.

primary negotiating concerns of the so-called Miami Group, consisting of GMO-producing countries such as the USA, Canada, Australia, Chile, Uruguay and Argentina. The Miami Group's main protagonist was the EU, which negotiated the Protocol largely from a GMO importer perspective. Also taking the position of current or potential future importers of GMOs was the Like-Minded Group of developing countries. These nations' concerns over the potentially wide-ranging impacts – including socio-economic consequences – of GMOs within their countries and their lack of technical and regulatory capacity to manage them saw them arguing for the greatest possible flexibility to control GMO imports under a global biosafety regime. Two other negotiating blocs formed during the negotiations: the Eastern European countries, which largely supported the EU position in light of their interest in future integration into the EU and its regulatory regime; and the Compromise Group consisting of countries such as Japan, Switzerland, New Zealand, Mexico, Norway, Singapore and South Korea that reflected a mix of interests aligned on different issues with either the Miami Group or the EU.[166] NGOs also played an important role in the negotiations by lobbying state participants. Supporting the Miami Group's position on trade issues was the Global Industry Coalition of agricultural, food and pharmaceutical companies. Taking up the Like-Minded Group's call for the Protocol to incorporate the greatest possible oversight over the international flow of GMOs were a range of environmental, development-oriented and consumer advocate groups.[167]

In the negotiation process for the Biosafety Protocol, the two most powerful diplomatic blocs – the EU and the Miami Group – each sought to 'internationalize and legitimize their domestic or regional approach to LMO regulation through the vehicle of the protocol'.[168] In the case of the EU, this involved fighting 'tooth-and-nail' in the final stages of the negotiations for the inclusion of the precautionary language that appears in the Protocol's decision-making provisions.[169] Not surprisingly, this language bears a strong resemblance to that found in the European Commission's Communication on the Precautionary

[166] *Ibid.*, pp. 62–4.

[167] For discussion see Stanley W. Burgiel, 'Non-state Actors and the Cartegena Protocol on Biosafety', in Michele M. Betsill and Elisabeth Corell (eds.), *NGO Diplomacy: The Influence of Nongovernmental Organizations in International Environmental Negotiations* (Cambridge, MA: MIT Press, 2007), p. 67.

[168] Gupta, 'Governing Trade in Genetically Modified Organisms', p. 26.

[169] Peter Andrée, 'The Cartagena Protocol on Biosafety and Shifts in the Discourse of Precaution', *Global Environmental Politics*, 5(4) (2005), 37.

Principle, discussed in Chapter 4. It also shares the same assumption as that document regarding the primary relevance of the precautionary principle to the decision-making stage of risk management (as opposed to scientific risk evaluation).[170]

Although conceding on the precautionary language in the Protocol, the end result of the negotiations from the Miami Group's perspective was not unfavourable. For example, the Protocol's decision-making procedures for LMOs take as their starting point the Group's favoured notion of scientifically sound risk assessment, rather than 'the much feared non-scientific criteria for decision-making'.[171] Vigorous objections by the Miami Group also saw the omission of the term 'principle' in provisions relating to the application of precaution, with the result that the controversial question of the international legal status of the precautionary principle (and its potential capacity to trump WTO obligations) was also left conveniently ambiguous.[172] On the pivotal issue of the relationship between the Protocol and its precautionary risk assessment provisions, and obligations under WTO rules such as the SPS Agreement, both the Miami Group and the EU were successful in having language inserted into the Protocol's preamble that preserved intact their respective positions.[173] Indeed, one EU representative remarked that the effect of the competing recitals – calling for the Protocol not to be interpreted 'as implying a change in the rights and obligations of a Party under any existing international agreements' and noting that the latter recital was 'not intended to subordinate [the] Protocol to other international agreements' – was to '"cancel each other out", leaving the legal relationship between the two regimes unclear and allowing both sides to claim a partial victory'.[174]

Accordingly, the version of precautionary risk regulation included in the Biosafety Protocol is a fairly narrow or weak notion of the

[170] See also Ryan Hill, Sam Johnston and Cyrie Sendashonga, 'Risk Assessment and Precaution in the Biosafety Protocol', *Review of European Community and International Environmental Law*, 13(3) (2004), 269.

[171] Gupta, 'Governing Trade in Genetically Modified Organisms', 30.

[172] Andrée, 'The Cartagena Protocol on Biosafety and Shifts in the Discourse of Precaution', 31 notes that the Miami group resisted the use of the term 'principle' because of their concern that this could give precaution formal international legal standing.

[173] Stewart and Johnson, 'A Nexus of Trade and the Environment', 22–3.

[174] Pollack and Shaffer, *When Cooperation Fails*, p. 154. The *GMO* case suggests this partial victory was pyrrhic for the EC given the panel's ruling that the Biosafety Protocol is not relevant to interpretation of the SPS Agreement.

precautionary principle, focused on scientific uncertainties 'due to insufficient relevant scientific information and knowledge' that pertain to 'the extent of the potential adverse effects of a living modified organism on the conservation and sustainable use of biological diversity ... taking also into account risks to human health'.[175] The reference to uncertainties arising as a result of 'insufficient relevant scientific information and knowledge' bears a strong resemblance to the language of Article 5.7 of the SPS Agreement. This suggests that precautionary action under the Protocol is likewise limited to cases where scientific evidence is quantitatively or qualitatively inadequate for the purposes of carrying out a technical risk assessment, rather than applying in situations where uncertainties call into question whether a risk assessment has or is capable of identifying all the hazards of concern to the community.[176] Further, the limitation to uncertainties regarding the *extent* of potential adverse effects of LMOs on biodiversity in the importing party, taking into account human health risks, assumes that the pathways by which LMOs could cause harm are not themselves in doubt.[177] This naturally leads to a focus on more well-defined ecological risk (and related human health risk) pathways, such as the potential for a GMO to transfer its genetic modification to a related organism and/or into gut bacteria of animals or humans consuming a GMO as food, rather than allowing for the fact that scientists may be ignorant of the manner in which a GMO might generate environmental or health hazards. Finally, precautionary action under the Protocol is established as an integral component of a risk-based decision-making framework such that the adoption of precautionary risk management measures is dependent upon a conclusion that uncertainties remain *following* a risk assessment 'carried out in a scientifically sound and transparent manner'.[178]

The dominance of intra-OECD (Organisation for Economic Co-Operation and Development) (and particularly US–EU) framings

[175] Aarti Gupta, 'Advanced Informed Agreement: A Shared Basis for Governing Trade in Genetically Modified Organisms', *Indiana Journal of Global Legal Studies*, 9 (2001), 277–8, arguing that the Protocol's precautionary language 'can be interpreted as privileging a quantitative risk assessment as the legitimate starting point for precautionary action'.

[176] In this respect, though, the Protocol's reference to deficiencies in scientific *knowledge* may be susceptible to a broader interpretation.

[177] Gupta, 'Advanced Informed Agreement', 278.

[178] Andrée, 'The Cartagena Protocol on Biosafety and Shifts in the Discourse of Precaution', 37.

of appropriate risk assessment in the Biosafety Protocol negotiations ultimately helped to achieve a result that minimised the potential for conflict between the Protocol's provisions and the WTO trade regime (a primary political objective of both the Miami Group and the EU). The model of risk assessment constructed during this process is one that is broadly compatible with the science-focused provisions of the SPS Agreement, though it can be said to favour GMO-importers interests by placing the burden of producing a risk assessment on GMO exporters, at least in the case of LMOs seen to pose the most significant risks to biodiversity (that is, LMOs deliberately introduced into the environment of an importing country).

Another, potentially significant, difference between the precautionary provisions of the Protocol and the equivalent allowance for precautionary action under the SPS Agreement (Article 5.7) is the time period over which precautionary measures can be maintained. Whereas the Protocol allows apparently indefinite bans on LMO imports in the face of 'lack of scientific certainty due to insufficient relevant scientific information and knowledge', only provisional measures are permitted under Article 5.7 of the SPS Agreement.[179] As Gupta has observed, precautionary risk assessment in the Protocol is thus best described not as operationalisation of *the* precautionary principle, but rather as 'a mix of existing understandings of precaution as articulated in other global fora, including Principle 15 of the Rio Declaration and Article 5.7 of the SPS Agreement'.[180]

Role of socio-economic factors in biosafety risk regulation

The particularity of the Protocol's version of precautionary risk assessment is highlighted if it is contrasted with the broader vision of biosafety that was advocated by the Like-Minded Group in the negotiations. As highlighted above, this Group pushed for an understanding of biosafety and scope of risk assessment in the Biosafety Protocol that would encompass the socio-economic impacts of LMO trade; an issue of particular salience for developing countries. On this question, the countries of the Like-Minded Group essentially put forward a broader notion of biosafety risk, arguing that decision-making about

[179] However, Article 12 of the Protocol narrows the differences with Article 5.7 of the SPS Agreement by providing for exporter-initiated reviews of import bans imposed by Protocol parties.

[180] Gupta, 'Advanced Informed Agreement', 276.

transboundary movements of LMOs 'transcended narrowly-defined conceptions of harm that were assessable and quantifiable through technical risk assessments (even those that could account for scientific uncertainties)'.[181] According to the Like-Minded Group, a variety of socio-economic concerns were also encompassed within the concept of biosafety, such as the problems of concentrating ownership over biological knowledge in a handful of multinational biotechnology companies, the impact of GMOs on traditional agricultural practices, and the lack of technical and regulatory capacity in developing nations necessary to ensure appropriate monitoring, evaluation and segregation of GMOs once introduced.[182]

In the initial phase of the Biosafety Protocol negotiations, some developing country representatives, together with their NGO allies, looked to the precautionary principle as 'an integral component of a comprehensive assessment of a wide range of social, environmental, health, and economic costs and benefits of LMOs that should be required before these organisms would be allowed to cross international borders'.[183] A strong precautionary approach was particularly advocated by the African member states of the Like-Minded Group, who were concerned over the potential for their territories to become dumping grounds for dangerous new technologies, in much the same way as had occurred with hazardous wastes and chemicals in the past.[184]

As an element of a global precautionary approach to LMO risk regulation, however, the Like-Minded Group's proposal attracted support neither from the Miami Group nor the EU. The latter maintained that precautionary decision-making 'should not be confused with decisions based on "socio-economic considerations"'[185] (although the EU's position, especially as regards the potential human health impacts of GMOs, has since evolved considerably in light of the transatlantic divide over trade in the products of biotechnology).[186] The Miami Group's opposition to the inclusion of socio-economic matters in biosafety risk assessment once again cited trade concerns. The Group contended that

[181] *Ibid.*, 269.

[182] Gupta, 'Advanced Informed Agreement', 72.

[183] Andrée, 'The Cartagena Protocol on Biosafety and Shifts in the Discourse of Precaution', 26.

[184] *Ibid.*, 29.

[185] Andrée, 'The Cartagena Protocol on Biosafety and Shifts in the Discourse of Precaution', 32, quoting from an interview with an EC negotiator.

[186] Gupta, 'Problem Framing in Assessment Processes', p. 73.

the country-specific nature of socio-economic considerations pertaining to GMOs would prevent the development of harmonised rules for LMO transfers under the international biosafety regime and would conflict with WTO rules.[187] The irony of this position was not lost on developing countries and their NGO supporters who noted that while the consideration of socio-economic harms from transboundary LMO movements was frowned upon, the legitimacy of the trade premise – that socio-economic benefits would flow from LMO trade – was not questioned.

While opposition from the Miami Group and the EU did not result in the outright exclusion of socio-economic considerations from the final text of the Biosafety Protocol, they are clearly segregated from matters of risk assessment and precautionary decision-making, which are limited to the consideration of the ecological risks of LMOs for biodiversity, taking also into account human health risks. Socio-economic impacts flowing from trade in LMOs are mentioned in a separate provision that was significantly watered down over the course of the negotiations. What now appears in Article 26 of the Biosafety Protocol is an authorisation that parties:

> in reaching a decision on import under this Protocol or under its domestic measures implementing the Protocol, may take into account, *consistent with their international obligations*, socio-economic considerations arising from the impact of living modified organisms *on the conservation and sustainable use of biological diversity*, especially with regard to the value of biological diversity to indigenous and local communities.[188]

Socio-economic considerations are hence potentially relevant to LMO risk decision-making pursuant to the Protocol, but are restricted in two important ways. The first is with respect to their scope, given the requirement that the socio-economic issues considered relate to environmental (biodiversity) risks. An even more critical limitation, however, is the thinly veiled reference to the need for consistency with parties' WTO obligations, a rider that may render the provision nugatory for those developing countries parties that are also members of the international trade organisation.[189] In the post-Protocol environment, the narrower notion of biosafety and LMO risk assessment reflected in

[187] *Ibid.*, p. 73.

[188] Biosafety Protocol, Article 26(1) (emphasis added).

[189] A point stridently made by the G-77 lead negotiator of the Like-Minded Group: Tewolde Egziabher, 'Safety Denied', Our Planet, 10(2) (1999), available at: www.unep.org/ourplanet/imgversn/102/viewpoint.html.

the treaty's provisions has helped to shift the political debate on GMOs to focus on scientific uncertainty and long-term (often hypothetical) environmental and health risks, even though this tends to devalue the immediate and often more readily apparent socio-economic consequences of the introduction of GMO technologies into agricultural and natural systems.[190] If the latter are to resonate within the Protocol's risk assessment framework, they 'may increasingly have to be articulated in the language of technical risk, harm, and safety'.[191]

As a case study of global requirements governing risk decision-making, the Biosafety Protocol illustrates a situation where solutions to the risk problem at hand had to be worked out against a background of 'persisting normative and scientific conflicts over potential harms'.[192] This context gave rise to a very particular notion of biosafety and LMO risk assessment that sought a reconciliation between regulatory models based on sound science and the precautionary principle advocated by leading negotiating blocs. While the language of 'technical risk' predominates in the Protocol,[193] it was not in fact conflicts over matters of scientific credibility that most heavily shaped the Protocol's provisions. The first-order concerns of negotiators were instead political in nature, though some matters (for instance, trade compatibility between the Protocol and WTO obligations) achieved dominance over others (such as the potential for GMOs to have broad socio-economic impacts in developing countries). As Gupta observes, one general lesson from the Biosafety Protocol experience is that a push for 'sound science' may carry little weight in the face of fundamental value conflicts relating to the framing of the risk issue concerned.[194]

The POPs Convention and its scientific advisory processes

The Biosafety Protocol is an unusual case among multilateral environmental agreements (MEAs) in that negotiations for the treaty were not preceded by a major scientific assessment, synthesising knowledge about the environmental safety issues posed by GMOs. More usually, negotiations for, and the conclusion of, an MEA take place following an initial

[190] Andrée, 'The Cartagena Protocol on Biosafety and Shifts in the Discourse of Precaution', 38.
[191] Gupta, 'Problem Framing in Assessment Processes', p. 80.
[192] Gupta, 'Advanced Informed Agreement', 265.
[193] *Ibid.*, 279.
[194] Gupta, 'Problem Framing in Assessment Processes', p. 79

identification of potential risks through scientific research and the conduct of some form of global-level technical evaluation of that body of research.[195] This global scientific assessment then feeds into political negotiation and policy formation processes, with different scientific assessments exhibiting varying levels of influence over such processes.

The POPs Convention, finalised soon after the Biosafety Protocol in 2001, and entering into force on 17 May 2004, is an example of an MEA where large-scale scientific assessments appear to have exercised significant influence over the nature of the risks dealt with under the treaty, and the manner of their regulation. Moreover, the Convention's principal scientific advisory body, known as the Persistent Organic Pollutants Review Committee (POPs Review Committee), plays an important role in ongoing implementation of global chemical regulation through its work in compiling risk profiles and risk management evaluations for new substances proposed for listing under the POPs Convention. On the basis of recommendations put forward by the POPs Review Committee, the state parties to the Convention agreed at their fourth conference held in May 2009 to list nine additional chemicals as POPs subject to the global regulatory regime.[196]

Some point to such achievements as an indication of the success of the Convention's processes for science–policy interaction and their potential to serve as a model for other MEAs.[197] However, the greatest test for these processes may still lie in the future as the Convention increasingly moves to regulate POPs whose toxicity is not uniformly accepted, and for which the socio-economic consequences of bans would be more acute for many countries.

International regulation of persistent organic pollutants

Chemicals posing risks to human health and the environment have long been a concern of the global environmental movement.[198] In 1962 Rachel Carson's book, *Silent Spring* – drawing attention to the toxic

[195] As was the case for the Montreal Protocol and the climate change treaties, discussed further below.

[196] Report of the Conference of the Parties of the Stockholm Convention on Persistent Organic Pollutants on the work of its fourth meeting, 8 May 2009, UNEP/POPS/COP.4/38.

[197] International Institute for Sustainable Development, 'Summary of the Fourth Meeting of the Persistent Organic Pollutants Review Committee of the Stockholm Convention: 13–17 October 2008', *Earth Negotiations Bulletin*, 15(161) (20 October 2008), 3, available at www.iisd.ca/chemical/pops/poprc4/.

[198] Declaration of the United Nations Conference on the Human Environment, Principle 21, U.N. Doc. A/CONF.48/14 (June 16, 1972) (Stockholm Declaration), Principle 6.

effects of the pesticide DDT for wildlife – sparked the introduction of pollution control laws in the USA and in other Western nations. DDT is a prominent example of a persistent organic pollutant (or POP): organic chemicals characterised by their capacity to persist in the environment, their tendency to accumulate in organisms up the food chain, and their ability to travel long distances in the atmosphere and in water posing risks to human health and the environment far from their site of production.[199] Other well-known POPs include polychlorinated biphenyls (PCBs), dioxins and furans. All are chemicals or chemical by-products of manufacturing processes that have been widely used in industrialised societies since the mid-twentieth century. For this reason, regulating the risks posed by POPs requires more than simply a ban on their use. In addition, there is a need to identify suitable substitutes for POPs in essential manufacturing processes, to remove stockpiles of the chemicals, to undertake the clean-up of contamination, to monitor their health and environmental effects and to initiate the implementation of cleaner technologies.[200] All such risk management measures may entail significant socio-economic consequences, particularly for less well-resourced developing countries.

While many POPs, such as DDT, have been the subject of domestic regulation for a decade or more, POPs only became a matter of international concern during the 1990s. The Agenda 21 action plan produced by the 1992 Rio Conference on Environment and Development was the first global instrument to call for risk reduction programmes focused on 'phasing out or banning of chemicals ... that are toxic, persistent and bioaccumulative and whose use cannot be adequately controlled'.[201] In 1995 the United Nations Environment Programme (UNEP) initiated a global scientific assessment process for twelve well-known POPs (described as the 'dirty dozen'), including DDT, PCBs, dioxins and furans.[202] This assessment was coordinated by the Intergovernmental Forum on Chemical Safety (IFCS) – a body operating under the auspices of the WHO, which consists of 'an alliance of all stakeholders concerned with the sound management of chemicals' (that is governments, international, regional and national organisations, industry groups, public

199 Noelle Eckley, 'Traveling Toxics: The Science, Policy, and Management of Persistent Organic Pollutants', *Environment*, 43(7) (2001), 26–7.

200 *Ibid.*

201 Commission on Sustainable Development, Agenda 21: the United Nations Programme of Action from Rio, (1992), 44.

202 Decision 18/32 Persistent Organic Pollutants, UNEP Governing Council, Nairobi, 25 May 1995.

interest associations, labour organisations, scientific associations and representatives of civil society).[203] The IFCS report identified the need for international action, including a global legally binding instrument, to reduce the risks to human health and the environment posed by the dirty dozen POPs.[204] International negotiations for the POPs Convention subsequently began in June 1998 and concluded in Stockholm in May 2001.[205]

An important precedent for both the IFCS assessment and negotiations for the POPs Convention was the 1998 POPs Protocol and associated scientific evaluations under the Long Range Transboundary Air Pollution Convention (LRTAP).[206] This long-standing treaty exists under the auspices of the United Nations Economic Commission for Europe, drawing its participants from Western Europe and North America. It is known for having a significant scientific component to its regulatory activities.[207] The issue of POPs was initially brought to the LRTAP forum by Canada, which had become increasingly concerned by the long-range, harmful effects of persistent, bioaccumulating chemicals but had been unable to interest global organisations in their regulation.[208] LRTAP presented a more receptive forum for the issues raised by Canada, resulting in scientific work being undertaken by various LRTAP task forces and working groups, and the initiation of negotiations for a protocol to deal with the long-range pollution issues posed by POPs.[209] As detailed further in the following section, the scientific assessment process for POPs in the LRTAP context would appear to have

[203] See the ICFS website at www.who.int/ifcs/page2/en/index.html.

[204] IFCS Ad Hoc Working Group on Persistent Organic Pollutants Meeting, Final Report, 21–22 June 1996, Manila, Philippines, IFCS/WG.POPs/Report.1, 1 July 1996, 4.1.

[205] Negotiations were initiated by Decision 19/13 C of 7 February 1997 of the Governing Council of UNEP.

[206] Convention on Long-Range Transboundary Air Pollution, 13 November 1979, 1302 UNTS 217, in force 16 March 1983 (LRTAP).

[207] See, generally, Jorgen Wettestad, 'The ECE Convention on Long-Range Transboundary Air Pollution: from Common Cuts to Critical Loads', in Steinar Andresen, Tora Skodvin and Arild Underdal (eds.), *Science and Politics in International Environmental Regimes: Between Integrity and Involvement* (Manchester University Press, 2000), p. 95; Rolf Lidskog and Göran Sundqvist, 'The Role of Science in Environmental Regimes: The Case of LRTAP', in Peter M. Haas (ed.), *International Environmental Governance* (Aldershot, Ashgate Publishing, 2008), p. 211.

[208] Noelle Eckley-Selin, 'From Regional to Global Information: Assessment of Persistent Organic Pollutants', in Ronald B. Mitchell *et al.* (eds.), *Global Environmental Assessments: Information and Influence* (Cambridge, MA: MIT Press, 2006), p. 178.

[209] Protocol to the 1979 Convention on Long-Range Transboundary Air Pollution on Persistent Organic Pollutants, 25 June 1998, 2230 U.N.T.S. 79, in force 23 October 2003.

played a major role in shaping risk regulation of POPs at the global level.[210]

Like the LRTAP POPs Protocol, the global POPs Convention focuses on controlling the use, production and trade in chemicals that are listed in one of the annexes to the Convention.[211] Initially, listed chemicals were confined to the dirty dozen POPs. However, the POPs Convention is intended to be a dynamic instrument for chemicals regulation, and thus incorporates a process for listing additional substances of concern based on scientific analyses of their persistence and accumulation.[212] Any party may propose a new chemical for listing under the Convention, though it must provide information with its proposal discussing the persistence of the chemical concerned, its bioaccumulation, its potential for long-range environmental transport and data on its adverse health effects or eco-toxicity.[213]

As discussed further below, a proposal that contains the requisite information is examined by the POPs Review Committee and, if it fulfils specified screening criteria, a risk profile and risk management evaluation are prepared by the Committee as the basis for a recommendation to state parties as to whether the proposal should proceed. Precaution is an explicit part of this decision-making process as 'lack of full scientific certainty' is not grounds for preventing a proposal from proceeding.[214] Moreover, the Committee's review and risk assessment process – while ostensibly scientific in nature – applies criteria that are in fact a mixture of scientific and policy judgments.[215] For instance, the purpose of the Committee's evaluation of a chemical proposed for listing is said to be 'whether the chemical is likely, as a result

210 See Eckley-Selin, 'From Regional to Global Information', p. 178.

211 There are three annexes to the Convention. Annex A lists chemicals that have been identified for elimination; Annex B lists chemicals the use of which has been restricted; Annex C lists chemicals for which special requirements apply when they are formed and released unintentionally from anthropogenic sources.

212 Noelle Eckley, 'Dependable Dynamism: Lessons for Designing Scientific Assessment Processes in Consensus Negotiations', *Global Environmental Change*, 12 (2002), 15.

213 POPs Convention, Annex D.

214 POPs Convention, Article 8(7)(a). Also see Article 1. Much as for the Biosafety Protocol, inclusion of precaution was one of the most controversial issues in the POPs negotiations, particularly between the USA (favouring a precautionary approach based on scientific assessments) and the EU (wanting explicit reference to the precautionary *principle*): Eckley, 'Traveling Toxics', 34.

215 Hence there are no scientifically verifiable thresholds that separate POPs from non-POPs: *ibid.*, 28–9.

of its long-range environmental transport, to lead to *significant adverse* human health and/or environmental effects, such that *global* action is warranted'.[216]

Influence of LRTAP scientific assessments on the POPs Convention

Having only been in force less than a decade, little detailed information exists regarding the operation of the POPs Convention and the performance of its expert Review Committee. One of the few studies to have been undertaken is an investigation by Noelle Eckley Selin, a researcher working with other Harvard University scholars to examine the influence of a range of global scientific assessments on environmental policy.[217] Eckley Selin's analysis of the POPs Convention, based on interviews and a survey of relevant assessments and documentary evidence, found an important influence in the POPs negotiations and in determining the shape of the eventual treaty was scientific assessments of POPs undertaken for the LRTAP regime.[218] She concluded that this occurred in three major ways.

First, the LRTAP assessment processes 'identified and defined the POPs problem, pushed the issue of POPs onto the global agenda, and set a dominant global framing of POPs as an international problem'.[219] Put another way, the LRTAP processes would appear to have played an important role in defining the nature of the risks of concern dealt with by the global chemicals regime. This was significant because the notion of a POP is not the only way of framing the problems or risks posed by chemicals such as DDT. As Eckley Selin points out, the global regime might equally have decided to focus on the hazards of pesticides or toxic chemicals in general, of chemical stockpiles, of the need for technical assistance in chemicals management or local problems associated with chemicals in international trade.[220] Moreover, the concept of a POP is not a purely scientific construct: scientific criteria can be used to determine values for traits such as persistence and bioaccumulation but cannot answer the broader value question of what levels give rise to 'significant adverse' effects, necessitating 'global' as opposed to domestic action.

[216] POPs Convention, Annex E (emphasis added).

[217] See the website for the Global Environmental Assessment Project: www.hks.harvard.edu/gea/.

[218] Eckley-Selin, 'From Regional to Global Information', pp. 176–7.

[219] *Ibid.*, pp. 177–8. [220] *Ibid.*, p. 179.

Second, Eckley Selin identified the function of assessment-related information on POPs as mobilising additional actors beyond the LRTAP domain around the issue at the global level. The core information used in the LRTAP process was the same as that used by the ICFS in compiling its report for UNEP. Powerful state actors, such as the USA and the EU, had thus already been exposed and agreed to this information. However, Eckley Selin notes the preparation of a separate assessment under the auspices of globally representative institutions was critical to other, non-LRTAP states' acceptance of the use of LRTAP information.[221] In addition, a series of subregional workshops was convened by UNEP during 1997–8 to consider the more localised problems posed by POPs, particularly in developing countries. This exercise also served to mobilise a greater range of actors, both 'by encouraging them to assess whether POPs were domestically regulated, and to conduct national scientific assessments of POPs', and also by facilitating the establishment of a trans-governmental network of chemicals regulators.[222]

Third, the existence of LRTAP and its assessments provided a 'road map' for the global POPs negotiations.[223] This was particularly important when it came to agreeing on the list of chemicals to be regulated under the POPs Convention, an issue that ordinarily might have been expected to be controversial.[224] However, the dirty dozen list identified by UNEP and IPCS was agreed fairly rapidly in the global POPs negotiations, facilitated by the scientific information on those chemicals that had been collected and compiled for the LRTAP discussions.[225] Eckley Selin also observes the 'large role' played by the LRTAP precedent in agreeing on a set of scientific criteria under the POPs Convention to be applied in the treaty's process for adding new chemicals to those already listed.[226] Although the LRTAP criteria were initially 'greeted with considerable controversy', especially by sceptical developing countries, discussion eventually 'coalesced around this precedent'.[227] Eckley Selin attributes this to the perceived scientific rationality of the criteria (given LRTAP's reputation for scientific credibility), their political rationality in that they resulted in 'a sensible policy outcome

[221] *Ibid.*, p. 181. [222] *Ibid.* [223] *Ibid.*

[224] This was the case in the LRTAP context: see Henrik Selin and Noelle Eckley, 'Science, Politics, and Persistent Organic Pollutants: The Role of Scientific Assessments in International Environmental Cooperation', *International Environmental Agreements: Politics, Law and Economics*, 3 (2003), 17.

[225] Eckley-Selin, 'From Regional to Global Information', pp. 182–3.

[226] *Ibid.*, p. 184. [227] *Ibid.*, pp. 188–9.

regarding which chemicals were included or not', and the fact that the criteria were subjected to a global review and reassessment, which de-emphasised the original LRTAP connection.[228]

While LRTAP scientific assessments of POPs were able to achieve significant influence over the shape of the global negotiations on the issue, a key challenge they faced was gaining acceptance with developing countries, tropical countries and southern hemisphere states.[229] POPs are significant local pollutants for many countries, hence different types of risks were seen to be of importance by these countries than the long-range transport and toxicological issues emphasised by the northern-focused LRTAP assessments.[230] The solution to this challenge lay in UNEP's activities to facilitate national and regional assessments of POPs problems.[231] This experience suggests that incorporating localised risk assessment data and ensuring broad participation in global scientific processes may be an important way in which 'universal' science can achieve general political acceptance.

Role of the POPs Review Committee in chemical risk assessment

There is some evidence to suggest that the dirty dozen list of chemicals initially included in the POPs Convention achieved broad global acceptance because these substances were well regulated and well known.[232] Indeed, many countries had already taken action to ban the dirty dozen POPs prior to the initiation of negotiations for the POPs Convention, leading to their designation as 'dead' chemicals in regulatory terms. By contrast, the procedures overseen by the POPs Review Committee for adding new chemicals to the lists under the Convention are likely over time to deal more with 'live' chemicals, that is those still in active use by countries around the world and traded internationally. A significant challenge thus facing the POPs Review Committee is whether its processes, which place an emphasis on scientific criteria and expert consensus, can achieve broad legitimacy where risk assessments are conducted against a backdrop of socio-economic conflicts.

As for other MEA and global scientific advisory bodies, questions over who makes up such bodies and how these experts are chosen are likely to be critical.[233] The POPs Convention sets out some standard, albeit fairly minimal, requirements for the selection of members of

[228] *Ibid.*, pp. 190–1. [229] *Ibid.*, p. 191. [230] *Ibid.*, pp. 191–2.
[231] *Ibid.*, p. 193. [232] *Ibid.*, p. 197.
[233] Winickoff and Bushey, 'Science and Power in Global Food Regulation', 17.

the POPs Review Committee, such as that they should be 'government-designated experts in chemical assessment or management' appointed 'on the basis of equitable geographical distribution'.[234] Given the significant economic consequences of listing new chemicals under the POPs Convention, conflict of interest rules and the extent of any affiliation between committee members and chemicals' manufacturers are also emerging as important considerations.[235]

The major function performed by the POPs Review Committee, as highlighted earlier, is the preparation of a risk profile and risk management evaluation for those chemicals proposed for listing by state parties that are assessed as meeting scientific criteria relating to persistence, bioaccumulation, long-range transport and adverse effects.[236] In carrying out this screening exercise the Committee is directed to act 'in a flexible and transparent way, taking all information provided into account in an integrative and balanced manner'.[237] If a state party disagrees with the Committee's assessment that a chemical does not meet the screening criteria, it can essentially seek a review of the Committee's decision by the Conference of the Parties (COP).[238]

For proposals for the listing of new chemicals that proceed past the screening stage, the Committee then prepares a draft risk profile in accordance with the requirements specified in Annex E of the POPs Convention. These requirements direct the conduct of a fairly standard, science-based chemical risk assessment process that combines a hazard assessment with an evaluation of the environmental fate of POPs and local exposure levels to come up with a characterisation of the risks in terms of 'whether the chemical is likely, as a result of its long-range environmental transport, to lead to significant adverse human health and/or environmental effects, such that global action is warranted'. The draft risk profile is circulated to parties and observers for their 'technical comments' before being finalised by the Committee.[239]

234 POPs Convention, Article 19(6)(a).

235 See Decision SC-1/7 of the Persistent Organic Pollutants Review Committee, in the Report of the Conference of the Parties of the Stockholm Convention on Persistent Organic Pollutants on the work of its fourth meeting, 8 May 2009, UNEP/POPS/COP.4/38, 13. The parties have also requested the Review Committee to propose to the COP at its fifth meeting in 2011 amendments, as appropriate, to the rules of procedure of the Committee set out in Decision SC-1/8 for preventing and dealing with conflicts of interest relating to the activities of the Committee.

236 Details of these criteria are set out in POPs Convention, Annex D.

237 POPs Convention, Article 8(3).

238 *Ibid.*, Article 8(5). 239 *Ibid.*, Article 8(6).

Based on the risk profile, if the Committee determines that there are global risks of concern posed by a chemical (applying a precautionary approach where appropriate), then it goes on to seek relevant risk management information from the parties and observers in order to prepare a risk management evaluation.[240] The information gathered should relate to 'socio-economic considerations associated with possible control measures', reflecting 'due regard for the differing capabilities and conditions among the Parties'.[241] It may include information on the efficacy and efficiency of possible control measures, their technical feasibility, available alternatives for the chemical concerned, waste and disposal implications, access to information and public education, and monitoring capacity. Also relevant is what might be described as a risk–risk trade-off evaluation of the positive or negative social aspects of implementing control measures (for example, the banning of DDT may be problematic for countries that use the chemical for controlling mosquito-borne malaria outbreaks).

The final step in the process involves the Committee submitting recommendations to the COP, on the basis of the risk profile and risk management evaluation, as to whether the chemical should be included in the lists under the Convention. The final decision on listing is thus that of the COP, which must take 'due account of the recommendations of the Committee, including any scientific uncertainty' and act 'in a precautionary manner'.[242]

At its May 2009 meeting the COP made decisions to list nine new chemicals that had been examined by the POPs Review Committee. The Committee's recommendations were developed on a consensus basis (as is urged by the Convention), and in accordance with the Committee's 'cooperative spirit ... grounded in both the principles of science-based evaluation and the common goal of protecting humans and the environment from the risks posed by POPs'.[243] While the report of the COP does not give much indication of the extent of the role played by the Committee's scientific advice and risk evaluations in the parties' decision-making process, the analysis prepared by

[240] *Ibid.*, Article 8(7). There are review rights specified for parties in the event of the Committee deciding to set a proposal aside at this stage: Article 8(8).

[241] *Ibid.*, Annex F.

[242] *Ibid.*, Article 8(9).

[243] International Institute for Sustainable Development, 'Summary of the Fourth Meeting of the Persistent Organic Pollutants Review Committee of the Stockholm Convention', 16.

the independent reporting service known as the Earth Negotiations Bulletin is more revealing. It highlights that agreement on the listing of some chemicals for which technical or economic issues were more acute (for instance, the substance perfluorooctane sulfonate used widely in a variety of industrial, fire-fighting and pest-control applications) proved more difficult than for others.[244] Agreement was eventually reached on the basis of a compromise package that tied agreement on new listings to greater provision for financial and technical assistance.[245] This result strongly suggests that where the social concerns surrounding health and environmental risks are more contested, scientific agreement alone will not be sufficient to generate political consensus.

As the POPs Convention moves to consider the listing of more 'live' chemicals, disputes over risks and the underlying scientific assessments are likely to become more common. In the case of 'live' chemicals, many are newer substances for which the human health and environmental impacts are less well understood. Scientific uncertainty over the hazards posed by potential POPs may lead to delays in the risk assessment process for proposed new listings as some experts call for more time in order for the scientific knowledge base to develop, whereas others stress the need for precautionary action to address potentially serious risks despite remaining uncertainties. The preference for consensus-based decision-making in the POPs Review Committee may also come under stress in such circumstances if several of the expert members object to deferring decisions to await the gathering of more data. Already, the consideration of listing proposals for some chemicals, such as endosulfan – a widely used agricultural insecticide – has generated significant disagreement within the Review Committee, necessitating a vote as to whether to continue with examination of the proposal.[246] In turn, if scientific consensus cannot be achieved on the listing of new chemicals, then this is likely to open up scope for greater political

244 See International Institute for Sustainable Development, 'Summary of the Fourth Conference of the Parties to the Stockholm Convention on Persistent Organic Pollutants: 4–8 May 2009', *Earth Negotiations Bulletin*, 15(174) (11 May 2009), available at www.iisd.ca/chemical/pops/cop4/.

245 Agreement on a non-compliance mechanism, which had been a key objection of developing countries such as China and India, was also sacrificed as part of the package deal.

246 See International Institute for Sustainable Development, 'Summary of the Fourth Meeting of the Persistent Organic Pollutants Review Committee of the Stockholm Convention'.

disagreement over risk management issues when decisions eventually come before the COP.[247]

Negotiated science-policy risk assessments in the IPCC

While socio-economic concerns are a source of dispute in the global regulation of chemical risks, such conflicts pale in comparison to those generated by the issue of climate change. Climate change regulation has been described as a field 'born in politics' given the important social, economic and value dimensions of the problem.[248] These relate to:

- the differing contributions of countries to the problem, with developed countries historically having emitted the highest levels of GHGs (although large developing countries such as China and India are set to exceed the emissions of the developed world in the near future);[249]
- the major changes in energy production and use necessary to reduce GHG (involving a switch from cheaper, dirtier sources such as coal and oil to cleaner, renewable technologies);
- the significant social ramifications of adapting to climate change, including the potential for large-scale migration of populations away from areas rendered uninhabitable by climate change in the future;[250]
- uncertainty over the precise effects of climate change on ecological and social systems, particularly at the local level;[251] and
- the long time horizons involved in responding to the problem, which extend far beyond the timeframe of political electoral cycles.

[247] See e.g., Report of the Conference of the Parties of the Stockholm Convention on Persistent Organic Pollutants on the work of its fourth meeting, 8 May 2009, UNEP/POPS/COP.4/38, pp. 10 and 13, discussing parties' concerns over the departure from consensus decision-making by the Review Committee in considering a proposal relating to endosulfan.

[248] Shardul Agrawala, 'Context and Early Origins of the Intergovernmental Panel on Climate Change', *Climatic Change*, 39 (1998), 614.

[249] Ross Garnaut et al, 'Emissions in the Platinum Age: the implications of rapid development for climate change mitigation', *Oxford Review of Economic Policy*, 24(2) (2008), 1.

[250] For discussion see Norman Myers, 'Environmental Refugees: An Emergent Security Issue' (13th Economic Forum, 2005); Christian Aid, *Human Tide: The Real Migration Crisis* (London: Christian Aid, 2007).

[251] This is a particular problem for developing countries because of the lack of data on the implications of climate change for these countries and their poor representation in global modelling: Paul J. Runci, 'Expanding the Participation of Developing Country Scientists in International Climate Change Research', *Environmental Practice*, 9(4) (2007), 225.

From an early stage, politicisation of the climate change issue brought concerns about legitimacy to the forefront in the design of global processes for assessing the scientific evidence of climate change and its impacts. As a consequence, the international body charged with this task – the IPCC – represents 'the careful crafting of a process of appointing scientists, reviewing reports, and producing policymaking summaries'.[252] The IPCC's risk assessment processes, which meld scientific data with policy input, have proved very successful in generating a high level of scientific consensus around the issue of climate change and widespread acceptance of this expert view of the nature and extent of climate change risks. Its remarkable achievements in this regard were recognised by the award of a Nobel Peace Prize in 2007 for its 'efforts to build up and disseminate greater knowledge about man-made climate change, and to lay the foundations for the measures that are needed to counteract such change'. Nonetheless, the challenge remains to translate such scientific consensus into global political agreement on stringent policy and legal measures for dealing with climate change.

Science–policy processes of the IPCC

The IPCC is an international, inter-governmental institution, drawing on scientific expertise from around the world. Its mandate is 'to assess on a comprehensive, objective, open and transparent basis the scientific, technical and socio-economic information relevant to understanding the scientific basis of risk of human-induced climate change, its potential impacts and options for adaptation and mitigation'.[253] While the kind of global risk assessments the IPCC generates are an increasingly common feature of multilateral environmental activity, the body is often regarded as a watershed institution given the size and comprehensiveness of its assessment processes, as well as its unique, inter-governmental structure especially designed to secure the credibility of its assessments with both scientific and political communities. Accordingly, the IPCC experience has been seen as one that promises

[252] Wendy E. F. Torrance, 'Science or Salience: Building an Agenda for Climate Change', in Ronald B. Mitchell *et al.* (eds.), *Global Environmental Assessments: Information and Influence* (Cambridge, MA: MIT Press, 2006), p. 51.

[253] Principles Governing IPCC Work, approved at the Fourteenth Plenary Session of the IPCC (Vienna, 1–3 October 1998) on 1 October 1998 and amended at the 21st Session (Vienna, 3 and 6–7 November 2003) and at the 25th Session (Mauritius, 26–28 April 2006), Principle 2.

to 'shed light on what may be fruitful ways to think about the role and status of scientific information used for policy purposes',[254] and is looked to as a possible model for global risk assessments in other fields.[255]

The IPCC was initially created in 1988 under the joint auspices of the World Meteorological Organisation (WMO) and UNEP. It produced its first report in 1990 in the lead-up to global agreement on the United Nations Framework Convention on Climate Change (UNFCCC),[256] and its second in 1995 prior to conclusion of the Kyoto Protocol in 1997.[257] Subsequent reports of the IPCC were released in 2001 and 2007,[258] with each report becoming progressively more sobering and less uncertain in its projections of climate change and the likely impacts on ecosystems and human societies.[259] The organisation is currently working towards its fifth assessment report with the aim of finalising it by the end of 2014.

The IPCC was not the first global body set up to investigate, and advise the international community on, the risk of climate change. In 1985 UNEP and the WMO, together with the International Council of Scientific Unions (now the International Council for Science),[260] established an Advisory Group on Greenhouse Gases consisting of a blue ribbon panel of experts with responsibility for assessing the available scientific information on atmospheric GHG levels and the likely impacts

254 Alison Shaw and John Robinson, 'Relevant but not Prescriptive? Science Policy Models Within the IPCC', *Philosophy Today*, 48(5) (2004), 84.

255 The IPCC process provided a blueprint for the Millennium Ecosystem Assessment completed in 2005 under UN auspices. Similar assessment processes have been called for for multidimensional global risk issues such as biodiversity loss and water scarcity: Bernd Siebenhüner, 'The Changing Role of Nation States in International Environmental Assessments – the Case of the IPCC', *Global Environmental Change*, 13 (2003), 117.

256 Intergovernmental Panel on Climate Change, 'Climate Change 1990: The Scientific Basis. Contribution of Working Group I to the Third Assessment Report of the Intergovernmental Panel on Climate Change', (1990).

257 Intergovernmental Panel on Climate Change, 'IPCC Second Assessment: Climate Change 1995. A Report of the Intergovernmental Panel on Climate Change', (1995).

258 Intergovernmental Panel on Climate Change, 'Climate Change 2001: Synthesis Report', (2001); Intergovernmental Panel on Climate Change, 'Climate Change 2007: Synthesis Report', (2007).

259 Runci, 'Expanding the Participation of Developing Country Scientists in International Climate Change Research', 225.

260 The ICSU is a scientific NGO, founded in 1931, and committed to the principle of the universality of science.

of predicted increases.[261] The Advisory Group structure was modelled on the ozone assessment processes that were widely regarded as highly successful in catalysing global action in the 1980s to control the production and consumption of ozone-depleting substances.[262] However, this model – essentially one that foresaw a linear transmission of high-quality scientific expertise into international policy and law-making processes – proved unworkable in the vastly more politically charged context of climate change. As Shardul Agrawala notes:

> Policy action on climate change needed to be *global*, would affect *entire economies* and hence widespread governmental support for any policy response from both developed and developing countries was a must.[263]

The defining feature of the IPCC, its *inter-governmental* status, can thus be seen as a play for gaining greater acceptance of scientific climate change assessments by policy-makers and governments by giving these actors a stake in the assessment process.[264]

Nonetheless, the initial stages of the IPCC's risk assessment process follow a more traditional sound science model, based upon expert review of the relevant peer-reviewed literature. Experts participating in each of the organisation's three working groups (dealing respectively with the physical scientific aspects of the climate system and climate change, the vulnerability and adaptation of socio-economic and natural systems to climate change, and options for mitigating climate change) contribute to reports of their working groups, with designated 'lead authors' responsible for coordinating the content of each chapter. The reports then undergo a process of expert peer review by independent scientists designed to check the scientific soundness of the information contained in the reports.

The review process for the reports does not end once the comments from scientific peer review have been incorporated. As was mentioned in Chapter 3, there are further review processes that take place which

[261] John Houghton, 'An Overview of the Intergovernmental Panel on Climate Change (IPCC) and Its Process of Science Assessment', *Issues in Environmental Science and Technology*, 17 (2002), 1.

[262] Richard Benedick, *Ozone Diplomacy: New Directions in Safeguarding the Planet*, 2nd edn (Cambridge, MA: Harvard University Press, 1998).

[263] Agrawala, 'Context and Early Origins of the Intergovernmental Panel on Climate Change', 612.

[264] Tora Skodvin, 'The Intergovernmental Panel on Climate Change', in Steinar Andresen *et al.* (eds.), *Science and Politics in International Environmental Regimes: Between Integrity and Involvement* (Manchester University Press, 2000), pp. 173–4.

involve the IPCC Panel, consisting of governmental delegations of all member countries.[265] All IPCC reports must be endorsed by the Panel during a working group or a plenary session, meaning that the scientific assessments they contain are subjected to scrutiny by hundreds of officials and experts from relevant government ministries, agencies and research institutions. In addition, the all-important summaries of the expert IPCC reports – known as the Summary for Policy-makers – require Panel approval. In this process, government representatives of the IPCC Panel engage in line-by-line (often word-by-word) discussion and agreement on the text of the summaries.[266]

The explicit blending of science and policy inherent in the IPCC's risk assessment processes has made the institution susceptible to critiques that the results of its assessments are politically biased.[267] Yet there is also a growing body of research that suggests that the involvement of governments in the IPCC's assessment process is a strength of the model, ensuring its continuing relevance and influence in the area of international climate change action.[268] The inclusion of governmental representatives seems not only to have played a role in educating a broad range of policy-makers about the problem of climate change, but has also aided scientists in tailoring their advice to address the questions of greatest concern to governments wishing to introduce measures that might address the problem. The IPCC experience hence indicates the importance of a global risk governance model 'that

[265] Membership of the IPCC is open to all member countries of the UN and currently stands at 194 members.

[266] 'Procedures for the Preparation, Review, Acceptance, Adoption, Approval and Publication of IPCC Reports', Appendix A, Principles Governing IPCC Work, adopted at the Fifteenth Session of the IPCC (San Jose, 15–18 April 1999) amended at the Twentieth Session (Paris, 19–21 February 2003) and Twenty-first Session (Vienna, 3 and 6–7 November 2003).

[267] Shaw and Robinson, 'Relevant but not Prescriptive?', 86; Sinclair Davidson and Alex Robson, 'Certainty Clouds the IPCC', *Institute of Public Affairs Review*, 59(1) (2007), 7. Critiques of the IPCC have also been undertaken from a social constructivist perspective, e.g., Sonja Boehmer-Christiansen, 'Global Climate Protection Policy: The Limits of Scientific Advice, Part 1', *Global Environmental Change*, 4(2) (1994), 140; Sonja Boehmer-Christiansen, 'Global Climate Protection Policy: The Limits of Scientific Advice, Part 2', *Global Environmental Change*, 4(3) (1994), 185.

[268] Skodvin, 'The Intergovernmental Panel on Climate Change', pp. 173–4; Agrawala, Context and Early Origins of the Intergovernmental Panel on Climate Change', 611; Bernd Siebenhüner, 'Can Assessments Learn, and If So, How?', in Alexander Farrell and Jill Jäger (eds.), *Assessments of Regional and Global Environmental Risks: Designing Processes for the Effective Use of Science in Decisionmaking* (Washington DC: Resources for the Future, 2006), p. 166; Shaw and Robinson, 'Relevant but not Prescriptive?', 90.

encourages science–policy interaction, and manages the negotiation at the interface' in order to '(co-)produce better questions, formulations, assessments, and products than either independently'.[269]

Institutional reforms in the IPCC to enhance credibility

The IPCC's quest to ensure that its assessments have policy salience (and hence are more attuned to real world conditions than narrowly science-focused evaluations) has not been without cost to perceptions of whether its work is also scientifically rigorous. Over its twenty-one-year history, the IPCC has undergone a number of institutional evolutions designed to enhance its scientific credibility in the face of charges of political bias.[270] Overall these changes have resulted in the introduction of more transparent and systematised procedures for the gathering, evaluation and review of the knowledge summarised in the IPCC's assessment reports. However, these procedural innovations have also contributed to making the IPCC's structures more rigid and cumbersome, potentially limiting the organisation's capacity to respond quickly as new questions arise, to adapt flexibly to the changing demands of policy-makers, or to recognise the impact of different social and cultural influences on the framing of climate change risks.

The evolution of the IPCC's processes for expert peer review of its reports over the course of the last two decades is a case in point.[271] These processes were radically overhauled following the release of the IPCC's first assessment report in 1990 that had included peer review of some of the working groups' conclusions but only on an ad hoc, informal basis. Following the release of its first assessment report, the IPCC came under intense media scrutiny and, at times, strident political attack, as the climate change issue gained increasing public and governmental attention, culminating in a decision of the United Nations General Assembly to launch negotiations for a climate change convention as part of the Rio Summit. Many of the challenges to the IPCC's conclusions came from US fossil fuel interests,

[269] Shaw and Robinson, 'Relevant but not Prescriptive?', 90.

[270] See generally, Bert Bolin, *A History of the Science and Politics of Climate Change: the Role of the Intergovernmental Panel on Climate Change* (Cambridge University Press, 2007). Bolin was the IPCC Chairman from 1988 to 1997.

[271] A detailed analysis is provided by Shardul Agrawala, 'Structural and Process History of the Intergovernmental Panel on Climate Change', *Climatic Change*, 39 (1998), 621.

such as the Global Climate Coalition, who brought to bear similar, adversarial strategies to those well rehearsed in domestic judicial review actions.[272]

The response of the IPCC, particularly under its second chairman, Robert Watson (a former, prominent scientific advisor to the US Clinton administration), bore many similarities to reforms applied by American risk regulatory agencies throughout the 1970s and 1980s.[273] For example, a formalised two-stage process of external review was introduced for all working groups which requires, first, formal review of draft chapters of reports by a large number of independent scientific experts, followed by a second review process in which both governments and experts are involved. Following the release of the IPCC's second assessment report, a further refinement to this process was introduced (again to counter science-based challenges to some of its methodologies). This reform saw the appointment of review editors for each chapter of the working group reports who take responsibility for ensuring that reviewers' comments are considered and dealt with appropriately in revising the text of the reports.

At the other end of the IPCC's risk assessment process – the line-by-line approval of the Summary for Policy-maker documents by governmental representatives – procedures have also been formalised in an effort to maintain the scientific integrity of the results. For instance, lead authors of the various chapters sit in during plenary sessions of the IPCC Panel and play an influential role given the requirement that all statements in the summaries have to be consistent with the bulk of the underlying technical reports.

Changes to the IPCC's peer review process have done much to shore up its international scientific credibility and to insulate it against

[272] Holly Doremus, 'Lots of Science, Not Much Law: Why Knowledge Has Not (Yet) Been Power Over Greenhouse Gas Emissions', in William H. Rodgers, Jr, Jeni Barcelos, Anna T. Moritz and Michael Robinson-Dorn (eds.), *Global Warming: A Reader* (Durham, NC: Carolina Academic Press, forthcoming 2010) (copy on file with the author). The Coalition suffered large membership losses following the issue of the IPCC's Third Assessment Report and has since been deactivated.

[273] Clark Miller, 'The Design and Management of International Scientific Assessments: Lessons from the Climate Change Regime', in Alexander Farrell and Jill Jäger (eds.), *Assessments of Regional and Global Environmental Risks: Designing Processes for the Effective Use of Science in Decisionmaking* (Washington DC: Resources for the Future, 2006), p. 193. For discussion of the evolution of US risk regulatory processes see Chapter 4.

science-based challenges; a remarkable achievement given the extensive involvement of government representatives in its assessment process.[274] Yet it is unlikely that the IPCC science–policy processes could be applied as a general model for the use of science in global risk governance. At a practical level, the IPCC's complex assessment and extended peer review process – involving thousands of scientists worldwide – would be unwieldy for all risk problems and, indeed, may be unnecessarily involved where the issues at hand are less controversial from a scientific or political perspective.[275] Some have also argued that it is not the involvement of governments and policy-makers in the IPCC's processes that is problematic but the institution's rigorous abstention from policy prescription given that the IPCC purports to confine itself to assessments that are 'neutral with respect to policy'.[276] This is said to reduce the impact of its risk assessment findings and limit the possibilities for stimulating research in areas where knowledge is most lacking.[277]

[274] The US National Research Council, reviewing the IPCC's Third Assessment Report, concluded that the analyses on which it was based were 'scientifically credible' and the conclusions in the policy-makers' summaries 'consistent with the main body of the report': National Research Council, *Climate Change Science: An Analysis of Some Key Questions* (Washington DC: National Academies Press, 2001), p. 22. More recently, the Netherlands Environmental Assessment Agency's review of the IPCC's Fourth Assessment Report - *Assessing an IPCC assessment, An analysis of statements on projected regional impacts* (2010) - confirmed its most important conclusions, albeit identifying a need for greater transparency and a tendency to over-emphasise worst-case scenarios in some cases. However, the 'Climategate' scandal that came to light in November 2009, demonstrates the continuing potential for challenges to the credibility of climate science and the IPCC's assessment reports relying on such science. Climategate arose after hackers uncovered and leaked emails written by scientists at the Climatic Research Unit of the University of East Anglia, many of whom contributed to IPCC reports. The emails appeared to suggest that the scientists concerned withheld, deleted or manipulated data to exaggerate the case for global warming. Subsequently, an Independent Climate Change Email Review, headed by former senior British public servant Sir Muir Russell, backed the integrity of the scientists and found they had not undermined the conclusions of the IPCC. Nonetheless, in the wake of Climategate, senior climate scientists have conceded the need to be more upfront, open and explicit about uncertainties in their research: Paola Totaro, 'Climate scandal a 'game changer'', *The Age* (Melbourne), 6 July 2010, at www.theage.com.au/world/climate-scandal-a-game-changer-20100705-zxjw.html.

[275] The IPCC's Fourth Assessment Report was the work of 130 lead authors, with contributions from more than 800 scientists and the oversight of more than 2,500 expert reviewers.

[276] Principles Governing IPCC Work, Principle 2.

[277] Agrawala, 'Structural and Process History of the Intergovernmental Panel on Climate Change', 631; A. Barrie Pittock, 'What Next for IPCC?', *Environment*, 44(10) (2002), 20.

In addition, the institutional reforms that have been necessary to improve the perceived scientific credibility and policy salience of the IPCC's work could restrict the organisation's capacity to take account of the risk perspectives of other, broader audiences who are increasingly affected by its work. Early in the IPCC's history, critiques of the organisation's legitimacy focused mainly on poor participation in its assessment processes by experts and governmental representatives from developing countries. Significant reforms have since taken place to improve developing countries' participation.[278] As a result, the numbers of participants from developing countries involved in IPCC assessments, or attending its meetings, have been greatly boosted (although a number of persistent obstacles to full developing country involvement remain).[279]

Beyond the issue of developing country participation, recently emerging critiques of the IPCC suggest that in the future it will need to secure even broader legitimacy for its findings through demonstrating capacity to adapt to, and incorporate, risk framings that depart from dominant scientific (and policy) understandings of the climate change problem. Responding to such concerns may require attention not only to national perspectives, but also to those of sub-national groups – indigenous peoples, communities, and NGOs – who are often most directly affected by any strategies put in place to adapt to, and mitigate, the risks of climate change.[280]

IPCC and the international climate change regime

There is much evidence to suggest that the IPCC's assessment reports have had important and growing influence over global policy and legal development in the area of climate change.[281] Particularly since the release

[278] For example, the IPCC has established a trust fund that sets aside funding to facilitate the participation of scientists from developing countries to attend its functions: see 'Procedures' at www.ipcc.ch/organization/organization_procedures.htm; Appendix B to the Principles Governing IPCC Work, Financial Procedures for the Intergovernmental Panel on Climate Change, adopted at the Twelfth Session of the IPCC (Mexico City, 11–13 September 1996), [3].

[279] These include contending policy priorities, financial constraints and lack of scientific capacity: Richard Moss *et al.*, *Towards New Scenarios for Analysis of Emissions, Climate Change, Impacts, and Response Strategies* (Geneva: IPCC, 2008).

[280] Miller, 'The Design and Management of International Scientific Assessments'; Myanna Lahsen, 'Transnational Locals: Brazilian Experiences of the Climate Regime', in Sheila Jasanoff and Marybeth Long Martello (eds.), *Earthly Politics: Local and Global in Environmental Governance* (Cambridge, MA: MIT Press, 2004), p. 151.

[281] Report of the Conference of the Parties on its Thirteenth Session, held in Bali from 3 to 15 December 2007, Annex 1, Decision 1/CP.13, FCCC/CP/2007/6/Add.1, preamble (Bali Action Plan), preamble.

of the organisation's fourth assessment report in 2007, there have been noticeable shifts in international climate policy that reflect broad acceptance of the view that there is now scientific consensus on the causes and likely impacts of climate change. In the Summary for Policymakers produced for its Fourth Assessment Report, the IPCC declares:

> Warming of the climate system is unequivocal, as is now evident from observations of increases in global average air and ocean temperatures, widespread melting of snow and ice and rising global average sea level.[282]

It also describes as 'very likely' (a term reflecting an assessed probability of occurrence of greater than 90 per cent),[283] that '[m]ost of the observed increase in global average temperatures since the mid-20th century is ... due to the observed increase in anthropogenic GHG concentrations'.[284] The IPCC projects continuing growth in global GHG emissions that 'would cause further warming and induce many changes in the global climate system during the 21st century that would very likely be larger than those observed during the 20th century'.[285] Even if global temperature increases are stabilised at a level of two degrees Celsius above pre-industrial levels – a figure that has been widely discussed as a tipping point for dangerous climate change – the IPCC has indicated that serious impacts on environmental and human systems are still likely to occur.[286]

The 'two degrees' limit has nonetheless become enshrined in the global policy sphere as the accepted 'magic number' for avoiding dangerous climate change.[287] In July 2009 the G-8 group of industrialised nations, together with the leaders of Australia, Brazil, China, India, Indonesia,

[282] Fourth Assessment Report, Summary for Policymakers, 2.

[283] See Fourth Assessment Report, Synthesis Report, 27.

[284] Fourth Assessment Report, Summary for Policymakers, 5.

[285] *Ibid.*, 7.

[286] Intergovernmental Panel on Climate Change, Climate Change 2007: Impacts, Adaptation and Vulnerability. Contribution of Working Group II to the Fourth Assessment Report of the Intergovernmental Panel on Climate Change (2007), chapter 19. Other scientists warn that limiting to 2 degrees warming will not be adequate to prevent irreversible and destructive climate change: Martin Parry, Jason Lowe and Clair Hanson, 'Overshoot, Adapt and Recover', Nature (online), 30 April 2009, 1102; Myles Allen *et al.*, 'Warming Caused by Cumulative Carbon Emissions toward the trillionth tonne', Nature (online), 30 April 2009, 1163; Malte Meinshausen *et al.*, 'Greenhouse-gas Emission Targets for Limiting Global Warming to 2°C', Nature (online), 30 April 2009, 1158.

[287] However, some countries – such as small island states and low-lying countries threatened by rising sea levels – pushed for a more stringent goal at the Copenhagen Conference of limiting global temperature rises to a maximum of 1.5 degrees Celsius.

Korea, Mexico and South Africa, recognised 'the scientific view that the increase in global average temperature above pre-industrial levels ought not to exceed 2 degrees C'.[288] These nations pledged to work together 'to identify a global goal for substantially reducing global emissions by 2050'.[289] In December 2009 the world's major emitters, including the USA and China, concluded the so-called 'Copenhagen Accord' at the fifteenth Conference of the Parties under the UNFCCC. In slightly stronger language than the G-8 declaration, this Accord records countries' recognition of 'the scientific view that the increase in global temperature should be below 2 degrees Celsius'.[290]

While the IPCC's work would appear to have been pivotal in generating a political consensus on the need to avoid a significant global temperature increase, this has so far not yielded a binding international agreement on the legal and policy measures necessary to reduce emissions and stabilise the world's climate. The Copenhagen Conference held out the promise of reaching a new international climate change pact to replace (or supplement) the Kyoto Protocol that expires in 2012. Under the so-called Bali Action Plan concluded in December 2007, state parties to the UNFCCC launched 'a comprehensive process to enable the full, effective and sustained implementation of the Convention through long-term cooperative action, now, up to and beyond 2012'.[291] This negotiation process was intended to reach an 'agreed outcome' on a post-2012 climate change treaty in time for adoption at the Copenhagen Conference in December 2009.

The disappointing results of the Copenhagen COP – which failed to put in place a binding post-2012 international climate change arrangement – do no bode well for efforts to translate a firm global scientific understanding of climate change risk into stringent legal and policy measures for addressing climate change.[292] Major sticking points at the Copenhagen COP and in negotiations since then are the question of targets for emissions reduction by 2050 and, more pertinently, by 2020, and the role of developing countries in global action to reduce

[288] The G-8 comprises the USA, the United Kingdom, France, Germany, Italy, Canada, Japan and Russia.

[289] Declaration of the Leaders, The Major Economies Forum on Energy and Climate, G8 Italia, L'Aquila, 8–10 July 2009.

[290] The Copenhagen Accord is a non-binding outcome of the fifteenth COP that was merely noted, but not adopted, by the Conference itself. It can be found at http://unfccc.int/files/meetings/cop_15/application/pdf/cop15_cph_auv.pdf.

[291] Bali Action Plan, Decision 1/CP.13, FCCC/CP/2007/6/Add. 1.

[292] For full coverage and analysis of the outcome of the Copenhagen COP see the Earth Negotations Bulletin website at www.iisd.ca/climate/cop15/.

GHG levels. In respect of these issues, the matters that split countries are largely non-scientific in nature. For instance, key developing countries such as China and India have insisted that developed countries make good on their commitment to 'take the lead in combating climate change and the adverse effects thereof'[293] by making or committing to deep cuts in their emissions before developing countries are asked to take on reduction targets.[294] Other developing countries have indicated that their acceptance of targets will need to be underpinned by substantial financial and technical assistance from developed countries to allow poorer nations to make the required changes in energy production and use.[295]

Some scientists now despair of the world ever being able to reach agreement on policy and legal measures to deal with climate change in time to prevent the most severe effects predicted by scientific risk assessments. In a number of countries, including the USA, the political debate over climate change has not progressed substantially beyond the issue of whether or not global warming is a real phenomenon. Climate change experts often express their frustration that their warnings of a climate emergency seem to fall on deaf ears. However, others, such as internationally recognised climate change scientist Graeme Pearman, are now coming to realise that they have been 'suffering under the delusion that as knowledge of the physical world improves, rationally based information would lead to rational responses to such threats as climate change'.[296] Pearman has instead looked to the social and behavioural sciences to understand what shapes people's perceptions of climate change risk. His findings – that rationality is circumstantially based and that people experience many alternative emotions and employ different coping strategies when confronted with the threat of climate change – point to the importance of socio-cultural factors in shaping risk perception and resonate with the understanding of risk presented by cultural theory (see Chapter 3).

293 United Nations Framework Convention on Climate Change, 9 May 1992, Rio de Janeiro, 1771 UNTS 164, in force 24 March 1994 (UNFCCC), Article 3(1).

294 Developing countries called on industrialised countries to reduce their emissions 40 per cent below 1990 levels by 2020 permitting developing countries to continue economic growth to better position them to take on reduction commitments and adapt to climate change.

295 See John Vidal, 'Bangkok climate talks end in recrimination', *The Guardian* (online), 9 October 2009, available at www.guardian.co.uk/environment/2009/oct/09/bangkok-climate-talks-end.

296 Quoted in Jo Chandler, 'Journey to a Hostile Climate', *The Age* (Melbourne), 13 June 2009, 1.

Pearman's work suggests that the future role of the IPCC in driving policy and legal change in the international climate change regime may be limited, at least if it continues to focus solely on developing scientifically sound, policy-ready assessments of climate change risk. Instead, the new frontier in confronting climate change may well lie in the realm of social science rather than the physical and natural sciences, requiring broader engagement with experts from these fields, and the public, in order to help people understand their responses to climate change risks and to move policy discussion to a new level.

Conclusion: influence of science in global risk governance

The case examples considered in this chapter – while they represent some of the most prominent and/or influential sites to have grappled with the question of the appropriate role for science in global risk governance – are only a cross-section of the wide variety of international policy processes and legal regimes that make use of science to inform judgments about health and environmental risk. The case studies, moreover, have been drawn from different institutional settings and deal with different types of risks. It is hence not possible to generalise from this experience *the* ideal configuration for scientific and risk assessment processes in international law in the sense of a one-size-fits-all prescription. Rather the case studies offer a source of ideas for the design of new risk governance arrangements, as well as for experimentation and institutional reform of existing processes.

In this regard, a common, useful lesson that the case studies present is the insufficiency of science alone to guide the assessment and management of risks in a way that will be broadly acceptable at the global level. Contrary to the prevalent notion that sound science and scientifically rigorous processes of risk evaluation are the necessary foundations for international decision-making on health and environmental questions,[297] the case examples emphasise that science is only ever one factor determining the influence and acceptance of assessments of risk. To summarise these insights:

- In the case of judicial review of national risk regulations under the auspices of the GATT, we saw how the relevant tests devised by the

[297] A linear science–policy model has been developed particularly in the work of Peter Haas. See Peter M. Haas, *Saving the Mediterranean: the Politics of International Environmental Cooperation* (New York: Columbia University Press, 1990).

WTO Appellate Body do not depend exclusively on examining the sufficiency of the scientific basis of the measures. Rather, in *Shrimp/Turtle* the Appellate Body turned to various process-based proxies to evaluate whether risk claims were genuine and legitimate, and in *Asbestos* it undertook an explicit weighing of value concerns in determining the necessity for trade-restrictive measures to implement the health objective at issue.

- Within the Codex Alimentarius Commission, and in the context of the WTO regime that utilises the former's standards as a benchmark for international harmonisation, the scientific soundness of Codex's work has been highly valued. At the same time, however, Codex has undertaken various institutional reforms since 1995 that reflect an acknowledgement of the political dimensions of its standard-setting exercise and the need to accommodate such concerns if its standards are to achieve broad international consensus.
- The negotiations for the Biosafety Protocol were examined as an illustration of the different contributions of science and factors outside of the scientific realm in framing the notion and treatment of biosafety risk in the international treaty. In the face of uncertainty over the health and environmental impacts of GMOs and consequent political differences over the way GMO risks should be defined and addressed, the relevant science and expert assessments played only a minimal role. Indeed, the experience of the international biosafety negotiations suggests that 'fundamentally normative and political conflicts regarding the nature of an environmental problem and its appropriate resolution cannot be resolved by reference to science that is merely technically credible'.[298]
- The international regulation of POPs chemicals is perhaps the only example considered in the chapter where science could be said to have played a predominant role, both in terms of determining the substances initially included under the POPs Convention and in the treaty's ongoing processes for adding new chemicals to the list of globally regulated substances. Even in this setting, however, we saw that scientific assessments by themselves were not enough to generate broadly acceptable risk regulation of chemicals. Instead, it was necessary to adapt ostensibly universal expert assessments to take account of local knowledge and policy concerns. The scientific and decision-making institutions of the POPs Convention are also paying increasing attention to socio-economic issues with the move to regulate chemicals still in widespread use that have more uncertain implications for human health and the environment.

[298] Ronald B. Mitchell, William C. Clark and David W. Cash, 'Information and Influence', in Ronald B. Mitchell *et al.* (eds.), *Global Environmental Assessments: Information and Influence* (Cambridge, MA: MIT Press, 2006), p. 312.

- Finally, the example of the IPCC is one where politicisation of the climate change issue has driven very explicit choices in the design of the institution to eschew a reliance on science alone in risk assessment and to embrace instead processes that blend scientific and policy inputs. This has allowed the IPCC to build government support for dealing with the issue of climate change, although the institution has also undertaken significant reforms to its peer review processes to ensure its assessments retain their perceived scientific credibility. The IPCC has achieved considerable success in generating a sense of international urgency for addressing climate change. Its future influence may be dependent on the institution continuing to display agility with regard to its assessment processes in order to take account not only of policy concerns but also socio-economic factors that underlie people's perceptions and response to climate change risk.

The conclusions that can be drawn from the case studies considered in this chapter are in line with the findings of other recent work undertaken by social scientists evaluating the influence of scientific assessments in determining how society deals with global environmental problems.[299] Research conducted by the Global Environmental Assessment Project group based at Harvard University has emphasised that the influence and effectiveness of scientific inputs into global risk governance does not just depend upon 'getting the science right'.[300] Rather, the researchers conclude that a combination of attributes of an assessment are important: the scientific credibility of the ideas, information and knowledge produced is one factor but assessments also need to be viewed by potential users as salient (that is responsive to local concerns and relevant for decision-making) and legitimate (that is produced via processes that are perceived as fair in the sense of having considered the values, concerns and perspectives of different stakeholders).[301] In addition, the researchers have found that different audiences will perceive issues of credibility, salience and legitimacy differently, depending to a large degree on the extent to which they have been able to participate in an assessment process.[302] This adds a

[299] William C. Clark, Ronald B. Mitchell and David W. Cash, 'Evaluating the Influence of Global Environmental Assessments', in William C. Clark *et al.* (eds.), *Global Environmental Assessments: Information and Influence* (Cambridge, MA: MIT Press, 2006), p. 1.

[300] *Ibid.*, p. 15.

[301] *Ibid.* See also Mitchell *et al.*, 'Information and Influence', pp. 314, 320.

[302] Mitchell *et al.*, 'Information and Influence', p. 309. As the authors stress, this may entail capacity building to enable actors to participate in scientific assessments.

layer of complexity where scientific assessments are used in global risk governance since decision-makers are then speaking to much broader and more diverse audiences. These audiences may extend beyond the traditional international domain of states to sub-national entities, NGOs and communities, and are likely to have differing, if not conflicting, risk perceptions and priorities.

To the attributes of credibility, salience and legitimacy we might add two other factors that the case studies considered in the chapter suggest are important in shaping the use and influence of science in international risk assessment. The first is the nature of the particular risk being assessed and managed. The case examples indicate a difference in the part science plays where governance exercises are directed to controversial risks as opposed to those where there exists greater scientific and political consensus. A stable scientific consensus regarding particular risks – such as the cancer-causing properties of asbestos fibres or the 'dirty dozen' list of well-known toxic organic pollutants – has provided the platform for achieving broad political consensus on the necessity for global risk regulation. On the other hand, the role of science has been more marginal in the international regulation of risks that attract greater controversy; for instance the health and environmental impacts of GMOs, 'live' chemicals proposed for listing under the POPs Convention, and health risks associated with very small quantities of contaminants present in foods. That said, the IPCC experience suggests that for some controversial risks, such as climate change, scientific assessment processes can build credibility over time as uncertainties lessen and efforts are made to bring in a diverse range of risk perspectives (for example those of developing countries).

The second factor that the case studies suggest is pivotal to how science is used in risk regulation in international fora is the relevant institutional context. Institutional settings will invariably differ in terms of their histories, mandate and powers, flexibility and capacity for adaptation, degree of transparency and openness to outside (for example NGO) participation. For instance, Codex's relatively long-established practices of engagement with international NGOs have allowed that body to incorporate a greater role for non-state perspectives in standard-setting in response to concerns over the legitimacy of its food safety standards. In the case of the IPCC, it has demonstrated a significant capacity to evolve its processes in order to buttress the scientific credibility of the organisation's climate change risk assessments. At the same time, the efforts to strengthen the perceived soundness of the IPCC's science through extended peer review have made the institution more cumbersome, leading some to criticise its assessments for being already out-of-date by the time they are released.[303]

The characteristics of the institution or legal regime within which risk governance takes place assume particular importance when it comes to calls for reforms to processes in order to embrace broader forms of knowledge that extend beyond science, narrowly conceived.[304] Supplementing science with local knowledge, information drawn from experience or even public opinion surveys can help to remedy the reductions and constraints inherent in scientific forms of knowledge that produce a limited framing of risk problems.[305] The implementation of such reforms depends upon the existence of (or capacity to establish) appropriate mechanisms for facilitating participation by those with other risk knowledges or perspectives beyond the expert views usually relied upon in risk regulation. As discussed further in the next chapter, at the international level such avenues for participation or broader involvement in risk decision-making are not always readily available, limiting the scope for achieving a 'democratisation' of global scientific assessment processes. Moreover, although the democratisation of science may help to improve the salience or legitimacy of risk assessments for some audiences, there is generally always a trade-off involved: such benefits may only come at the expense of decreased credibility of the information for other audiences.[306] Hence, as the researchers of the Global Environmental Assessment Project have remarked, 'a tension exists between the desire for science to be simultaneously well informed and well analyzed and also to be democratic.'[307]

303 David Leary, 'From Bali to Poznan: An Assessment of Australia's Response to Climate Change in 2008', *Environmental And Planning Law Journal*, 26 (2009), 194–5. See also Chris Mitchell, 'The Role of Science in the Analysis of Climate Change: A Perspective Based on Recent Research', in Wayne Gumley and Trevor Daya-Winterbottom (eds.), *Climate Change Law: Comparative, Contractual and Regulatory Considerations* (Sydney: Thomson Reuters, 2009), p. 1, discussing developments in climate science since the release of the IPCC's Fourth Assessment Report.

304 One manifestation of the move to embrace knowledge rather than just science is provisions in global environmental treaties for the inclusion of indigenous and traditional knowledge. See generally Sheila Jasanoff and Marybeth Long Martello (eds.), *Earthly Politics: Local and Global in Environmental Governance* (Cambridge, MA: MIT Press, 2004).

305 Sheila Jasanoff and Marybeth Long Martello, 'Conclusion: Knowledge and Governance', in Sheila Jasanoff and Marybeth Long Martello (eds.), *Earthly Politics: Local and Global in Environmental Governance* (Cambridge, MA: MIT Press, 2004), p. 338.

306 Farrell *et al.*, 'Overview: Understanding Design Choices', p. 10.

307 Clark *et al.*, 'Evaluating the Influence of Global Environmental Assessments', p. 16.

7 Democratising global risk governance

Introduction

As risk decision-making has moved increasingly from sites of national regulation to global governance structures, science and experts – aided by shared positivist and universalist traditions – have been readily able to relocate. They occupy a central place in the processes of international standard-setting organisations, under treaty provisions governing the assessment and management of risks and as part of advisory bodies to international adjudicators like those of the WTO dispute settlement system. Even in an international institution such as the Inter-governmental Panel on Climate Change – deliberately designed to create shared science–policy understandings of climate change risks – much attention has been paid to the best ways of bringing scientific expertise into its processes and to retaining credibility with the broader scientific community despite governmental involvement in the production of risk assessments. However, as the examples considered in the previous chapter illustrated, recognition of the importance of scientific input into global risk governance sits alongside a growing acknowledgement that science and/or technical risk assessments alone will generally offer inadequate foundations for effective and legitimate risk regulation at the international level.

Reflecting this, questions about the legitimacy of international risk decision-making are becoming an important topic of debate, both in the literature and in the practice of global institutions. Perspectives from many different fields have been brought to bear on the issue of the use of science in international risk regulation, including those of

An earlier version of this chapter was published as Jacqueline Peel, 'International Law and the Legitimate Determination of Risk: Is Democratising Expertise the Answer?', *Victoria University of Wellington Law Review*, 38 (207), 106.

post-normal science, constructivist perspectives on risk, and science and technology studies of the co-production of scientific knowledge and risk policy. A common finding is the need to meld science with a broader array of perspectives or values in order to generate a legitimate basis for international decisions about risk.[1] Calls for greater participation in global risk governance extend beyond the academic literature. As Christian Joerges and Jürgen Neyer note:

> Even though the epistemological reasons for the fragility of the knowledge base of regulatory politics are hardly common knowledge, the public has nevertheless learned to mistrust experts and administrators who claim to act on an incontestable objective knowledge basis.[2]

Consequently, in the wake of debates over the contingency and uncertainty of scientific knowledge, together with greater public awareness of prior failures of regulatory science, 'a democratic impulse has emerged as a counterweight to the "technocratic view"'.[3]

It is this 'impulse' for the democratisation of science in global risk decision-making and governance that is the focus of this chapter. The initial part of the chapter introduces readers unfamiliar with the relevant social scientific literature to concepts and models of democratised expertise. However, the primary concern of the chapter is with an aspect of the democratisation of global risk governance that has received less attention to date; namely, how democratised science might be institutionalised in global risk decision-making and associated international legal structures. This is a critical issue yet, as the chapter emphasises, one that raises difficult challenges for international law and institutions. In the pursuit of broader visions of science and pluralist risk assessment processes it seems that '[f]or some time to come, we shall live with a confused and rather uncomfortable mix of highly imperfect attempts to democratize global decision-making'.[4]

[1] Ronald B. Mitchell, William C. Clark and David W. Cash, 'Information and Influence', in Ronald B. Mitchell *et al.* (eds.), *Global Environmental Assessments: Information and Influence* (Cambridge, MA MIT Press, 2006), p. 324.

[2] Christian Joerges and Jürgen Neyer, 'Politics, Risk Management, World Trade Organisation Governance and the Limits of Legalisation', *Science and Public Policy*, 30(3) (2003), 220.

[3] Joanne Scott, 'On Kith and Kine (and Crustaceans): Trade and Environment in the EU and WTO', in J. H. H. Weiler (ed.), *The EU, the WTO, and the NAFTA: Towards a Common Law of International Trade?* (Oxford University Press, 2000), p. 158, citing Dorothy Nelkin, 'The Truth About Law's Truth', (EUI Working Paper No. 90/1).

[4] Ralf Dahrendorf, 'Can European Democracy Survive Globalization?', *The National Interest*, Fall(65) (2001), 22.

Whereas the previous chapter canvassed the use of science in a range of global risk governance settings, this chapter adopts a narrower focus, looking closely once more at WTO dispute settlement under the SPS Agreement. Most proposals put forward in the legal scholarship that could be seen as advocating the democratisation of international risk regulation have focused on this institutional setting. Commentators in this vein acknowledge the central importance of scientific input and its benefits for SPS risk regulation, but have argued that the science that informs global risk decision-making should take account of a broad range of uncertainties, value considerations and public risk framings. Focusing on democratisation proposals in the context of SPS dispute settlement allows a more detailed examination of their potential to enhance the legitimacy of risk determination in this forum, as well as the potential hurdles presented by underdeveloped participatory structures and the lack of modes of democratic representation equivalent to those in domestic systems. The concluding part of the chapter seeks to distil the insights from various proposals in the legal literature to provide guidance on practical measures for democratising the global governance of risk regulation applicable in the SPS setting. It is hoped this analysis will encourage further research into the possibilities for democratising science in other areas of international law concerned with risk regulation.

Democratisation of science in global risk governance

The notion of democratising science in global risk governance represents the confluence of two streams of thought that have gained prominence in recent years: those that challenge positivist constructs of science and risk assessment as a value-free process, and calls for international decision-making with an inherent value dimension and broad socio-economic effects to incorporate the views of those affected. However, the marriage of democratic theory with science is not straightforward. Conventional notions of science, as was highlighted in Chapter 3, conceive it as a body of knowledge produced via particular methodologies, whose nature and modes of application are only comprehensible to experts. By contrast, democracy (albeit having many and varied forms) is seen to rest on an alternative premise of all members of a community having an equal right to engage in political discussion and debate.[5] These notions of science and democracy

[5] For a discussion of different models of democracy see David Held, *Models of Democracy*, 3rd edn (Cambridge: Polity Press, 2006).

have lent themselves to dichotomies between facts and values, with experts assigned responsibility for discovering the former, while political processes are restricted to dealing with the latter. In the field of risk regulation this dichotomy is often reproduced in the form of a distinction (and as rigorous a separation as possible) between processes of risk assessment – as the domain of scientific experts – and procedures for risk management, which might be a forum for democratic politics.[6]

Democratic legitimacy for global risk governance

As we saw in Chapter 2, the possibilities for global governance to gain some form of democratic legitimacy have been a matter of intense debate in the academic literature and institutional practice. The consensus view is that global democratic processes – in the conventional representative democratic form – are not achievable at this time given the lack of an identifiable global public. This has not deterred some authors from exploring other ways of achieving democratically legitimate governance at the international level. The concepts of democracy that tend to be drawn on for this purpose are ones in the liberal 'proceduralist' tradition, which emphasise processes for ensuring transparency, deliberation and public participation in decision-making as the basis of legitimate authority.[7]

One prominent commentator to explore this approach is Gráinne de Búrca. In her 2008 Article, 'Developing Democracy Beyond the State', de Búrca argues that 'we should not jettison democratic values when we attempt to shape more legitimate governance structures beyond the state'. Rather we should isolate the building blocks of democracy – such as the fullest possible participation and representation of those affected – and strive to translate them to the international context in order to 'start to provide the legitimating democratic dynamic of non-state sites of governance'.[8] De Búrca's 'democratic-striving' approach proposes that global governance processes be designed 'to strive continuously to develop the best possible degree of participation' while recognising the possibility of 'incomplete success or possible failure and the expectation of the need for regular revision of the process as a

[6] Angela Liberatore and Silvio Funtowicz, '"Democratising" Expertise, "Expertising" Democracy: What Does this Mean, and Why Bother?', *Science and Public Policy*, 30(3) (2003), 148–9.

[7] Robert Howse, *The WTO System: Law, Politics and Legitimacy* (London: Cameron May, 2007), p. 218.

[8] Gráinne de Búrca, 'Developing Democracy Beyond the State', Colum. J. Transnat'l L., 46 (2008), 226–7, 237.

whole'.[9] She suggests that participation in global governance activities be structured in a way that keeps the circle of potential participants continuously open and sets incentives to generate the fullest degree of participation possible. Initially this would involve identifying and providing for participation by relevant stakeholders most likely to be concerned by the decisions or policies in question. However, the process would need to be open to revision at the end of a regulatory cycle in order to include any new actors or interests who identify themselves as having a potential claim to be included.[10]

De Búrca emphasises the need to strive for the democratic legitimacy of global governance given that its outcomes – often complex and opaque – have significant effects on the lives of many people. In the area of risk regulation, the translation of core democratic values, such as transparency and participation, into relevant global governance practices could facilitate external monitoring of these processes and their outcomes, making them more accountable to the people affected by such decisions. Democratic participation in global risk regulation might also play a substantive role as a mechanism for airing concerns about harm to valued environments, which could then inform more broadly framed risk assessments.

Democratised expertise: social science concepts

The social scientific literature on democratic science supports the need for standard, technically oriented, expert-dominated risk assessment processes to be opened up to a greater range of participants. Its underlying premise is the impossibility of separating science and politics in risk decision-making, which challenges the idea of scientific expertise 'as a self-referential system in which only peers can recognise and judge each other'.[11] In place of positivist understandings of science, authors in this tradition have proposed new notions of science or ways of doing risk assessment, a number of which were reviewed in Chapter 3. In general, these alternative notions of science seek to strengthen its utility as a resource for risk decision-making by making it more cognisant of areas of uncertainty and hence more aware of the potential for human activities to have unforeseen consequences.[12] Another common feature of such broader visions of science is the call for greater attention to, and openness regarding, 'the normative that lurks within

[9] *Ibid.*, 252. [10] *Ibid.*, 253. [11] *Ibid.*, 147.

[12] Sheila Jasanoff, 'Technologies of Humility: Citizen Participation in Governing Science', *Minerva*, 41(3) (2003), 240.

the technical' so that points at which experts move beyond their areas of direct competence are made explicit.[13] Proposals for democratising expertise involve drawing together such extended notions of science with core democratic values such as participation.[14] The purpose of democratisation is to expose conventional forms of scientific research and assessment to a greater range of perspectives and knowledges, thereby bringing science 'into the public debate along with all the other issues affecting our society'.[15]

For those with a strong allegiance to positivist scientific traditions, calls for democratisation are often misconstrued as a challenge to the central role of science in health and environmental regulation or as a threat to the credibility of risk assessment. However, far from 'turning over the research labs to untrained persons',[16] the underlying rationale of democratising expertise is one of better equipping science for the policy and decision-making settings in which it now finds itself. For some, adoption of this approach can be justified on a purely pragmatic basis as a means for 'producing social consensus around public decisions and ... defusing controversy – something which the application of scientific reason has conspicuously failed to do'.[17] More often the democratisation of expertise is called for in order to improve the quality of information available for risk decision-making by:

> extend[ing] the amount and types of information incorporated into decisions, ... ensur[ing] that experts alone are not charged with making value-laden decisions ... serv[ing] to expose and communicate uncertainties and information often not considered during decision-making processes and [bringing] forth unrecognized alternatives and solutions to problems.[18]

Hence, democratising expertise can be viewed as a means for ensuring the continuing relevance of science for risk regulation, even in complex or contested situations. As Helga Nowotny observes, the expectation that science can adapt to the new demands imposed by democratised

[13] *Ibid.*

[14] Literature on the democratisation of science and expertise shares much in common with that on transdisciplinarity in scientific research. See further, Chapter 3.

[15] Silvio Funtowicz and Jerome Ravetz, 'Three Types of Risk Assessment and the Emergence of Post-Normal Science', in Sheldon Krimsky and Dominic Golding (eds.), *Social Theories of Risk* (Westport, CT: Praeger Publishers, 1992), p. 254.

[16] *Ibid.*

[17] Scott, 'On Kith and Kine', p. 158.

[18] Joel A. Tickner and Sara Wright, 'The Precautionary Principle and Democratizing Expertise: a US Perspective', *Science and Public Policy*, 30(3) (2003), 213.

visions and so form a part of a comprehensive response to uncertain, multifaceted risk problems 'is an expression of confidence in its potentiality, not a loss of trust'.[19]

Models for democratising expertise

Reflecting the developing nature of scholarship on the democratisation of science, many and varied approaches have been proposed in the social scientific literature regarding the way in which expertise used in risk decision-making might enter the public arena. The most conservative proposals in this regard depart little from the technical perspective on risk described in Chapter 3, contending that enhancing the accountability of decision-makers for the outcomes of risk decision-making depends as much upon science disciplining politics, as politics disciplining science.[20] For instance, Cass Sunstein prescribes the increased use of technocratic tools, such as quantitative risk assessment and cost–benefit analysis, as a means for systematising risk regulation and rendering it transparent to the general public.[21] Along similar lines are proposals for improved risk communication, which are designed to mediate divergences between scientific and public perspectives on risk (most commonly) through experts educating the public about the 'true' nature of risks.[22]

While such proposals are at one end of a possible 'continuum' of approaches for democratising expertise, they still tend to depend heavily on conventional notions of science conceived as a body of knowledge capable of sharp demarcation from the political domain.[23] In circumstances where that boundary is blurred – for example, where there are intractable uncertainties that undermine the reliability of scientific knowledge or value conflicts that influence the way risks are framed for assessment – more transparent processes might ameliorate some of the deficiencies of science-based decision-making but cannot make it value neutral.

19 Helga Nowotny, 'Democratising Expertise and Socially Robust Knowledge', *Science and Public Policy*, 30(3) (2003), 151.

20 E.g., Jeremy D. Fraiberg and Michael J. Trebilcock, 'Risk Regulation: Technocratic and Democratic Tools for Regulatory Reform', *McGill Law Journal*, 43 (1998), 835.

21 Cass Sunstein, *Risk and Reason: Safety, Law, and the Environment* (Cambridge University Press, 2002), p. 294.

22 Carlo Jaeger *et al.*, *Risk, Uncertainty, and Rational Action* (London: Earthscan Publications Ltd, 2001), p. 127.

23 Liberatore and Funtowicz, '"Democratising" Expertise, "Expertising" Democracy', 149.

Those approaches which, by contrast, seek to make science a more 'socially robust' contributor to risk regulation recognise that its knowledge claims need to demonstrate 'validity outside as well as inside the laboratory', are 'most likely to be achieved by involving an extended group of experts' and will generally result 'from having been repeatedly tested, expanded and modified'.[24] Producing such knowledge requires 'genuine debate on the way a problem is formulated, knowledge is developed and uncertainties are dealt with' within a framework that envisages a more extensive role for participants beyond the traditional domains of expertise.[25]

The precautionary principle, discussed in Chapter 4, has been advocated by some commentators as a regulatory model capable of housing such an approach. Commentators who advance this view align implementation of the precautionary principle with the goals of democratising expertise by emphasising the utility of the principle in generating a greater cognisance of uncertainty, awareness of the normative dimensions of risk assessment and openness to collective decision-making.[26]

Other models for democratising expertise also take as their departure point the difficulties of risk decision-making under conditions of uncertainty. For example, Silvio Funtowicz and Jerome Ravetz's model of 'extended peer review' proposes the use of 'an extended peer community' in risk assessment to augment the work of technically trained scientists by employing 'extended facts', such as anecdotal evidence and statistics gathered by the community.[27] Funtowicz and Ravetz maintain that this mode of risk assessment is most appropriate in situations where 'hard decisions' must be made based on 'soft science'[28] because 'facts are uncertain, values in dispute, stakes high, and decisions urgent'.[29]

For others, however, the way that questions are selected and framed for risk assessment is at least as important, if not more important, than the way such questions are subsequently assessed.[30] Here the critical

[24] Nowotny, 'Democratising Expertise and Socially Robust Knowledge', 155.

[25] Liberatore and Funtowicz, '"Democratising" Expertise, "Expertising" Democracy', 147.

[26] E.g., Theofanis Christoforou, 'The Precautionary Principle and Democratizing Expertise: a European Legal Perspective', *Science and Public Policy*, 30(3) (2003), 205.

[27] Funtowicz and Ravetz, 'Three Types of Risk Assessment and the Emergence of Post-Normal Science', p. 254.

[28] *Ibid.*, p. 259. [29] *Ibid.*, pp. 253–4.

[30] E.g., Brian Wynne, 'Risk and Social Learning: Reification to Engagement', in Sheldon Krimsky and Dominic Golding (eds.), *Social Theories of Risk* (Westport, CT: Praeger Publishers, 1992), p. 275.

focal point for democratisation is the scoping exercises undertaken at the outset of risk assessment that determine what harms and uncertainties are of importance; processes which are often dominated by experts despite the normative content of these questions. Broadening out these exercises so as to take account of a greater array of information and views on the risks of concern is seen as a means to redress the constraints inherent in technical forms of knowledge that produce a limited framing of risk problems.[31]

As Angela Liberatore and Silvio Funtowicz point out, in practice there are many overlaps and areas of coexistence between these various models of democratised expertise.[32] Reflecting their democratic aspirations, a common feature of many is a reliance on improving the transparency of risk regulatory processes and avenues for participation by those outside the conventional spheres of regulatory agencies and their expert advisors. Depending on the particular model adopted, this might serve to allow a broad initial framing of risk questions or a more comprehensive response to uncertainty, or provide a means for obtaining public input into value-laden questions raised by risk decision-making regarding harms of concern and their relative significance.

Proposals for democratising WTO SPS risk governance

Outside of the social scientific literature explicit discussion of democratised expertise models is rare. One consequence of this is that while significant attention has been paid to some issues – such as the way processes of risk assessment may 'co-produce' scientific knowledge about risks – less analysis has been directed to institutional questions regarding the implementation of more democratic forms of science and expertise in global governance and international legal structures. There is, however, a growing body of commentary in the legal literature proposing reforms to international risk governance – especially in the context of WTO dispute settlement in SPS cases – that could be seen as sharing the same intellectual inspiration as the social scientific scholarship on democratised expertise. Like their

[31] Sheila Jasanoff and Marybeth Long Martello, 'Conclusion: Knowledge and Governance', in Sheila Jasanoff and Marybeth Long Martello (eds.), *Earthly Politics: Local and Global in Environmental Governance* (Cambridge, MA: MIT Press, 2004), p. 338.

[32] Liberatore and Funtowicz, '"Democratising" Expertise, "Expertising" Democracy', 149.

social science counterparts, these legal commentators start from the premise that science as a sole basis for risk regulation is problematic in conditions of uncertainty or where there exist divergent perspectives on risk drawn from different socio-political or cultural traditions. The emphasis on scientific justification in the SPS Agreement and SPS jurisprudence is thus viewed with concern since it might restrict the scope for national risk measures to respond to situations of more complex, uncertain risks.

Suggestions for reforms or different modes of operation of WTO dispute settlement processes in SPS disputes in order to take account of a broader range of risk perspectives have taken a variety of forms. However, a common source of inspiration is innovations in domestic risk regulatory schemes, especially that of the USA. A particular focus in this regard is the use of participatory mechanisms in the US system (such as notice and comment procedures, *amicus curiae* briefs and the involvement of public-interest organisations in regulatory decision-making) as well as deferential review processes developed by the US federal courts in dealing with science-based risk regulation (see further, Chapter 4). Reference to such domestic models is used to emphasise the importance of risk regulation having a broader basis than just science in order to garner credibility and legitimacy in democratic societies.

Despite their frequent domestic inspiration, the proposals that have been developed are generally more than a simple plea for national regulatory autonomy in an age of globalisation. Authors thus pay explicit attention to, and attempt to reconcile, the competing concerns of the national and the international in the mechanisms they put forward for enhancing democratic input into WTO risk decision-making. It is this attempt that makes the proposals an interesting focus for analysis as they highlight the kinds of institutional obstacles that will need to be grappled with if international risk governance is to give effect to notions of democratised expertise. Far more so than calls for global risk decision-making to be based only on sound science, these proposals also expose the hard decisions thrown up by the frequent need to rely on soft science in health and environmental risk assessment. In this respect, they share in common with de Búrca's democracy-striving approach (discussed above) a desire to explore the potential for greater democratisation of global risk governance processes, albeit in the knowledge that the results may be imperfect and in need of ongoing reassessment and revision.

Strategies of deference

When processes for international supervision of risk decision-making were first introduced into instruments such as the SPS Agreement, forward-thinking commentators, such as David Wirth, remarked on the potential for highly intrusive scrutiny of risk regulations taking insufficient account of the fact that 'science may inform the regulatory process but cannot, by itself, determine the result with particularity'.[33] In response, Wirth called for WTO decision-makers reviewing domestic risk measures to 'be highly deferential to scientific determinations of national authorities that underlie regulatory measures to protect the environment and public health'.[34] In essence, Wirth saw questions of the appropriate balance between science and values in risk regulation as being best worked out at the domestic level rather than by international decision-makers guided only by expert advice.

Deference to the risk determinations of national authorities

A decade or so later, proposals for deferential review of disputed SPS measures by WTO adjudicators continue to find favour in the literature, although different views have been expressed as to how much deference ought to be accorded to the risk determinations of national authorities.[35] At one end of the spectrum are those who call for complete deference by international adjudicators to domestic risk assessments. Andrew Guzman, for example, has argued that WTO review of SPS measures is 'inappropriately intrusive' and that 'substantive review' should be forgone in favour of deference 'to the implementing state's evaluation of the level of risk it is willing to tolerate, the relevant scientific evidence, and the relationship between the measure and the risk assessment'.[36]

33 David Wirth, 'The Role of Science in the Uruguay Round and NAFTA Trade Disciplines', Cornell Int'l. L.J., 27 (1994), 833.

34 *Ibid.*, 859.

35 E.g., Theofanis Christoforou, 'Settlement of Science-Based Trade Disputes in the WTO: A Critical Review of the Developing Case Law in the Face of Scientific Uncertainty', N.Y.U. Environmental Law Journal, 8 (2000), 622; J. Martin Wagner, 'The WTO's Interpretation of the SPS Agreement has Undermined the Right of Governments to Establish Appropriate Levels of Protection Against Risk', *Law and Policy in International Business*, 31 (2000), 855.

36 Andrew Guzman, 'Food Fears: Health and Safety at the WTO', Virginia J. Int'l L., 45 (2004), 4.

Ilona Cheyne also suggests that the stage of 'risk evaluation' – where national authorities assess 'the significance and acceptability of a risk' – is one at which an external evaluation is least appropriate. She argues that the significance of a risk 'is only measurable in the society in which the risk poses a threat, and therefore the decision-making power must lie with the institutions and procedures of that society'.[37] Cheyne is also sceptical of the scope for WTO decision-makers to engage in objective review of domestic assessments of whether a risk exists, its probability and magnitude given the subjective component of decisions about how evidence is to be weighed and interpreted and competing opinions evaluated. She thus recommends that although 'a baseline test of objectivity' should be applied, review of SPS measures 'must inevitably be a light touch'.[38]

At one level, proposals such as these for deference seem to have much to recommend them, not least being that they largely avoid the problems that might occur in seeking to democratise scientific inputs to *international* risk governance processes by assuming that uncertainty and broader risk concerns are adequately taken into account in the domestic regulatory system. For some commentators, another virtue of a strategy of deference towards the risk determinations of domestic regulatory authorities is said to be its consistency with judicial review practices in member states (such as the USA and EC) that avoid second-guessing the scientific conclusions of more expert, regulatory agencies.[39] Accordingly, the practice of courts such as the US federal courts or the European courts reviewing risk regulatory measures devised by, or with the advice of, expert agencies is often put forward in support of the adoption of a similarly deferential stance on the part of international adjudicators within the WTO dispute settlement system. However, as the discussion in Chapter 4 highlighted, a policy of deference by the courts to risk assessments conducted by responsible government agencies has not prevented the development of judicial practices – most particularly in the USA – that subject the science underlying risk regulatory measures to intensive review.

In the WTO SPS context, as was discussed in Chapter 5, members' arguments for complete deference to national risk assessments have

[37] Ilona Cheyne, 'Risk and Precaution in World Trade Organisation Law', *Journal of World Trade*, 40(5) (2006), 842.

[38] *Ibid.*

[39] Wirth, 'The Role of Science in the Uruguay Round and NAFTA Trade Disciplines', 843.

not been met with a favourable response before the Appellate Body. In a trade forum this reaction might be attributed partly to the suspicion with which strict health and environmental (and especially precautionary) regulatory measures are frequently viewed given assumptions of regulatory capture. However, there may be a legitimate foundation for caution given that the conditions under which science alone is a deficient tool for risk assessment – prevalent uncertainties, disputed risk framings and so on – are also those which may promote its exploitation by governments for instrumental purposes, either to erect trade barriers, or to further an anti-regulatory agenda.[40] Moreover, where domestic risk regulation is heavily influenced by a particular cultural world-view, there may be little justification for preferring one member's risk perspective to those of other WTO members by way of application of a strategy of deference.

A further, fundamental difficulty with proposals for absolute deference to national risk determinations in WTO SPS review is that they relinquish possibilities for encouraging national authorities to be more cognisant of the potential for differing risk perspectives and evaluations by communities in other countries. As Chapter 2 highlighted, instilling 'other-regarding' practices that force national authorities to be more accountable for the effects of their decisions beyond borders is often the *raison d'être* for global forms of risk regulation and governance more broadly. There is also a strong argument that, in a globalised context, efforts to integrate the concerns of actors outside the national polity is 'democracy-reinforcing' because it serves to enhance the accountability of state-based risk regulation to all those who might be affected by it.[41]

Deferential reasonableness standard

While there are strong counter-arguments against the adoption of complete deference by WTO reviewers to national risk assessments in *all* SPS proceedings, there may nonetheless be some categories of health or environmental risk (or at least risks at certain stages in the course of knowledge and experiential evolution regarding them) which are best excluded from international review and

[40] Sheila Jasanoff, '(No?) Accounting for Expertise', *Science and Public Policy*, 30(3) (2003), 159.

[41] Joanne Scott, 'European Regulation of GMOs: Thinking About "Judicial Review" in the WTO', in Michelle Everson and Ellen Vos (eds.), *Uncertain Risks Regulated* (Milton Park: Routledge-Cavendish, 2009), pp. 304–5.

decision-making processes, and left instead to the judgment of different, domestic regulatory authorities. This was effectively the argument put by the EC in the *GMO* dispute when it highlighted the political, scientific, social and legal complexities of regulating risks from GMOs that underlay the 'prudent approach' taken in its regulatory framework.[42] The EC went on to declare that '[t]here is a serious question as to whether the WTO is the appropriate international forum for resolving all the GMO issues that the Complainants have raised in these cases'.[43] This might be read as an appeal for greater deference, exercised on a case-by-case basis, in response to factors such as the novelty of the risks at issue, the level of uncertainty and the degree of public concern.

The basis for a differentiated, deferential approach might be found in the standard of review applied by WTO panels and the Appellate Body in SPS cases. Taking this approach, some authors have suggested that under the SPS Agreement the most appropriate balance between science and normative goals of risk protection can be achieved by the adoption of a deferential reasonableness standard of review.[44] For instance, in her 2004 book, *The Power to Protect: Trade, Health and Uncertainty in the WTO*, Catherine Button proposes WTO decision-makers adopt a 'standard of review hinging on reasonableness'. She describes this standard as one designed to ensure that:

> a defending Member is meeting its substantive commitment only to take regulatory action where there is scientific justification without unnecessarily intruding into the national regulator's assessment of the significance of various pieces of scientific evidence or its integration of social and cultural factors into the regulatory process.[45]

As Button points out, the principal value of such an approach is that it may illuminate the multidimensional, complex nature of risk questions and invite a more contextualised assessment by 'concentrating

[42] *EC – Measures Affecting the Approval and Marketing of Biotech Products*, WTO Docs WT/DS291, WT/DS292, WT/DS293 (2004) [1] (First Written Submission by the EC), [2]–[3].

[43] *Ibid.*, [10].

[44] See also Vern Walker, 'Keeping the WTO from Becoming the "World Trans-science Organisation": Scientific Uncertainty, Science Policy and Fact-Finding in the Growth Hormones Dispute', Cornell Int'l. L.J., 31 (1998), 280–5.

[45] Catherine Button, *The Power to Protect: Trade, Health and Uncertainty in the WTO* (Oxford: Hart Publishing, 2004), p. 235. Button develops her proposal from statements of the panel in the *Asbestos* case: see Chapter 6.

the panel's mind on the fact that it is reviewing a regulatory action and not a disembodied set of facts'.[46]

Alexia Herwig also advocates the use of 'an unreasonableness standard' in WTO SPS cases, which she describes as one that examines 'whether scientific knowledge claims fail to cohere with larger belief systems'.[47] Accordingly she advises that WTO dispute settlement bodies ought not to determine whether the regulating member has used the 'best' form of scientific evidence applying technical criteria such as specificity and reliability. Rather they 'should only ask the experts to evaluate whether the evidence relied on is not *directly* contradicted by other knowledge claims taking into account experience with potential hazards and methods of assessment and general background knowledge about the specific and closely related substances'.[48]

Left unclear by such proposals for a standard of review 'hinging on reasonableness' is how WTO decision-makers might evaluate this criterion with respect to particular uses of science underlying divergent risk assessments. What is considered reasonable is likely to depend on WTO decision-makers' perception of the purpose of their review exercise (risk protection or trade liberalisation?); a matter regarding which the institutional structures of the multilateral trading system do not provide clear normative guidance.[49] In this context, a standard of review based on what is reasonable might well devolve in practice to relatively intense scrutiny of the science underlying a risk regulatory measure.[50] It is notable, for instance, that the Appellate Body's attempt in *Hormones II* to craft a middle-of-the-road review standard based on a panel's determination of whether a risk assessment 'is supported by coherent reasoning and respectable scientific evidence' is underpinned by a methodology that still retains a substantial emphasis on scientific factors and expert advice.[51]

46 *Ibid.*, 221.

47 Alexia Herwig, 'Whither Science in WTO Dispute Settlement?', *Leiden Journal of International Law*, 8, 21 (2008), 842.

48 *Ibid.*, 842–3.

49 Button, *The Power to Protect*, p. 211.

50 Herwig's example of the application of an unreasonableness standard still requires WTO decision-makers to undertake complex evaluations of scientific and other forms of evidence: 'Whither Science in WTO Dispute Settlement?', 843.

51 See the discussion of this case in Chapter 5.

Proceduralist approaches

The drawbacks of a simple deference approach, and the imprecision of review standards that call for an appraisal of 'reasonableness', have spurred the development of more nuanced proposals for SPS dispute settlement reform that attempt to strike a balance between 'localized science policy decision-making' and global assessments of risk.[52] Common to such proposals is the view that faith in science as an objective arbiter of risk is misplaced and, in any event, poses difficult problems for international decision-makers charged with evaluating the truth of divergent risk claims.[53] Nonetheless, it is recognised that, left to themselves, domestic risk regulatory authorities, even (or especially) in democratic societies, are likely to accord insufficient regard to the risk perspectives of those outside of the national polity. Hence a compromise is proposed whereby domestic authorities would retain competency over the substantive evaluation of SPS risks, while the role of WTO reviewers would be to focus on the regulatory process followed by those authorities and whether their procedures allow for an adequate consideration of different views about risk.

Applying these ideas in the context of an SPS dispute would mean that a national risk determination could not be evaluated simply on the basis of its consistency with a narrowly focused, science-based risk assessment. To do so would deny a 'pluralistic understanding' of knowledge and experience with regard to risk.[54] Rather the task of the WTO dispute settlement organs would be to assess the way in which scientific and other inputs were gathered and evaluated in reaching decisions on SPS risk. The adequacy of this process might be judged against criteria of inclusiveness (were all relevant risk perspectives – scientific, economic, and public – considered?), transparency (were the reasons for the ultimate decision explained?) and intellectual rigour (were scientific inputs peer-reviewed and subjected to open and critical discussion?).[55] To this list some would add a requirement for

[52] David Winickoff *et al.*, 'Adjudicating the GM Food Wars: Science, Risk, and Democracy in World Trade Law', Yale J. Int'l L., 30 (2005), 111.

[53] Oren Perez, *Ecological Sensitivity and Global Legal Pluralism: Rethinking the Trade and Environment Conflict* (Portland: Hart Publishing, 2004).

[54] *Ibid.*, 152.

[55] E.g., Robert Howse, 'Democracy, Science, and Free Trade: Risk Regulation on Trial at the World Trade Organization', Michigan L. Rev., 98 (2000), 2330; Jan Bohanes, 'Risk Regulation in WTO Law: A Procedure-Based Approach to the Precautionary Principle', Colum. J. Transnat'l L., 40 (2002), 365–70.

'institutional reflexivity', embracing both an openness of the risk assessment process 'to speakers with various institutional and ideological affiliations', and 'a capacity for self-assessment or self-critique'.[56]

A procedural approach in practice

Indications of how a procedurally oriented model might operate in global risk disputes are offered by recent international case law. One such example is the 2005 decision of an arbitral tribunal under the North American Free Trade Agreement (NAFTA) in the *Methanex* dispute. The *Methanex* case was brought by a Canadian corporation, Methanex, under the auspices of NAFTA's Chapter 11 dispute settlement procedures, which allow investors from NAFTA states to bring actions directly against government members of NAFTA where the latter are alleged to have acted inconsistently with investors' rights guaranteed under the treaty.[57] Methanex was (and remains) the world's largest producer of methanol, a feedstock for the fuel additive MBTE. The central argument underlying Methanex's case against the USA was that Californian regulations banning MBTE on the basis of its potentially detrimental health and environmental effects were based on a 'sham' risk assessment. Consequently, the company sought substantial compensation from the USA for resultant economic losses to its methanol production business.

Before the NAFTA tribunal, detailed scientific evidence regarding issues of health and environmental risk was produced by both parties to the dispute. Methanex attempted to demonstrate that the Californian risk assessment underlying the MBTE ban was technically unsound, while the USA defended the risk assessment report as one reflecting 'substantial scholarship'.[58] The NAFTA arbitral tribunal, however, eschewed a detailed review of the scientific merits of the Californian risk assessment report, or the competing views presented by the parties' experts respecting matters of environmental risk. Instead it pointed to the fact that the assessment – albeit open to the possibility 'for other scientists and researchers to disagree in good faith with certain of its methodologies, analyses and conclusions' – had been subjected to 'open and informed debate', including 'public hearings, testimony and

56 Oren Perez, *Ecological Sensitivity and Global Legal Pluralism*, pp. 154–5.

57 North American Free Trade Agreement, 17 December 1992, in force 1 January 1994, (1993) 32 ILM 605, Articles 1115–38.

58 *Methanex Corporation* v. *United States of America*, NAFTA Chapter 11 Arbitral Tribunal (2005) 44 ILM 1345 (*Methanex*), Part IIIA, [37]–[40].

peer-review'.[59] According to the tribunal, the assessment's 'emergence as a serious scientific work' from such a process was 'the best evidence that it was not the product of a political sham engineered by California'. In any case, the tribunal also went on to hold that it was not persuaded that the risk assessment was scientifically incorrect, having been 'much impressed by the scientific expert witnesses presented by the USA and tested under cross-examination by Methanex'.[60]

The kind of review of risk regulatory measures suggested by the tribunal's decision in *Methanex* – deferential with respect to matters of scientific substance and instead focusing on the way scientific information about risk is used and debated in the public arena – is one that an increasing number of commentators would see as being appropriate also in the realm of WTO SPS review. In this light, the panel's pursuit of process-based review in the *GMO* case is instructive. The *GMO* panel's attempt to review the regulatory process of the EC for the assessment and approval of biotechnological products was described in Chapter 5. The panel applied procedural criteria in its evaluation, such as transparency and timeliness, which might have been thought to offer greater scope for regulatory diversity in the biotechnological field than a detailed scientific review would have done.[61] However, the panel came to the conclusion that the EC measures failed these standards, notwithstanding evidence concerning scientific uncertainty surrounding GMOs and the regulatory diversity present in international and domestic GMO frameworks. The panel's report stands as a salutary reminder that international review focused on issues of regulatory process will not always provide an approach that is sensitive to localised science–policy decision-making.

That procedurally based international review of risk regulation has the potential to yield mixed results is something that should not surprise, given similar experience in domestic systems such as that of the USA. As we saw in Chapter 4, procedural review of agency risk measures in the USA, in the name of preserving a diversity of risk perspectives, has as often served to promote regulatory 'ossification'.[62] Moreover, a commitment to reviewing only the procedures followed

[59] *Ibid.*, [101]. [60] *Ibid.*

[61] For a detailed, step-by-step discussion of the panel's methodology and reasoning see Gregory Shaffer, 'A Structural Theory of WTO Dispute Settlement: Why Institutional Choice Lies at the Centre of the GMO Case', *New York University Journal of International Law and Politics*, 41 (2008), 1, Annex.

[62] Thomas McGarity, 'The Courts and the Ossification of Rulemaking: A Response to Professor Seidenfeld', *Texas Law Review*, 75 (1997), 525.

in regulating risks has not necessarily answered hard questions about where the boundary between issues of substance and those of process lies in risk decision-making.[63]

The difficulty facing a review of risk regulatory process, as much as for an assessment of the scientific justification for the resulting measures, is that normative judgments are still required in order to determine what amounts to an adequate process. For the *Methanex* tribunal, it seemed that its decision to uphold challenged risk regulations, provided they were 'non-discriminatory' and 'enacted in accordance with due process',[64] was underpinned by a preference for the normative claims of public interest environmental protection over those of private investors. By contrast, in assessing the justifiability of the EC regulatory process in the *GMO* case, the panel appeared to weight more heavily the goal of minimising the adverse trade effects of risk regulatory measures. A commitment on the part of international institutions to ensuring democratic legitimacy – or at least 'striving' for such – might supply an alternative normative basis for upholding domestic risk regulatory measures that are underpinned by assessments that are produced via a transparent process and embrace wide-ranging participation (including from outside the polity concerned).

Risk situation continuum

While simply shifting the focus of international review from matters of scientific substance to those of regulatory process will not do away with the need for difficult value judgments about risk, it may be possible to construct a framework around the decision-making process that allows decision-makers to better justify their decisions as legitimate ones. In this regard, the proposal put forward by a quintet of distinguished social science professors in an article published in the lead-up to the panel's decision in the *GMO* dispute provides important insights as to how democratisation of WTO SPS risk review might be effected in a way that both addresses concerns over the credibility of science-based evaluations in uncertain or contested risk scenarios, and also buttresses the legitimacy of international risk decision-making.

David Winickoff and his co-authors, in common with other proponents of process-based approaches, believe that WTO decision-makers conducting SPS review should generally 'steer away from adopting any

[63] Button, *The Power to Protect*, p. 153.
[64] *Methanex*, Part IVD, [7].

member state's conclusions as scientific truths' and instead 'act more as an administrative tribunal searching for transparency and procedural adequacy'.[65] Key to their proposal is the notion that the balance between substantive and process-based review – and so between competing normative perspectives regarding risk regulation – can be determined in light of the characteristics of the risk situations at issue. They see two aspects of risk situations as critical in this respect: the level of 'certainty' surrounding a risk, which can be determined in light of the available knowledge base and the analytic methods to be applied in assessing risks; and the degree of social 'consensus' regarding risk, which relates to the framing of the scientific issues to be addressed and the values to be protected through public policy.

Using these criteria, the authors argue that risk situations can be conceptualised on a 'continuum', running from 'low certainty and low consensus' at one end, to 'high certainty and high consensus' at the other.[66] Factors that would allow decision-makers to locate risk situations on this certainty–consensus continuum would include the relative novelty of the technologies or activities with which risks may be associated, whether it has been possible to agree upon international standards relating to those risks, the degree to which unknowns affect current knowledge of, or methodologies for, assessing risks, and the extent to which evidence sourced in the literature, regulatory experience or public dialogues suggests a lack of consensus and certainty as to the nature, sources and extent of the risks involved.[67]

Winickoff and his co-authors argue that, in cases of low certainty and low consensus, public input into the risk decision-making process assumes both social and scientific importance in order to 'frame risk in ways that make regulation more relevant and effective' and 'present the relevant questions that need to be answered before risks are assumed'.[68] Hence, in low (or possibly even medium) certainty and consensus situations, they believe this should invite a 'more deferential approach to the science-based decision-making of members' on the part of WTO reviewers, giving national regulators greater room 'to take public value choices into consideration when setting appropriate regulatory standards'.[69] On the other hand, the authors acknowledge that in cases where consensus and certainty are high, 'the range of rational measures to

[65] Winickoff *et al.*, 'Adjudicating the GM Food Wars', 107. Winickoff's co-authors were Sheila Jasanoff, Lawrence Busch, Robin Grove-White and Brian Wynne.

[66] *Ibid.*, 104. [67] *Ibid.*, 115–16. [68] *Ibid.*, 105–6. [69] *Ibid.*, 117–18.

address the risk situation should be more limited'.[70] Accordingly, they contend that 'in situations of high consensus and high certainty, a heavier burden [should] be placed on [WTO] members to establish that their [SPS] measures stem from non-protectionist values'.[71]

While some might still contend that international deference is never justified in respect of national risk regulatory processes with beyond-border impacts (particularly where detrimental effects are likely to be borne predominantly by exporters in developing countries), what is significant about the approach put forward by Winickoff and his co-authors is that it emphasises the need to assess such competing normative claims in light of the characteristics of the risk situation at hand. In the framework they propose, the normative claims of locally based risk concerns would only be given greater weight via mechanisms of process-based review in circumstances of low certainty and significant social disagreement over health and environmental risks. Put another way, this approach uses characteristics of the particular risk situation being addressed as a means for determining, on a case-by-case basis, the balance struck between competing considerations of scientific credibility and political legitimacy in decision-making, thereby defining what is included within the scope of the notions of science and risk assessment that are applied. As discussed further in the final section of the chapter, the risk continuum notion is a useful device for identifying different categories of risk situations that should elicit different approaches on the part of WTO decision-makers engaged in the review of domestic SPS measures.

Enhancing transparency and participation

While those advocating the adoption of process-based review in SPS decision-making have focused on the transparency and openness of the domestic risk regulatory procedures under scrutiny, other authors have argued for similar standards to apply at the international level through opening up WTO dispute settlement to a wider range of participants. In its broadest sense, public participation in WTO dispute settlement would include all institutionalised forms of interaction in the decision-making process between the dispute settlement organs and external actors who are independent of WTO member governments.[72] Participation in WTO dispute settlement proceedings might be enhanced in a modest way

[70] *Ibid.*, 118. [71] *Ibid.*, 123.

[72] Yves Bonzon, 'Institutionalizing Public Participation in WTO Decision Making: Some Conceptual Hurdles and Avenues', J. Int'l Economic Law, 11(4) (2008), 753.

by improving the transparency of the process to outsiders. More extensive reforms would involve allowing for outside participation in dispute settlement proceedings through the active engagement of non-state actors in the decision-making process.

Proposals for improving participation in WTO dispute settlement have been extensively canvassed, both within the institution itself and in associated literature. In recent times such proposals have gained a new prominence as possible means for improving the legitimacy of WTO governance and decision-making.[73] Based upon notions of participatory democracy, it is argued that opening up the WTO to a greater range of participants will improve the accountability of this forum of global governance for the outcomes of its decision-making.

From the specific perspective of democratising the science relied upon in WTO SPS disputes, improving the transparency of the decision-making process and opening up opportunities for outside participation could also serve to augment the available information base for risk decision-making. In this way, participation could enhance the quality of the decision-making process by bringing to bear a wider range of information and knowledges to broaden the framing of risk assessments and highlight areas of uncertainty.[74] In addition, research concerning the influence of global scientific assessments suggests that the extent to which such assessments take account of and accommodate stakeholders' viewpoints is critical to their perceived legitimacy.[75] Hence the use of science and risk decision-making by panels and the Appellate Body in SPS disputes might gain added legitimacy if opportunities were provided for external participation in these processes. Such added legitimacy would be particularly important where WTO decision-makers are seeking to rely on uncertain science to justify a particular risk regulatory approach. In the absence of 'strict transparency and participation requirements', the WTO's dispute settlement organs are unlikely to be seen to have 'sufficient credibility' to exercise the level of discretion inherent in decision-making undertaken in circumstances of scientific uncertainty.[76]

[73] *Ibid.*, 760–4.

[74] Michelle Everson and Ellen Vos, 'The Scientification of Politics and the Politicisation of Science', in Michelle Everson and Ellen Vos (eds.), *Uncertain Risks Regulated* (Milton Park: Routledge-Cavendish, 2009), p. 8.

[75] Mitchell *et al.*, 'Information and Influence'.

[76] Veerle Heyvaert, 'Facing the Consequences of the Precautionary Principle in European Community Law', European L. Rev., 31(2) (2006), 202–4.

Transparency in WTO SPS dispute settlement

Enhancing the transparency of WTO dispute settlement in SPS (and other) cases represents a relatively straightforward means for improving the perceived legitimacy of the institution's risk governance. Transparency in decision-making can be achieved by simple, readily implemented measures such as the timely public release of dispute settlement documentation. Indeed, procedures instituted by the WTO General Council provide for all dispute settlement documentation to be derestricted following the conclusion of a dispute – a process which generally sees documents made publicly available within a six- to twelve-week timeframe. However, subsequent release of the documents pertaining to a dispute is not adequate to allow interested persons to follow the course of the dispute while in progress, as a true commitment to transparency would demand. For this purpose, real-time release of documents via the internet (a practice some governments have initiated with respect to their submissions as parties to SPS and other WTO disputes)[77] would better achieve the goal of ensuring transparent processes.

Advocates for greater transparency in WTO dispute settlement have long called not just for the public release of dispute settlement documentation but also the opportunity for public access to WTO hearings before panels and the Appellate Body. This reform has been sought for a number of reasons, including to demonstrate that the WTO has 'nothing to hide' when it comes to its dispute settlement process by allowing the public to see that the process 'in reality [is] highly professional, engaged, impartial, and objective', as well as to strengthen public trust and confidence in WTO judicial decision-making, thereby making its results more acceptable to those affected.[78] Moreover, open hearings can benefit the WTO membership itself, particularly developing country parties, by giving members the opportunity to observe a dispute closely and to attain familiarity with the process.[79]

The *Hormones II* dispute marked a watershed moment in WTO/SPS dispute settlement history with the panel for the first time agreeing to

[77] E.g., this was the practice of the parties in the *GMO* case. More recently in the *Hormones II* dispute and the trans-Tasman *Apples* case, the parties have also made their submissions publicly available via the internet.

[78] Lothar Ehring, 'Public Access to Dispute Settlement Hearings in the World Trade Organization', *Journal of International Economic Law*, 11(4) (2008), 1023.

[79] *Ibid.*, 1024.

open its proceedings, including the consultations with experts, to the public. This took the form of a live broadcast of the proceedings to a separate room in the WTO building where there were members of the public who had registered (on a first-come, first-served basis) to attend the hearings.[80] In subsequent disputes, such as the trans-Tasman *Apples* case between Australia and New Zealand, this practice has been continued and expanded to allow the public to observe the presentations made by third parties in the dispute.[81] In July 2008 another landmark was achieved with the Appellate Body also agreeing to hold open hearings in the *Hormones II* appeal.[82]

While public attendance at open hearings in the WTO has been patchy (especially after the novelty wore off) and continues to generate some opposition within the WTO membership, it seems that this practice has had positive flow-on effects for WTO dispute settlement. Lothar Ehring points out that for the purpose of ensuring transparency in international governance, the opportunity for the public to observe WTO hearings is more important than the actual levels of public attendance in any one case.[83] He also notes that as WTO members have become more familiar with the practice, seen that it does not involve great institutional outlays or disrupt the proceedings in any significant way, many more have dropped their objections to open hearings, including some members, such as Brazil, which had previously maintained strong opposition.[84] Ehring concludes that there is thus the potential for open hearings to become the norm in the majority of future SPS and other WTO dispute settlement proceedings.

External participation in WTO SPS dispute settlement

Greater transparency in WTO dispute settlement is an important first step for enhancing public confidence in the outcomes of this

[80] *United States* and *Canada – Continued Suspension of Obligations*, Communication from the Chairman of the Panels of 1 August 2005, WT/DS320/8, WT/DS321/8, 2 August 2005.

[81] Ehring, 'Public Access to Dispute Settlement Hearings in the World Trade Organization', Appendix provides a table summarising the modality and degree of openness of all open hearings to date.

[82] *United States* and *Canada – Continued Suspension of Obligations in the EC-Hormones Dispute*, Reports of the WTO Appellate Body, WT/DS320/AB/R, WT/DS321/AB/R, 16 October 2008, Annex IV.

[83] Ehring, 'Public Access to Dispute Settlement Hearings in the World Trade Organization', 1027.

[84] *Ibid.*, 1032.

international decision-making process. However, it remains a long way from notions of democratic legitimacy built upon ensuring the fullest possible participation by those affected. In the SPS context, a transparent dispute settlement process might allow observers to monitor the risk information and perspectives put before a panel, but offers no opportunity to correct inaccurate representations or to contribute alternative views or new information on the risks concerned. It cannot be assumed that the risk assessments undertaken by domestic authorities will have considered all relevant risk perspectives and information as not all members provide for wide-ranging public participation in risk regulation, and even those that do rarely invite participation by those outside the polity. For these reasons, the democratisation of SPS risk governance in most cases calls not only for greater transparency but also external (that is, non-party mediated) participation in the dispute settlement process.

Adopting a broader understanding of expertise and information

In improving external participation in WTO dispute settlement – and SPS cases in particular – there are a number of available options that can be accommodated (with more or less controversy) within the existing institutional framework. At the more conservative end of the scale are proposals to expand the types of expert advice provided to a panel as a proxy for the consideration of a greater range of risk perspectives. Such proposals rely on existing provisions under the WTO SPS Agreement and the Dispute Settlement Understanding (DSU) that permit panels to engage in information-gathering in order to assist them in the task of dispute settlement. Article 11.2 of the SPS Agreement directs panels to 'seek advice from experts chosen by the panel in consultation with the parties to the dispute' where the dispute raises 'scientific or technical issues'. The relevant provision of the DSU is broader, conferring on a panel 'the right to seek information and technical advice from any individual or body which it deems appropriate', including experts.[85]

Pursuant to these provisions, panels in SPS disputes have so far limited themselves to seeking technical and scientific information from a fairly limited range of 'experts'. However, there is nothing to prevent WTO panels making more extensive and imaginative use of these provisions to solicit expert advice regarding 'other bodies of knowledge, such

[85] *Agreement on the Application of Sanitary and Phytosanitary Measures*, 15 April 1994, in force 1 January 1995, 1867 UNTS 493 (SPS Agreement), Article 13.

as sociological and anthropological studies of the communities closely related to the risks in question (e.g. farmers, and veterinarians)'.[86] The powers granted to panels to engage in information-gathering might also provide an avenue for greater inter-institutional cooperation and multilateral engagement, say by encouraging WTO decision-makers to seek advice from broadly based international organisations in the United Nations system, such as the United Nations Environment Programme or the World Health Organization.[87] While this would not equate to the democratic ideal of the fullest possible participation and representation of those affected, it might nonetheless enhance the quality and perceived inclusiveness of a panel's risk decision-making process, thereby making its results more acceptable to a wider range of audiences.

Broadening the notion of who is an 'expert' or what is a 'relevant source' of information as a means for democratising WTO dispute settlement in SPS cases has the advantage that it is relatively easy to accommodate within existing legal structures. Of course, simply providing panels and the Appellate Body with a broader range of expert opinion or information is no guarantee that this material will actually be taken into consideration. The panel decision in the *GMO* case illustrated this only too well, with the panel making few references to the extensive expert advice compiled over the course of the dispute, which included perspectives from experts in a range of disciplinary fields. The panel also declined to consider the information put forward in *amicus* briefs filed by a range of NGOs and academic experts on risk assessment and risk perception because it 'did not find it necessary to take the *amicus curiae* briefs into account'.[88]

Involving more experts in WTO SPS risk decision-making – while likely to highlight a greater variety of discipline-specific ways of 'framing' the risks at issue – may yet not be sufficient to capture knowledge existing outside the expert domain (for example, that held by industry, NGOs or community members) nor to prevent the imagination block that may be engendered by always looking at risk problems through an expert lens.[89] Expert deliberations on risk are often value-laden

[86] Perez, *Ecological Sensitivity and Global Legal Pluralism*, p. 153.

[87] *Ibid.*, p. 157.

[88] *European Communities – Measures Affecting the Approval and Marketing of Biotech Products*, Reports of the Panel, WTO Docs WT/DS291/R, WT/DS292/R, WT/DS293/R, 29 September 2006 (*GMO* case), [7.11].

[89] David Fisk, 'Environmental Science and Environmental Law', J. Envt'l L., 10(1) (1998), 3, 5–6.

exercises, involving the demarcation of areas within the competence of the experts involved (which are considered to be relevant matters for assessment) and areas that fall outside (deemed irrelevant).[90] There is thus a danger that reliance solely on bodies of experts to better inform WTO panels in relation to relevant risks and uncertainties may simply become another layer of 'technocracy', albeit one that canvases a greater range of views.[91]

In addition, a consensus achieved by experts – even those from multiple disciplines – may not necessarily be indicative of a wider social consensus on the most appropriate ways of framing the risks concerned. Sheila Jasanoff maintains that the 'boundaries drawn by experts, and the resulting analytical frames, therefore need to be continually interrogated; otherwise experts are in danger of over-extending their capacities ... or overlooking potentially crucial inputs from interested and affected parties'.[92] This argues for making available an even broader information base to those undertaking risk decision-making in governance institutions such as the WTO; one that includes public views on a given risk issue, which can then be used to contest expert framings of risk and uncertainty.

Public participation in WTO risk governance

Increasingly, broad participation is viewed as a 'vital element' of democratic legitimacy in supranational risk governance, as well as 'a positive response' to the complex task of regulating risk under conditions of uncertainty.[93] At the same time it poses difficult operational issues concerning who should participate (for example stakeholders, NGOs and civil society and/or the general public), how participants should be defined and what degree of representativeness they should have.[94] In addition, proposals for participation (and particularly public participation) raise significant institutional challenges, given the need to identify mechanisms and procedures that

[90] Jasanoff, '(No?) Accounting for Expertise', 160.

[91] Steve Rayner, 'Democracy in the Age of Assessment: Reflections on the Roles of Expertise and Democracy in Public-Sector Decision Making', *Science and Public Policy*, 30(3) (2003), 169.

[92] Jasanoff, '(No?) Accounting for Expertise', 160–1 (some text omitted).

[93] Everson and Vos, 'The Scientification of Politics and the Politicisation of Science', p. 8.

[94] Ellen Vos, 'The EU Regulatory System on Food Safety: between Trust and Safety', in Michelle Everson and Ellen Vos (eds.), *Uncertain Risks Regulated* (Milton Park: Routledge-Cavendish, 2009), pp. 259–60.

reconcile potentially irreconcilable rationalities of science and ethics, and which further guard against [regulatory] capture by inappropriate interests, or even an utter collapse in the [regulatory] function as plural processes of knowledge creation descend into meaningless and fruitless conflict between varied views and interests.[95]

Developing modes of public participation in the WTO and international risk governance is complicated by the fact that there is a lack of well-developed participatory models that could be adapted from the domestic context. While public participation is a feature of the risk regulatory systems of Western democracies – with some procedures in place for a number of decades – there remains a limited understanding of what makes an effective public participation mechanism. Consequently, 'the efficacy of public participation remains largely a matter of faith and of what model of society and citizenship one is committed to'.[96]

In addition, public participation exercises in domestic risk governance settings are dogged by practical problems, including that members of the public lack the resources to take advantage of formal procedures and that participation frequently occurs too late in the regulatory process to identify alternative risk framings and management options.[97] The experience of domestic regulatory systems, such as that of the USA, which have placed a premium on transparency in risk decision-making, points to other potential perils. For instance, Sheila Jasanoff notes consistent findings in empirical research that 'transparency may exacerbate rather than quell controversy, leading parties to deconstruct each other's positions instead of deliberating effectively'.[98] Transposing participatory mechanisms to international risk governance structures, such as those of the WTO, may seem only to invite an exacerbation of the problems uncovered domestically, given that greater divergences of interests and values are likely. It is not surprising then that for every potential mechanism of public participation in global risk decision-making that can be identified, problems or objections remain.

This is even so for more modest participatory reforms that rely upon the acceptance of public opinion data by panels as a source of information to guide judgments about acceptable risk. Under the GATT,

[95] Everson and Vos, 'The Scientification of Politics and the Politicisation of Science', p. 9.

[96] Rayner, 'Democracy in the Age of Assessment', 168.

[97] Jasanoff, 'Technologies of Humility', 237.

[98] *Ibid.*

consumer survey information may be taken into account when panels are evaluating the likeness of products.[99] It would not seem to be too much of a step from this to a consideration of public opinion about the significance of particular risks in determining, for example, whether an SPS risk assessment is adequate. The Appellate Body has left the door open for such an approach with its ruling that the factors specified in Article 5.2 of the SPS Agreement, which are to be taken into account in risk assessment, comprise a non-exhaustive list.[100]

A more radical version of this proposal would involve WTO decision-makers in SPS disputes recognising a specific and explicit public opinion style defence. This would involve endorsing as SPS-compliant, measures implemented 'precisely, and conceivably exclusively, because the public considers them to be so'.[101] Such situations would generally arise in respect of risks for which uncertainties abound, corroborating scientific evidence is minimal, yet public concern remains high.

Although she raises a public opinion defence as a possible mode of accommodating public views within the WTO dispute settlement process, Joanne Scott nonetheless foresees objections 'so numerous and so intense that it may be misconceived to even contemplate travelling down this road'.[102] After all, public opinion is notoriously difficult to survey accurately, both because of its constantly evolving nature, and due to the difficulties of designing mechanisms that capture the diversity of views that exist in any society. In a trade context there are also significant concerns as to the possibilities for manipulation of consumer opinion, not least through the creation of regulatory distinctions between different products, which can contribute to public perceptions of riskiness. Accordingly, Scott argues that if public opinion is to play a role in the WTO dispute settlement process, the institution would need 'to instantiate with care the conditions according to which [the entry of public opinion] into the justification equation may be mediated'.[103] This might include requirements pertaining to the way public opinion is surveyed and substantiated, transparency as to its

[99] European Communities – Measures Affecting Asbestos and Asbestos-Containing Products, Report of the WTO Appellate Body, WT/DS135/AB/R, 12 March 2001, [113].

[100] *European Communities – Measures Concerning Meat and Meat Products*, Report of the WTO Appellate Body, WT/DS26/AB/R & WT/DS48/AB/R, 16 January 1998 (*Hormones*), [187].

[101] Scott, 'European Regulation of GMOs', p. 317.

[102] *Ibid.* [103] *Ibid.*, p. 319.

use, and an acknowledgement of the volatile nature of public views by enacting risk measures on a provisional basis only.[104]

Another avenue for public opinion to play a role in WTO risk decision-making would be via mechanisms that allow for direct participation by individuals or public interest groups in the dispute settlement process. This proposal immediately raises the question of who should be recognised as an appropriate participant. Given the large number of people impacted indirectly as a result of SPS rulings (for instance, consumers wishing to adopt their own 'risk management' procedures by not purchasing GMO-derived products), extending participatory rights to all those potentially affected by a WTO dispute would be unwieldy and might only serve to expose decision-makers to an overwhelming array of conflicting viewpoints. On the other hand, limiting participation simply to 'stakeholders' who are clearly and directly affected by a decision may present decision-makers with a range of risk perspectives that approximate poorly those of the public at large. Indeed, enhancing stakeholder involvement at the expense of true public participation may only serve to polarise the process since stakeholder groups often represent those with views at the extremes of the spectrum of public opinion.[105]

Given the hurdles facing participation by the public at large in WTO dispute settlement, it has tended to be other actors better capable of operating at the global level, such as environmental and other NGOs, which have purported to voice the concerns of the public internationally.[106] In recognition of their public interest role, reforms to allow greater NGO participation in international fora have gained significant momentum in the last few decades, not least so in trade and economic institutions where some of the most innovative developments have taken place.[107] In WTO dispute settlement modest moves have been made in this direction via the acceptance (albeit rarely the substantive

[104] *Ibid.*

[105] Jan McDonald, 'Mechanisms for Public Participation in Environmental Policy Development – Lessons from Australia's First Consensus Conference', *Environmental and Planning Law Journal*, 16 (1999), 258.

[106] Andreas Klinke, 'Inclusive Risk Governance through Discourse, Deliberation and Participation', in Michelle Everson and Ellen Vos (eds.), *Uncertain Risks Regulated* (Milton Park: Routledge-Cavendish, 2009), p. 407.

[107] For discussion see Kal Raustiala, 'The "Participatory Revolution" in International Environmental Law', Harv. Envtl. L. Rev., 21 (1997), 537; Peter Van den Bossche, 'NGO Involvement in the WTO: A Comparative Perspective', J. Int'l Economic Law, 11(4) (2008), 717.

consideration) of *amicus curiae* briefs submitted by NGOs and other organisations.[108] Potentially this mechanism could be used by WTO decision-makers to gather additional information on risks and areas of uncertainty, and to gain an understanding of alternative risk perspectives in high-profile SPS disputes. However, strong opposition on the part of many government members to the acceptance of *amicus curiae* briefs by WTO dispute settlement bodies has seen panels and the Appellate Body eschew explicit reliance on such briefs in any case to date.[109]

Moves to allow any greater NGO involvement in WTO dispute settlement, for example, as advisors to, or complainants before, panels, also attract a barrage of criticisms. The most intractable relate to NGOs' representativeness of wider public concerns, especially those of the developing world, where organisations seeking involvement come primarily from Western countries or advocate a narrow viewpoint.[110] Quite apart from these concerns, there is also little guarantee that institutional reforms to enhance NGO participation would favour increased exposure of decision-makers to public risk perspectives. As Peter Van den Bossche notes, the NGOs seeking access to the WTO are often professional groups or business organisations that represent special interests rather than the views of the general public.[111] If NGOs were to play a more extensive role in the WTO dispute settlement process as advocates of the public interest, this would seem to necessitate procedures for NGO selection. Selection criteria could be targeted to matters such as the objectives of an organisation, its mode of financing and its capacity to 'add value' to the deliberative process of a panel or the Appellate Body by offering information or views not already presented by the parties to the dispute.[112]

Undoubtedly the theoretical and practical difficulties in enhancing external participation in WTO risk decision-making remain serious.

108 For an outline of the Appellate Body's practices with regard to the acceptance of *amicus* briefs see Mary Footer and Saman Zia-Zarifi, 'European Communities – Measures Affecting Asbestos and Asbestos-Containing Products: The World Trade Organization on Trial for its Handling of Occupational Health and Safety Issues', *Melbourne Journal of International Law*, 3 (2002), 1200.

109 Van den Bossche, 'NGO Involvement in the WTO', 741.

110 For a critical perspective on NGO representativeness see Tim Forsyth, 'Social Movements and Environmental Democratization in Thailand', in Sheila Jasanoff and Marybeth Long Martello (eds.), *Earthly Politics: Local and Global in Environmental Governance* (Cambridge, MA: MIT Press, 2004), p. 195.

111 Van den Bossche, 'NGO Involvement in the WTO', 721.

112 *Ibid.*, 743–7.

However, the challenges posed by public opinion or local knowledge inputs are not ones which can or should be avoided.[113] Where openness to such inputs is vital to the legitimacy and ongoing public acceptance of WTO risk decision-making, less than perfect procedures may have to be tolerated though always on the basis of striving to ensure that they are the best possible. In this regard it should not be expected that grafting public participation procedures onto risk decision-making mechanisms will lead to an automatic democratisation of the underlying science or encourage decision-makers to take into account a broader range of views about risks. Hence, participatory reforms, if they occur, will need to be accompanied by a wider change to the '*culture* of governance' in international institutions such as the WTO, which pays attention not just to 'the mechanics, but also to the substance of participatory politics'.[114]

Adaptive governance

One proposal that focuses explicitly on changing WTO modes of *governance* under the SPS Agreement as a means of responding to situations of uncertain, complex risks is Rosie Cooney and Andrew Lang's call for the institution of 'adaptive governance' in the trade regime. Cooney and Lang draw their notion of adaptive governance from the well-established literature and practice of adaptive management in the ecological field. Adaptive management in that context involves 'learning by doing' based on information garnered from post-decision monitoring of the environmental impacts of an activity as it progresses.[115] Essentially adaptive management is a form of real world experimentation where the results of studies are progressively fed into decision-making. The idea is to establish 'feedback mechanisms' 'so that management experience could inform system understanding and, thus, improvements in management'.[116] In natural resource management, where there is

[113] Scott, 'European Regulation of GMOs', p. 320.

[114] Jasanoff, 'Technologies of Humility', p. 238. See also Anne Orford, 'Beyond Harmonization: Trade, Human Rights and the Economy of Sacrifice', *Leiden Journal of International Law*, 18(2) (2005), 208, who suggests that increased transparency and openness in governance that does not challenge 'the form of law mandated by international economic agreements' may simply mean that calculation and decisions in response to the demands of market integration are 'made in public, rather than in secrecy'.

[115] Carl Walters, 'Is Adaptive Management Helping to Solve Fisheries Problems?', *Ambio*, 36 (2007), 304.

[116] Warwick Gullett, 'The Precautionary Principle in Australia: Policy, Law and Potential Precautionary EIAs', *Risk: Health, Safety and Environment*, 11 (2000), 96.

often pervasive scientific uncertainty, such strategies allow ongoing management arrangements to be adjusted and improved as more environmental information comes to light.

Cooney and Lang see significant potential for the ideas and mechanisms of adaptive management to be adopted in WTO/SPS risk governance given the scientific uncertainties frequently encountered in decision-making regarding SPS risks. They suggest that an adaptive governance framework in the WTO would involve five key parameters:

1. **Continuous learning** to develop better solutions to defined problems but also in order to redefine the problem to be addressed, to revisit what constitutes relevant knowledge about a problem and to develop critical awareness of the inherently limited nature of human knowledge.[117]
2. **Policy-making as experimentation** so that action can be taken despite a high level of uncertainty. The purpose of this 'learning by doing' process is to produce critical information that may help to reduce uncertainty and broaden the base of knowledge and experience for risk decision-making.[118]
3. **Avoiding irreversible harm** by ensuring that policy interventions are highly provisional and reversible, as well as being subject to strict oversight mechanisms. Cooney and Lang emphasise that this requirement means that policy-making by experimentation will only be appropriate 'where the system in question has some resilience, that is, where the changes induced by adaptive management interventions do not risk unacceptable and/or irreversible outcomes'.[119]
4. **Monitoring and feedback** so that the substantive outcomes of a policy are subject to an iterative process of review and revision.[120]
5. **Open, transparent and plural processes** that allow alternative knowledges to be marshalled, uncertainties to be mapped out and a disciplined process for decision-making to occur despite areas of uncertainty.

It is particularly this last element of Cooney and Lang's proposed governance framework that aligns with the ideas of democratised expertise. Like authors in the social science field, they emphasise the importance of involving a range of knowledge perspectives, beyond the traditional domain of scientific expertise, in order to reach sound decisions on risks in circumstances of scientific uncertainty.

[117] Rosie Cooney and Andrew T. F. Lang, 'Taking Uncertainty Seriously: Adaptive Governance and International Trade', European J. Int'l Law, 18(3) (2007), 534–5.
[118] *Ibid.*, 535–6. [119] *Ibid.*, 536. [120] *Ibid.*, 537.

What is most innovative about Cooney and Lang's proposal is their evaluation of how an adaptive governance model might work in the specific institutional context of WTO dispute settlement taking place under the SPS Agreement. Examining the current processes of WTO review, they see Article 5.7 of the SPS Agreement as the best avenue for implementation of an adaptive governance model, coupled with a procedurally based approach to review, such as that discussed previously. They argue that 'the expanded application of Article 5.7-type procedures to many (perhaps most) SPS measures could provide a beneficial mix of relaxed substantive supervision, combined with forms of context-specific ongoing supervision aimed at encouraging continuous learning in particular domestic environments'.[121]

As a means of improving the capacity of WTO risk governance to cope with uncertain risks, Cooney and Lang's proposal has great merit, although it might work best in less formal arenas, such as the multilateral policy processes of the SPS Committee. Indeed, the authors themselves seem to favour the SPS Committee as the best site for the operation of adaptive governance mechanisms such as continuous learning, policy experimentation, monitoring and feedback loops.[122] Less obvious is whether the adaptive governance idea could work effectively in the dispute settlement context, even to address environmental risks plagued by significant uncertainties, such as the risks associated with the introduction of invasive alien species that are the focus of Cooney and Lang's discussion.

In contrast to the conventional adaptive management scenario where careful risk-taking is endorsed as a means of resolving uncertainty, at issue in SPS dispute settlement are measures specifically put in place to guard against risks that might be introduced with the import of a particular product. Such measures may be ostensibly provisional in nature, giving the impression that they will be removed or revisited as knowledge about the risks at issue improves. However, often the uncertainties that are of concern to regulatory authorities (and the public) in these cases are ones that cannot be resolved by normal scientific work, since science itself lacks appropriate techniques for this purpose. For instance, the cancer-causing potential of very low residues of a chemical in food may not be measurable or able to be distinguished from the risk created by other substances also present. If a ban on foods containing the chemical is adopted as a risk management approach in

[121] *Ibid.*, 546. [122] *Ibid.*, 548–9.

this situation, it will generally be on the basis of socio-cultural preferences for risk aversion where the potential consequences could be catastrophic or irreversible. Such cases offer little scope for an adaptive management strategy of 'learning by doing' as gathering more scientific data and experience is unlikely to resolve the uncertainties at issue.

Conclusion: democratising SPS risk governance

The notions and models of democratised expertise being developed in the social scientific literature offer a means for responding to many of the limitations posed by an over-reliance on science and science-based risk assessment in complex and uncertain risk situations. Applied in international contexts such as WTO decision-making in SPS disputes, such notions could provide a more credible and legitimate basis for determinations of what the Appellate Body has referred to as 'risk in human societies as they actually exist'.[123] Particularly where uncertainties or disputed risk framings are an issue, models for democratising expertise present ways of supplementing scientific knowledge used in risk assessment, so that it can better take account of unknowns and the different values at stake in health and environmental regulation. Importantly, they also provide a means for strengthening science to make it more socially robust in an environment where it is continually asked 'to transgress the boundaries between specialised knowledge and its multiple, many-layered (and often unforeseeable) context of implication'.[124]

Moving from theory to the practice of international risk governance is the difficulty that now presents itself. In this respect, proposals for reform to WTO dispute settlement processes in SPS cases in order to create avenues for the introduction of a greater diversity of risk knowledges and perspectives present a microcosm of the broader institutional challenges likely to face efforts to democratise science in global risk regulation. In the setting of WTO/SPS dispute settlement this chapter has reviewed a variety of proposals that seek more transparent and participatory processes in the institution's risk governance. They include calls for deference to national or local risk preferences in WTO review; proceduralisation of WTO review in SPS disputes eschewing

[123] *Hormones*, [187].
[124] Nowotny, 'Democratising Expertise and Socially Robust Knowledge', 152.

substantive review of the underlying science but maintaining international controls over the process by which risk measures are created and implemented; enhancing the transparency and opportunities for participation in WTO dispute settlement to expose decision-makers to a range of alternative risk perspectives and knowledges; and instituting processes of adaptive governance that emphasise continuous learning and policy experimentation.

Looking at this array of proposals for the democratisation of SPS dispute settlement, it is important – as Gregory Shaffer has recently emphasised – to note that we are presented with a series of imperfect alternatives.[125] As Shaffer argues, when selecting among these alternatives, we are making institutional choices, each with their own advantages and disadvantages. For instance, a policy of deference allocates the responsibility for risk decision-making to national regulatory bodies, which may be responsive to the risk concerns of their own publics but take little account of those of affected outsiders. On the other hand, deference to risk measures that emerge from an international political process, such as the standard-setting procedures of Codex, might be inclusive of affected stakeholders but more remote from citizens and less responsive to local risk concerns.[126] Attempts to create more direct avenues for involvement of non-standard knowledges and non-expert participants in WTO SPS decision-making also encounter difficulties. Reasonable questions may be raised in relation to the legitimacy and authority of WTO decision-makers to judge the importance of different value concerns, the relevance of different framings of risk, or the plausibility of various policy responses in the face of scientific uncertainty. Moreover, if WTO panellists and the Appellate Body seek to inform themselves more widely as to relevant risk perspectives and value concerns, the institutional structures within which they must operate are not well adapted to gathering such information directly from individuals or groups, rather than from national governments.

In highlighting these difficulties with institutional alternatives to the current mode of SPS risk decision-making, it is worth emphasising

[125] Shaffer, 'A Structural Theory of WTO Dispute Settlement'.

[126] *Ibid.*, 54–5. See also Martin Shapiro, '"Deliberative", "Independent" Technocracy v. Democratic Politics: Will the Globe Echo the E.U.?', IILJ Working Paper 2004/5 (Global Administrative Law Series, 2004), available at www.iilj.org, questioning the adequacy of procedural safeguards such as transparency, especially where interested actors have unequal capacities to monitor and participate in the vast array of contemporary risk decision-making processes.

that this does not confer on the default option - application of narrowly based scientific standards by an international judicial body - any greater credibility or legitimacy. Indeed, continuation of the status quo may be the more problematic alternative if by restricting the scope of uncertainties considered or discarding 'novel viewpoints, radical critiques, or considerations lying outside the taken-for-granted framing of the problem'[127] the WTO risk governance process (at best) produces overly narrow assessments of risk or (at worst) misses the warning signs of future harms to health or the environment. Recognising the relative imperfections of all institutional alternatives, we are thus faced with a choice between real world options; a choice that will inevitably be dependent upon the decision-making context.[128]

Arguably one of the most important contextual factors dictating the choice of institutional forum is the risk situation at issue in an SPS dispute. The risk situation continuum developed by Winickoff and his co-authors provides useful criteria for distinguishing different types of SPS risk concerns based on associated levels of uncertainty and social consensus. In situations corresponding to conditions of high certainty-high consensus (for example asbestos-related health risks), science is likely to serve as an adequate proxy for the normative dimensions of risk regulation given plausible assumptions that expert understandings of risk would be considered credible and legitimate by a range of audiences. Most such risk situations will be amenable to resolution in multilateral, technically oriented bodies, such as the expert committees of standard-setting organisations such as Codex, or in political fora that operate on the basis of consensus decision-making, such as the SPS Committee.

Risk situations involving medium-to-high uncertainty and high social consensus are those where SPS dispute settlement is most likely to play a constructive role as part of a broader WTO risk governance system. Relevant examples might be measures taken to address risks from diseases such as bovine spongiform encephalopathy (mad cow disease) or swine influenza where there is broad public acceptance of the potential for harm despite many gaps in relevant scientific knowledge. In dealing with these types of risk situations, the SPS Agreement already has a variety of tools at decision-makers' disposal that could promote WTO review that is science-based yet also cognisant of the values

[127] Jasanoff, 'Technologies of Humility', 237.
[128] Shaffer, 'A Structural Theory of WTO Dispute Settlement', 6.

inherent in risk regulation. These include focusing review on the risk assessment supporting measures (pursuant to Article 5.1) rather than on the relevant scientific evidence (pursuant to Article 2.2) in order to allow for risk assessment that takes account of non-scientific factors; giving weight to international, consensus-based standards on the risks concerned (via Article 3.1) as an indicator of the best-practice global regulatory approach; permitting risk measures on a precautionary basis until scientific knowledge improves (pursuant to Article 5.7); or evaluating the proportionality of the measures adopted in light of the perceived seriousness of the risks or the measures' consistency with those taken to address similar risks (in line with Articles 5.5 and 5.6). These tools might be used in combination with some of the proposals discussed in the chapter, for instance allowing provisional measures under Article 5.7 subject to adaptive management-type conditions for continuous scientific research into the risks concerned, or conducting a review of a supporting risk assessment under Article 5.1 primarily on the basis of procedural criteria provided there is some reputable scientific opinion indicating the potential for harm.

At the other end of the risk situation continuum, disputes over risks characterised by low certainty and low consensus will generally be unsuited to resolution via international decision-making processes. Indicators of such risks might be the persistent failure of political negotiation processes in standard-setting organisations or in treaty settings to reach consensus on suitable risk management measures, the existence of divergent regulatory approaches in different member countries, or public opinion data showing high levels of concern about particular risks. Examples include disputes over the risks posed by hormone residues in meat and GMOs. At issue in these cases is a clash of regulatory approaches that represent different risk management choices in the face of intractable uncertainties.

In dealing with a case involving this kind of risk situation, WTO decision-makers might choose to devolve decision-making back to local authorities via a deferential review approach that recognises the need to preserve scope for a plurality of risk regulatory approaches. Where uncertainty is a significant factor affecting scientific knowledge regarding the risks in question (as for GMOs), deference might take the form of permitting the implementation of provisional measures under Article 5.7, subject to requirements for ongoing review of the underlying science and the need for the measures. Alternatively, the maintenance of some level of procedural supervision over the domestic regulatory process via requirements for transparency and broad participation could offset, at least partially, the potential for measures to be adopted for a protectionist purpose. A more radical strategy would involve WTO decision-makers declining jurisdiction in normatively charged cases, leaving the resolution of the risk dispute to political processes of negotiation.[129] In the *Hormones* saga, for instance, it looks likely that a political deal between the USA and the EU will do more to resolve the dispute than the rulings from international judicial interventions over the past two decades.

Over time, if uncertainty concerns abate and risk framings converge, low certainty–low consensus risk situations may move further along the continuum towards high certainty–high consensus. Allowing scope for the pursuit of a diversity of local or regional risk management approaches in the interim period may facilitate this, especially in the case of new technologies where a longer history of use in some parts of the world without the materialisation of harm may convince other countries that the benefits of the technology outweigh any potential risks. This is, of course, a more modest vision for international risk regulation than living in an interdependent, globalising world might seem to demand.[130] Yet if trust and legitimacy take time to build as societies adjust to the new reality of an international level of risk governance, the drive for science-based global risk regulation may need to slow its pace accordingly so as not to outstrip their evolution.

[129] See, e.g., the proposal put forward by Jeffrey Dunoff, 'The Death of the Trade Regime', *European Journal of International Law*, 10(4) (1999) ,733.

[130] Subsidiarity, regionalisation and similar approaches necessarily compete with goals of harmonisation, which have been a primary concern of the WTO. For an exploration of these tensions see Daniel Esty, 'Good Governance at the Supranational Scale: Globalizing Administrative Law', Yale L.J., 115 (2006), 1490.

8 What role for science in international risk regulation?

Introduction

In an age of globalisation, when health and environmental problems are increasingly being identified on a global scale, it may seem paradoxical that one of the most intractable questions facing international law is how to ensure the regulation of risks in ways that will also accommodate local concerns and diverse risk perspectives.[1] That the international law of risk regulation should have reached this point is a testament to the rapidity of its development from a system of primarily inter-governmental negotiation, implementation and adjudication to one embracing regimes of global governance with the capacity to penetrate deeply into national regulatory orders. The importance of risk management in contemporary societies, and the ease with which many health and environmental issues can be framed as global problems, seem set to ensure a central place for issues of risk in international law and associated governance structures. Concepts of risk and procedures for assessing and managing risk are already an established feature of international law in diverse fields such as international trade law, international environmental law, laws governing hazardous pollutants and food contaminants, and international climate change law.

Expectations for this system of global risk governance are high. On the one hand, it is anticipated that risk governance should deliver effective responses to globalised health and environmental risk problems that are developed on the basis of credible information. In this

[1] Marybeth Martello and Sheila Jasanoff, 'Introduction: Globalization and Environmental Governance', in Sheila Jasanoff and Marybeth Martello (eds.), *Earthly Politics: Local and Global in Environmental Governance* (Cambridge, MA: MIT Press, 2004), p. 1.

respect, science seems to offer optimal tools for international risk regulation. The claims of scientific knowledge to universal, value-free knowledge sit well with a system of risk governance that is global in its ambition. Moreover, as a practical matter, science is an essential input into decision-making dealing with the kinds of 'invisible' risk problems that dominate the international agenda – climate change, ozone depletion, biodiversity reduction, carcinogenic pollutants and the long-term toxic effects of hazardous chemicals.[2]

But expectations of global risk governance increasingly extend much further. Alongside effectiveness, legitimacy has become an important criterion against which global risk regulation is measured. In addition to credibility, there are also demands that risk regulation accommodate a plural understanding of risks and have the capacity to take account of uncertainties in available knowledge to protect against 'surprises' over the long term.[3] In meeting these additional expectations, conventional science and scientific risk assessment processes have struggled. Accordingly, a constant theme of this book has been that despite the common representation of risk governance as a technical endeavour, in most, if not all, settings science alone is insufficient – to legitimate the outcomes of global structures of risk governance; to assess risks accurately under conditions of uncertainty; and to reach decisions on the management of risks that will achieve broad acceptance with a range of different audiences.

In this final chapter the task is first to summarise the problems with too heavy a reliance on science in international law governing risk regulation. We then turn to the difficult challenge this raises; namely if science by itself is an inadequate resource for global risk governance, how can it be supplemented or strengthened to give it the necessary 'robustness' to tackle the task of decision-making on uncertain risks?[4] Given the contextualised nature of risk assessment, it is argued that differentiating risk problems into different categories of risk situations is a necessary first step in shaping a response to this question. While it is not possible to be prescriptive about the role of science in international risk regulation given the many different decision-making settings and

[2] Ulrich Beck, *Risk Society: Towards a New Modernity* (London: SAGE Publications, 1992), pp. 21–3.

[3] European Environment Agency, *Late Lessons from Early Warnings: the Precautionary Principle 1896–2000* (Luxembourg: European Union, 2001).

[4] Helga Nowotny, 'Democratising Expertise and Socially Robust Knowledge', *Science and Public Policy*, 30(3) (2003), 151.

applicable international rules, this chapter offers guidelines about how science might be blended with other non-scientific inputs in the different categories of risk situations faced in global risk governance.

The dilemma of science in international risk regulation

Ever since risk management became a preoccupation of modern societies, risk issues have been treated primarily as a technical concern, seen as requiring the input of science and expertise for their assessment and resolution. In the international context, the seemingly neutral and authoritative standards of science and technical expertise have also held out the potential for overcoming otherwise intractable differences in the self-interested bargaining of governments on risk issues. However, as Helga Nowotny has observed, '[e]xpertise has never before been so indispensable, while being simultaneously so contested'.[5] Various regulatory failures of science in predicting and managing technological risks from Chernobyl to the Challenger disasters have exposed the uncertainties underlying many areas of scientific knowledge and the critical role that social and institutional assumptions play in framing science-based risk assessment. As a consequence, although science remains an important resource for contemporary risk decision-making, its knowledge claims no longer command the same respect and deference as in the past.

The 'dilemma' presented by science in international risk regulation is that while the desirability of scientific input is generally recognised, so is its inability to answer the questions involved in purely scientific terms.[6] Throughout the book we have seen numerous manifestations of this tension. For instance, it underlies the divergence between the dominant risk regulatory paradigms of sound science and the precautionary principle – both science-based regulatory approaches – but ones that prescribe different management responses in conditions of uncertainty in light of differing value concerns and socio-cultural risk attitudes. An inability to select between the competing normative claims of precautionary and sound science-based approaches at the international level has led to uncomfortable pairings of scientific risk assessment requirements with precautionary decision-making provisions in treaties such as the Biosafety Protocol. The disputes under the SPS Agreement considered in Chapter 5 also present many examples

[5] *Ibid.*, 151–2. [6] *Ibid.*

of the limitations of science as a basis for risk decision-making. While experts are consulted by WTO decision-makers to determine whether there is 'sufficient' scientific evidence to support the existence of a particular risk pathway or whether a member's risk assessment is adequate, technical experts are unable to answer these questions fully as they ultimately depend on value judgments about the acceptability of the risks at issue.

In essence the problem that faces risk regulation in international law is that it cannot rely exclusively on scientific assessments, which purport to be free of normative content,[7] to reach what are ultimately normative decisions about risk-taking and prevention. This has not prevented normatively charged questions continually being put to scientists or referred to processes of science-based risk assessment, as in the case of the expert review committee under the POPs Convention which must assess whether a chemical 'is likely, as a result of its long-range environmental transport, to lead to *significant* adverse human health and/or environmental effects, such that global action is warranted'.[8] Certainly 'the skilful employment of scientific discourse' can go a long way in some contexts towards securing agreement on regulatory action to address globally identified risks.[9] However, in the end, the discourse of science is no replacement for political contestation over the appropriate goals of international risk regulation. If instead of engaging in such debate, global risk governance turns exclusively to science – thereby extending scientific knowledge beyond the bounds of its accepted competence and representing it as definitive in circumstances where it is subject to uncertainty – science will eventually lose all credibility as a resource for global risk decision-making.

The inadequacy of science alone as a basis for global risk regulation is a lesson that has been accepted, to varying degrees, in a number of international settings outside the SPS arena.[10] Most prominently, the

7 As Chapter 3 discussed, in practice scientific risk assessment always has a degree of normative input; however, this only becomes problematic if science is deployed in policy and legal settings as if this were not the case.

8 Convention on Persistent Organic Pollutants, 23 May 2001, Stockholm, in force 17 May 2004 (2001) 40 ILM 532 (POPs Convention), Article 8(6).

9 Karen T. Litfin, *Ozone Discourses: Science and Politics in Global Environmental Cooperation* (New York: Columbia University Press, 1994), p. 198.

10 William C. Clark *et al.*, Global Environmental Assessments: Information and Influence (Cambridge, MA: MIT Press, 2006); Noelle Eckley, Designing Effective Assessments: The Role of Participation, Science and Governance, and Focus – Environmental Issue Report No. 26 (Copenhagen: European Environment Agency, 2001); Alexander E. Farrell and Jill Jäger (eds.), *Assessments of Regional and Global*

Inter-governmental Panel for Climate Change (IPCC) employs processes that explicitly set out to generate shared science–policy understandings of climate change risks that are more likely to garner acceptance with the institution's primary audience of government policy-makers. Even in more technically oriented international organisations, such as the Codex Alimentarius Commission, there has been recognition that science-based evaluation of the risks posed by food contaminants and food-borne diseases needs to be tempered by political considerations instilled in the standard-setting process via a requirement for consensus decision-making.

Differentiating risk situations in global risk governance

In designing a more appropriate role for science in global risk governance processes an important first step is to recognise that there is considerable variation in the nature of the risk situations dealt with by international law. The politically charged atmosphere in which the IPCC operates is quite different from that attending meetings of bodies such as the POPs Review Committee or the SPS Committee. Likewise, decision-making about risks for which the science is relatively settled (such as the carcinogenic effects of exposure to asbestos fibres) presents a very different task from evaluations of measures taken to deal with uncertain environmental risks (such as those posed to local biodiversity by GMOs).

Key parameters in respect of which risk situations differ are the degree of scientific certainty with respect to the knowledge base to be relied upon and the methods to be applied, and the level of social consensus with respect to the framing of the risk issues to be addressed and the values to be protected.[11] Combining these parameters, social scientists such as David Winickoff and his co-authors have produced a framework for distinguishing different risk situations raised in disputes under the SPS Agreement as either high certainty–high consensus, high certainty–low consensus, low certainty–low consensus or low certainty–high consensus. This framework can be applied more generally to discern different risk situations encountered in global risk

Environmental Risks: Designing Processes for the Effective Use of Science in Decisionmaking (Washington DC: Resources for the Future, 2006).

[11] David Winickoff *et al.*, 'Adjudicating the GM Food Wars: Science, Risk, and Democracy in World Trade Law', Yale J. Int'l L., 30 (2005), 104.

Table 8.1. *'Traffic light' spectrum of risk situations*

Risk situation	Description
Red	'Hot' risks where normative conflicts overshadow science
Amber	Some dispute over uncertainties or appropriate risk framing which may be ameliorated through more research, risk communication or participation
Green	Technical understandings of risk have broad social acceptance

governance. An alternative terminology or metaphor we might apply in describing the different categories of risk situations so derived is a risk situation 'traffic light' displaying green, amber and red risks (Table 8.1).[12]

The green risk category encompasses situations where the relevant scientific evidence is reasonably settled, areas of uncertainty are minimal (or do not give rise to significant public concern) and there is alignment of technical and social understandings of the risks involved. These situations will generally be susceptible to science-based assessment and management at the global level. Under the SPS Agreement some types of quarantine risk may fall in this category, for instance where SPS measures are taken to address well-known pests or diseases, readily transmitted via trade, that have the potential to cause severe environmental or agricultural impacts in the country of import.[13] Another example that we encountered in Chapter 6 is so-called 'dead' chemicals regulated under the POPs Convention. Stringent risk regulatory measures for these chemicals have attracted little political opposition given

[12] In identifying these categories of risk situations guidance has been derived from both Winickoff *et al.*, *ibid.*, and the tripartite framework of normal science, professional consultancy and post-normal science put forward in the work of Silvio Funtowicz and Jerome Ravetz, 'Three Types of Risk Assessment and the Emergence of Post-Normal Science', in Sheldon Krimsky and Dominic Golding (eds.), *Social Theories of Risk* (Westport, CT: Praeger Publishers, 1992), p. 251. For discussion of the latter, see Chapter 3.

[13] The Appellate Body has implicitly recognised the difference between health and quarantine risk categories in specifying different stringencies of risk evaluation for each: see further, the discussion in Chapter 5.

the fact that they are usually well-studied pollutants that are heavily regulated in many nations already.

By contrast, red risk situations – characterised by high levels of uncertainty and a lack of social consensus on risk framing and management measures – present a difficult prospect for regulatory efforts at the global level. Cass Sunstein describes these situations as ones involving 'hot' risks for which normative or cultural commitments play a central role in people's assessments of risk.[14] When dealing with red or hot risk situations, normative conflicts may overshadow the role of science, as was the experience in negotiations for the Biosafety Protocol governing the risks associated with GMOs in international trade.[15] Alternatively, disputes over the scientific evidence and areas of uncertainty may simply serve as cover for more fundamental differences over the risks and benefits of the technology or product involved (for example in the *Hormones* SPS cases).

The final category of risk situations involving amber risks lies in between the two extremes represented by green and red risk situations. For risks in the amber category levels of scientific uncertainty may still be relatively high or a broad consensus may not yet have been achieved on appropriate ways of framing the risks concerned, perhaps because different risk perspectives are put forward by different groupings of countries. However, in amber risk situations these problems are not intractable as they are in red or hot risk situations. For instance, there might be the capacity for further scientific research to resolve and allay concerns over areas of uncertainty. Alternatively, social risk framings might converge over time as global assessment processes are adjusted to take on board local concerns and experience.

From a legal perspective, another aspect of risk situations of importance besides levels of uncertainty and social consensus is the relevant institutional context for decision-making and the applicable legal rules. These factors will be critical in determining the capacity of different global risk governance settings to respond to the risk situation at hand. In some institutional contexts a transparent decision-making process, participation by non-state actors or mechanisms for the ongoing review of decisions over time may be easier to achieve than in others. This might be the result of a greater capacity to interpret applicable

[14] Cass R. Sunstein, 'Misfearing: A Reply', *Harvard Law Review*, 119 (2006), 1115.

[15] Aarti Gupta, 'Problem Framing in Assessment Processes: The Case of Biosafety', in Ronald B. Mitchell *et al.* (eds.), *Global Environmental Assessments: Information and Influence* (Cambridge, MA: MIT Press, 2006), p. 57.

legal instruments in a flexible manner; an institution's history of openness to, and accommodation of, perspectives beyond those of conventional participants such as government representatives and experts; or the evolution of structures well suited to post-decision monitoring and review.[16]

Guidelines for the use of science in international risk regulation

In designing mechanisms for the use of science in international risk regulation (or evaluating existing ones), one might hope that it would be possible to find a reliable formula of universal application. This, after all, has long been touted as the appeal of sound science, although, as we have seen in practice, dependence on science as a universal arbiter and legitimator places unrealistic demands on scientific knowledge and experts.[17] Instead what we face are real world situations where contextual factors are important and institutional constraints limit the possibilities for the implementation of ideal models. Consequently, the task at hand is one of devising the best possible mechanisms for the use of science in global risk governance that are responsive to different types of risk situations and which take account of the inherent limitations of international law and institutions in securing broad participation in decision-making.

Preparing science for this new role in international risk regulation will involve recasting science (or our expectations of scientific knowledge) from the role of truth giver to a supplier of what Sheila Jasanoff has called 'serviceable truths'. Jasanoff uses the concept of serviceable truth to signify knowledge that satisfies tests of scientific acceptability and supports reasoned decision-making, but which also assures those exposed to risk that their interests have not been sacrificed on the altar of impossible scientific certainty.[18] In essence this means that the science used in global risk decision-making should have both technical credibility and broad social acceptance.

[16] These resources might be concentrated in NGOs associated with a particular regulatory regime rather than the institution itself. For instance, the Convention on International Trade in Endangered Species of Wild Fauna and Flora (CITES) has had a long association with the NGO TRAFFIC which undertakes the lion's share of compliance monitoring under the treaty. For details see www.traffic.org/.

[17] See also Vern Walker, 'The Myth of Science as a "Neutral Arbiter" for Triggering Precautions', B.C. Int'l & Comp. L. Rev., 26 (2003), 197.

[18] Sheila Jasanoff, *The Fifth Branch: Science Advisors as Policymakers* (Cambridge, MA: Harvard University Press, 1990), p. 250.

For science to serve as a source of serviceable truth will generally involve melding scientific knowledge with other inputs that overcome science's deficiencies as a resource for international risk regulation. These inputs might range from anecdotal evidence of experience with the management of similar risks (useful where novel risk situations are at issue with little specific scientific knowledge available) to local knowledge (in cases where the impact of particular risks will be differentiated given different environmental conditions) and public opinion (an important input where the socio-political dimensions of risk regulation are salient). In each case solutions will need to be worked out on a case-by-case basis and are likely to involve an element of compromise between the desirability of garnering credibility for global risk decision-making (an area in which conventional scientific knowledge still plays an important part) and securing its legitimacy (which will require attention to matters of risk politics). The end result will be negotiated agreements about acceptable ways of balancing conventional and non-standard knowledges, expertise and politics, science and democracy, that are targeted to dealing with the risk situations faced by a particular governance mechanism and sensitive to its institutional constraints. The same task has confronted many domestic risk regulatory systems that have sought to incorporate participatory reforms as part of risk decision-making processes. In some ways, international law may be well placed to engage with this challenge given its perpetual quest to find ways of living with, and accepting, uncertainty, anxiety and instability.[19]

In some risk decision-making settings (for example. those involving green risks), it may still be possible to rely heavily on technical risk assessment because low levels of uncertainty and high levels of social consensus make legitimacy concerns less salient in agreeing on an appropriate science–politics boundary. In other cases, more pressing problems surrounding how to respond to uncertainty or define the nature of the risks involved might promote an approach where there is a division of labour between international and national (or regional) levels of regulation. In this scenario, the latter would take on primary responsibility for the deliberation of risk questions and devising risk management responses, whereas the former would merely exercise a supervisory function over the way risk determinations are reached.

[19] Anne Orford, 'The Destiny of International Law', *Leiden Journal of International Law*, 17 (2004), 441.

Rethinking the use of science in SPS dispute settlement

As a major focus of analysis throughout the book, and in the legal literature generally, WTO dispute settlement under the SPS Agreement is a good place to start when thinking about how general guidelines regarding the use of science in international risk regulation might be implemented in a particular institutional setting. A necessary first step in any WTO review of national SPS measures would be to determine the type of risk situation at issue, bearing in mind that green risk situations are unlikely to be those that generate legal disputes in most cases.[20]

At the other extreme, cases that raise red risk situations are ones for which WTO dispute settlement is unlikely to produce a long-lasting resolution of the underlying dispute. Ideally, WTO members should exercise restraint so as to avoid referring these kinds of cases to legal dispute settlement, instead remaining within fora that allow more scope for negotiating political differences over risks such as the SPS Committee or international bodies such as the Codex Alimentarius Commission. If, however, disputes involving red risks are referred by members to the WTO dispute settlement system, decision-makers might respond with the deployment of legal tools that – while by no means perfect – are the most permissive of a diversity of risk regulatory approaches.[21] As Chapter 7 highlighted, these might include:

- Favouring softer forms of review that give substantial deference to the risk judgments of national authorities. Rather than examining the sufficiency of the science on which SPS measures are based, reviewers would look to domestic regulatory processes, their transparency and openness to a broad range of participants (including those outside the polity).
- Focusing review under Article 5.1, which allows consideration of whether disputed measures are warranted in light of a risk assessment that may take into account non-scientific factors as well as the available scientific evidence. By contrast, pursuing review under Article 2.2 tends to lead WTO reviewers to heavy scrutiny of the scientific evidence only and its capacity to support particular risk conclusions.

[20] Gavin Goh, 'Tipping the Apple Cart: The Limits of Science and Law in the SPS Agreement after *Japan-Apples*', J. World Trade, 40(4) (2006), 655. The *Asbestos* case (decided under the GATT) might be regarded as an exception to this rule.

[21] Alan O. Sykes, 'Domestic Regulation, Sovereignty, and Scientific Evidence Requirements: A Pessimistic View', *Chicago Journal of International Law*, 3 (2002), 353.

- Applying proportionality or 'balancing'-style tests, such as those under Articles 5.5 and 5.6, which seek to discern whether the stringency of disputed measures matches the severity of risk concerns.

Similar strategies may be appropriate in amber-type risk situations, depending upon whether the key problem facing risk decision-making is one of scientific uncertainty or a lack of social consensus on optimal ways of framing risks. Whereas Article 5.1-led review or the application of balancing tests might best deal with the latter type of situation, circumstances where uncertainty is the major issue due to a lack of scientific knowledge of a risk problem may benefit more from process-based review coupled with the acceptance of risk measures on a provisional basis under Article 5.7. Such an approach would provide some discipline over the adoption of risk measures in conditions of uncertainty but without denying WTO members the necessary flexibility to act in a precautionary manner until scientific knowledge of risks improves.

Both red and amber risk situations might also invite greater openness on the part of WTO reviewers to outside observation of the hearing process and external participation. As others have argued, in risk situations where uncertainties abound or there is intense socio-political debate over potential harms, public input into the risk decision-making process assumes both social and scientific importance in order to 'frame risk in ways that make regulation more relevant and effective' and 'present the relevant questions that need to be answered before risks are assumed'.[22] While it will not be possible – at least within existing institutional constraints – for WTO decision-makers to obtain views from all those (most) affected by the measures in dispute, the sensitivity of decision-making to a range of risk perspectives might still be improved by consulting a greater range of experts (including international institutions with relevant expertise like the World Health Organization or the Biodiversity Convention secretariat), holding open hearings and giving substantive consideration to arguments raised in NGO *amicus* briefs.

Conclusion

This book began with an observation about the dominance of questions of risk, and the scientific understandings of them, in contemporary international law and governance. This situation reflects the considerable faith

[22] Winickoff *et al.*, 'Adjudicating the GM Food Wars', 105–6.

governments around the world have placed in the capacity of science to identify global risks and provide a basis for developing acceptable solutions. As we have seen throughout the book, however, these tasks place substantial demands on science and scientific experts; expectations which will generally not be able to be met given uncertainties in scientific knowledge or the inherent normative aspects of risk regulation. Nevertheless, science remains an important resource for global risk decision-making that retains significant power to persuade governments and peoples of the need for risk regulatory measures.

Clearly the solution to the challenge posed by the use of science in international risk regulation is not the jettisoning of science altogether; science-based tools such as risk assessment play a crucial role in curbing the excesses of political debate on risk issues and can help to systemise and make transparent processes of risk decision-making.[23] On the other hand, the 'technical and normative frailties'[24] besetting scientific risk assessment require the intervention of non-scientific inputs, including public views, to improve both its credibility and broader social acceptance as the basis of global risk decision-making. Hence, in global risk governance 'science must discipline politics and politics must discipline science'[25] with the ideal balance between the two determined, as far as possible, in light of the characteristics of the risk situation at hand.

For international risk governance institutions, such as the WTO, to move towards such an approach will require much greater acknowledgement of the softness of science as a basis for global risk assessment, as well as increased courage on the part of governments to take on the hard political decisions that this acknowledgement requires. This will necessitate attention not only to the mechanics of using science in new ways – for example, by increasing the transparency of risk determinations or by broadening the information base available to decision-makers – but also to the need for wider changes to the

[23] Climate change illustrates this point with the scientific assessments produced by the IPCC apparently having been effective in ameliorating some of the ideologically charged political debates over the issue. Nonetheless, the 'Climategate' affair in late 2009 illustrates the continuing need for scientific studies and risk assessment to be undertaken in a transparent manner if they are to retain their broader legitimacy.

[24] Jeremy D. Fraiberg and Michael J. Trebilcock, 'Risk Regulation: Technocratic and Democratic Tools for Regulatory Reform', *McGill Law Journal*, 43 (1998), 835.

[25] *Ibid.*, 835.

culture of governance itself.[26] To date in some of the most prominent sites of global risk governance too much emphasis has been placed upon a fruitless quest for objectivity in the face of risk problems whose framing and assessment are strongly influenced by factors of context, culture and politics. The challenge for international risk regulation is to find ways of creating space for, and facilitating genuine debate, over the values that are ultimately crucial for developing responsible risk policy, rather than allowing these to be subsumed within a discourse of sound science.

[26] Sheila Jasanoff, 'Technologies of Humility: Citizen Participation in Governing Science', *Minerva*, 41(3) (2003), 238; Anne Orford, 'Beyond Harmonization: Trade, Human Rights and the Economy of Sacrifice', *Leiden Journal of International Law*, 18(2) (2005), 208.

Index

Locators with 'n' are notes; with 'fig' are figures.

CAMBRIDGE STUDIES IN INTERNATIONAL AND COMPARATIVE LAW

Books in the series

Science and Risk Regulation in International Law
Jacqueline Peel

Vicarious Liability in Tort: A Comparative Perspective
Paula Giliker

Legal Personality in International Law
Roland Portmann

Legitimacy and Legality in International Law: An Interactional Account
Jutta Brunnée and Stephen J. Toope

The Public International Law Theory of Hans Kelsen: Believing in Universal Law
Jochen von Bernstorff

The Concept of Non-International Armed Conflict in International Humanitarian Law
Anthony Cullen

The Challenge of Child Labour in International Law
Franziska Humbert

Shipping Interdiction and the Law of the Sea
Douglas Guilfoyle

International Courts and Environmental Protection
Tim Stephens

Legal Principles in WTO Dispute
Andrew D. Mitchell

War Crimes in Internal Armed Conflicts
Eve La Haye

Humanitarian Occupation
Gregory H. Fox

The International Law of Environmental Impact Assessment: Process, Substance and Integration
Neil Craik

The Law and Practice of International Territorial Administration: Versailles, Iraq and Beyond
Carsten Stahn

Cultural Products and the World Trade Organization
Tania Voon

United Nations Sanctions and the Rule of Law
Jeremy Farrall

National Law in WTO Law Effectiveness and Good Governance in the World Trading System
Sharif Bhuiyan

The Threat of Force in International Law
Nikolas Stürchler

Indigenous Rights and United Nations Standards
Alexandra Xanthaki

International Refugee Law and Socio-Economic Rights
Michelle Foster

The Protection of Cultural Property in Armed Conflict
Roger O'Keefe

Interpretation and Revision of International Boundary Decisions
Kaiyan Homi Kaikobad

Multinationals and Corporate Social Responsibility: Limitations and Opportunities in International Law
Jennifer A. Zerk

Judiciaries within Europe: A Comparative Review
John Bell

Law in Times of Crisis: Emergency Powers in Theory and Practice
Oren Gross and Fionnuala Ní Aoláin

Vessel-Source Marine Pollution: The Law and Politics of International Regulation
Alan Tan

Enforcing Obligations Erga Omnes in International Law
Christian J. Tams

Non-Governmental Organisations in International Law
Anna-Karin Lindblom

Democracy, Minorities and International Law
Steven Wheatley

Prosecuting International Crimes: Selectivity and the International Law Regime
Robert Cryer

Compensation for Personal Injury in English, German and Italian Law: A Comparative Outline
Basil Markesinis, Michael Coester, Guido Alpa, Augustus Ullstein

Dispute Settlement in the UN Convention on the Law of the Sea
Natalie Klein

The International Protection of Internally Displaced Persons
Catherine Phuong

Imperialism, Sovereignty and the Making of International Law
Antony Anghie

Necessity, Proportionality and the Use of Force by States
Judith Gardam

International Legal Argument in the Permanent Court of International Justice: The Rise of the International Judiciary
Ole Spiermann

Great Powers and Outlaw States: Unequal Sovereigns in the International Legal Order
Gerry Simpson

Local Remedies in International Law
C. F. Amerasinghe

Reading Humanitarian Intervention: Human Rights and the Use of Force in International Law
Anne Orford

Conflict of Norms in Public International Law: How WTO Law Relates to Other Rules of Law
Joost Pauwelyn

Transboundary Damage in International Law
Hanqin Xue

European Criminal Procedures
Edited by Mireille Delmas-Marty and John Spencer

The Accountability of Armed Opposition Groups in International Law
Liesbeth Zegveld

Sharing Transboundary Resources: International Law and Optimal Resource Use
Eyal Benvenisti

International Human Rights and Humanitarian Law
René Provost

Remedies Against International Organisations
Karel Wellens

Diversity and Self-Determination in International Law
Karen Knop

The Law of Internal Armed Conflict
Lindsay Moir

International Commercial Arbitration and African States: Practice, Participation and Institutional Development
Amazu A. Asouzu

The Enforceability of Promises in European Contract Law
James Gordley

International Law in Antiquity
David J. Bederman

Money Laundering: A New International Law Enforcement Model
Guy Stessens

Good Faith in European Contract Law
Reinhard Zimmermann and Simon Whittaker

On Civil Procedure
J. A. Jolowicz

Trusts: A Comparative Study
Maurizio Lupoi

The Right to Property in Commonwealth Constitutions
Tom Allen

International Organizations Before National Courts
August Reinisch

The Changing International Law of High Seas Fisheries
Francisco Orrego Vicuña

Trade and the Environment: A Comparative Study of EC and US Law
Damien Geradin

Unjust Enrichment: A Study of Private Law and Public Values
Hanoch Dagan

Religious Liberty and International Law in Europe
Malcolm D. Evans

Ethics and Authority in International Law
Alfred P. Rubin

Sovereignty Over Natural Resources: Balancing Rights and Duties
Nico Schrijver

The Polar Regions and the Development of International Law
Donald R. Rothwell

Fragmentation and the International Relations of Micro-States: Self-determination and Statehood
Jorri Duursma

Principles of the Institutional Law of International Organizations
C. F. Amerasinghe

For EU product safety concerns, contact us at Calle de José Abascal, 56–1°, 28003 Madrid, Spain or eugpsr@cambridge.org.

www.ingramcontent.com/pod-product-compliance
Lightning Source LLC
LaVergne TN
LVHW011530200926
848484LV00001B/222